Process
Dynamics
and Control

PRENTICE-HALL INTERNATIONAL SERIES
IN THE PHYSICAL AND CHEMICAL ENGINEERING SCIENCES

NEAL R. AMUNDSON, EDITOR, *University of Minnesota*

ADVISORY EDITORS

ANDREAS ACRIVOS, *Stanford University*
JOHN DAHLER, *University of Minnesota*
THOMAS J. HANRATTY, *University of Illinois*
JOHN M. PRAUSNITZ, *University of California*
L. E. SCRIVEN, *University of Minnesota*

AMUNDSON   *Mathematical Methods in Chemical Engineering*
ARIS   *Elementary Chemical Reactor Analysis*
ARIS   *Introduction to the Analysis of Chemical Reactors*
ARIS   *Vectors, Tensors, and the Basic Equations of Fluid Mechanics*
BALZHISER, SAMUELS, AND ELIASSEN   *Chemical Engineering Thermodynamics*
BOUDART   *Kinetics of Chemical Processes*
BRIAN   *Staged Cascades in Chemical Processes*
CROWE, HAMIELEC, HOFFMAN, JOHNSON, SHANNON, AND WOODS   *Chemical Plant Simulation*
DOUGLAS   *Process Dynamics and Control, Vol. 1, Analysis of Dynamic Systems*
DOUGLAS   *Process Dynamics and Control, Vol. 2, Control System Synthesis*
FREDRICKSON   *Principles and Applications of Rheology*
FRIEDLY   *Dynamic Behavior of Processes*
HAPPEL AND BRENNER   *Low Reynolds Number Hydrodynamics*
HIMMELBLAU   *Basic Principles and Calculations in Chemical Engineering, 2nd ed.*
HOLLAND   *Multicomponent Distillation*
HOLLAND   *Unsteady State Processes with Applications in Multicomponent Distillation*
KOPPEL   *Introduction to Control Theory with Applications to Process Control*
LEVICH   *Physicochemical Hydrodynamics*
MEISSNER   *Processes and Systems in Industrial Chemistry*
NEWMAN   *Electrochemical Engineering*
PERLMUTTER   *Stability of Chemical Reactors*
PETERSEN   *Chemical Reaction Analysis*
PRAUSNITZ   *Molecular Thermodynamics of Fluid-Phase Equilibria*
PRAUSNITZ AND CHUEH   *Computer Calculations for High-Pressure Vapor-Liquid Equilibria*
PRAUSNITZ, ECKERT, ORYE, O'CONNELL   *Computer Calculations for Multicomponent Vapor-Liquid Equilibria*
WHITAKER   *Introduction to Fluid Mechanics*
WILDE   *Optimum Seeking Methods*
WILLIAMS   *Polymer Science and Engineering*

PRENTICE-HALL, INC.
PRENTICE-HALL INTERNATIONAL, INC., UNITED KINGDOM AND EIRE
PRENTICE-HALL OF CANADA, LTD., CANADA

Process Dynamics and Control

Volume 2
Control
System Synthesis

R T HATCH

J. M. Douglas
Department of Chemical Engineering
University of Massachusetts

PRENTICE-HALL, INC., Englewood Cliffs, New Jersey

PRENTICE-HALL INTERNATIONAL, INC., *London*
PRENICE-HALL OF AUSTRALIA, PTY. LTD., *Sydney*
PRENTICE-HALL OF CANADA, LTD., *Toronto*
PERNTICE-HALL OF INDIA PRIVATE LIMITED, *New Delhi*
PRENTICE-HALL OF JAPAN, INC., *Tokyo*

10 9 8 7 6 5 4 3 2 1

Library of Congress Catalog Card Number: 70-171041

ISBN: 0-13-723056-7

Printed in the United States of America

To
R. L. Pigford and A. B. Metzner
and to the memories of
Kurt Wohl and J. A. Gerster

Contents

*The asterisk indicates sections or chapters considered to provide a basis for an undergraduate course.

9. Optimal Control Theory   248

Part IV   Deliberate Unsteady State Operation   355

10. Periodic Processing   357

Preface

In Volume 1, *Analysis of Dynamic Systems*, of this two–volume text on
Process Dynamics and Control, we developed the tools for treating dynamic
systems. By considering the optimum steady state design and operation of a
plant, we were able to pinpoint areas where the dynamic features of a plant
might be significant. Thus we studied procedures for deriving dynamic models
of process equipment, as well as methods for solving these equations. Now we
are interested in establishing the techniques we can use to modify the dynamic
characteristics of a plant if our dynamic study reveals that they are undesirable.

We consider a control system to be anything attached to a process unit in
an attempt to improve the dynamic performance of that unit. This volume is
primarily devoted to a study of the techniques for synthesizing control systems.
In Chapter 7 we treat conventional, servomechanism control theory, which is
applicable to linear plants having a single input and a single output. Of course,
most chemical processes have a large number of input and output streams;
therefore we devote Chapter 8 to extensions of the servomechanism ideas to
multivariable plants. Unfortunately, the conventional design criteria normally
are not applicable to these multivariable processes; consequently, in Chapter 9
we explore the possibility of designing optimal, multivariable feedback controllers.

In the three chapters mentioned, the focus is on the design of regulating
feedback controllers. In other words, our basic definition of the control problem
includes the assumption that it is always desirable to operate as close as possible
to the optimum steady state design of the plant. The material presented in

Chapter 10 provides a challenge to this assumption. Thus a large number of examples are presented where periodic operation is superior to the optimum steady state performance. Similarly, we show that in some cases it is advantageous to reverse the "rules of thumb" we developed in Chapters 7 through 9, such as using positive feedback controllers to make stable plants become unstable. The purpose of including this material is to point out some of the limitations of our present control theory and, hopefully, to excite the imagination of the reader into exploring new approaches.

The book is intended for use in conjunction with Volume 1 as a senior-year undergraduate text, an introductory graduate text, or as a reference for self-study by practicing engineers. Therefore most of the required mathematical tools have been presented as part of the development. The material designated by an asterisk in the Table of Contents of Volumes 1 and 2 is intended to provide the basis for an undergraduate course. Also, the exercises at the end of each chapter prefixed by the letter A are on the undergraduate level. Those labeled B are of an intermediate level, and those with a C are more difficult. Problems marked with an asterisk appear in more than one chapter (several techniques are applied to the same physical system) and those marked with a double dagger (‡) are "open-ended" to some extent.

The excellent annual reviews on process dynamics and control published by *Industrial and Engineering Chemistry* made it unnecessary to include a complete bibliography in the text. Readers should definitely become familiar with these review articles. The references listed were selected on a somewhat arbitrary basis and many significant papers have not been mentioned. The references are presented as footnotes, which, like the figures and tables, have been numbered consecutively within each section.

Perhaps it should be emphasized that many of the methods for synthesizing multivariable control systems have not yet been tested on industrial plants. Moreover, the rapid growth of the control field undoubtedly will make some of the ideas presented in the text become obsolete in the near future. However, man seems to learn best by a trial-and-error procedure, and it is hoped that this book will be helpful in this endeavor.

Process Control

Design
of Simple Control
Systems

7

In the first volume we described a procedure for determining the optimum steady state design of a plant. The information required for the analysis included an expression for the profitability of the plant written in terms of the capital and operating costs as well as sales income, a specified value for the desired production rate, a set of steady state material and energy balances describing each piece of process equipment, and the specification of certain system parameters, together with the average values of the properties of the input streams (feed composition, feed temperature, steam temperature, etc.). Once the design has been accomplished and the plant constructed, we still want to achieve the optimum economic operation of the plant. We know that some of the inputs, such as feed temperature, actually will vary with time, and therefore we attempt to manipulate some of the other inputs—the control variables—in such a way that we maximize the profit produced by the system. If we use the steady state material and energy balances in this procedure, we call the result an optimum steady state control system.

Of course, it is always dangerous to use steady state equations to describe how a process changes with time. Thus it is necessary to develop dynamic

R T HATCH

models for the units. Even though we normally obtain coupled sets of non-linear, ordinary or partial, differential equations for these models, a first estimate of the dynamic response can be determined by linearizing the equations and using matrix methods to find solutions. Also, the range where the linearized models are valid can be established by using perturbation theory. This material was discussed in detail in Volume 1.

If the results of the dynamic analysis indicate that the process is in an unsteady state condition for only a small fraction of the normal operating time, an optimum steady state control system should be installed. However, if the dynamic study shows that unsteady operation is predominant, we must find some other way of ensuring that we obtain the most profitable operation. In other words, if we discover that the dynamic characteristics of the plant corresponding to the optimum steady state design are undesirable, we would like to find some way of modifying those characteristics. Procedures of this type provide the subject matter for most of this volume.

In Chapter 7 we consider the classical techniques for synthesizing control systems for plants with a single input and a single output. The extension of these methods to multivariable processes is discussed in Chapter 8, and the design of optimal, multivariable controllers is presented in Chapter 9. Before we discuss these various approaches, however, it is helpful to consider a specific example that demonstrates the importance of dynamic control.

SECTION 7.1   FEEDBACK CONTROL OF AN UNSTABLE STIRRED-TANK REACTOR

The operation of a nonisothermal stirred-tank reactor served as a vehicle for illustrating all the topics in Volume 1 of this text, and we will continue with this practice. One of the most interesting features of the reactor system is that a steady state design, and even an optimum steady state design, sometimes is unstable. For example, if we consider the process parameters given in Table 7.1–1 and refer to the optimum design procedure we described in Example 2.1–2 in the first volume, we obtain the optimum design conditions listed in the table. Now if we calculate the characteristic roots of the linearized system equations, using the method presented in Example 4.5–2 of Volume 1, we find that

$$\lambda_1 = 0.036 + j3.367 \qquad \lambda_2 = 0.036 - j3.367$$

Thus the real part of the characteristic roots is positive, which means that the system will always tend to leave the neighborhood of the optimum steady state design. Actually, the linearized dynamic equations indicate that the system output will become unbounded, but we know that this really implies that the outputs will grow to such an extent that the linearized equations no longer provide a valid description of the nonlinear system—that is, that the quadratic

TABLE 7.1–1
SYSTEM PARAMETERS

Cost data:

$C_A = \$0.081/(\text{cm}^2)(\text{hr})$	$C_f = \$0.01207/\text{g. mole}$
$C_H = \$0.00002/\text{gram}$	$C_V = \$0.0389/(\text{liter})(\text{hr})$

System constants:

$C_p \rho = 1.0 \, \text{cal}/(\text{cc})(°K)$	$C_{pH} = 1.0 \, \text{cal}/(\text{gram})(°K)$
$E = 60,352 \, \text{cal}/(\text{g. mole})$	$(-\Delta H) = 5888 \, \text{cal}/(\text{g. mole})$
$k_0 = 4.094 \times 10^{32} \, \text{sec}^{-1}$	

Design data and system inputs:

$A_f = 0.01 \, \text{g. mole/cc}$	$G = 2700 \, \text{g. mole/hr}$
$T_H = 373 \, °K, \quad T_f = 300 \, °K$	$U = 1.0 \, \text{cal}/(\text{cc})(\text{sec})(°K)$

Optimum design variables:

$A = 0.0025 \, \text{g. mole/cc}, \quad A_H = 794.7 \, \text{sq. cm.}, \quad k = 0.36 \, \text{hr}^{-1}$
$Q_H = 2384 \, \text{cal/sec}, \quad q = 100 \, \text{cc/sec}, \quad q_H = 596 \, \text{gram/sec}$
$T = 368 \, °K, \quad T_0 = 369 \, °K, \quad V = 3000 \, \text{liters}$

Optimum costs:

$C_A A_H = \$64.365/\text{hr}$	$C_f q A_f = \$43.456/\text{hr}$
$C_H q_H = \$42.912/\text{hr}$	$C_V V = .\$116.485/\text{hr}$
$C_T = \$267.218/\text{hr}$	

and higher-order terms in the Taylor series expansion become more important than the linear terms.

Obviously it is of great interest to determine the final state of an unstable system of this type. Although this problem is very complex, we will study the possibility of obtaining some alternate steady state output, which is not the optimum steady state, in this section. Also we will look for ways to use controllers to stabilize unstable steady states of various kinds. A more detailed discussion of the behavior of the unstable reactor will be presented in Chapter 10.

Multiple Steady State Solutions

For a first-order reaction, we showed that we could write the dynamic equations for a stirred-tank reactor in the form

$$V \frac{dA}{dt} = q(A_f - A) - kVA \qquad 3.3\text{--}2$$

$$VC_p \rho \frac{dT}{dt} = qC_p \rho(T_f - T) + (-\Delta H)kVA + \frac{UA_H Kq_H}{1 + Kq_H}(T_H - T) \qquad 3.3\text{--}3$$

where

$$k = k_0 \exp \frac{-E}{RT} \qquad K = \frac{2C_{pH} \rho_H}{UA_H} \qquad 3.3\text{--}4$$

At steady state conditions, the accumulation terms (the time derivatives) must both be equal to zero. Thus we can solve Eq. 3.3–2 for the steady state composition

$$A_s = \frac{qA_f}{q + k_s V} \qquad 7.1\text{-}1$$

and use this result to eliminate the composition from the energy equation. After some slight manipulation, we obtain

$$\left[1 + \frac{UA_H Kq_H}{qC_p\rho(1 + Kq_H)}\right]T_s - \left[T_f + \frac{UA_H Kq_H}{qC_p\rho(1 + Kq_H)}\,T_H\right]$$
$$= \frac{(-\Delta H)A_f}{C_p\rho}\left(\frac{k_s V/q}{1 + k_s V/q}\right) \qquad 7.1\text{-}2$$

The terms on the left-hand side of the equation are related to the heat removed by convective flow and the heat added through the coil, whereas the term on the right-hand side indicates the amount of heat generated by the chemical reaction. If we plot both sides of this transcendental equation against temperature, the points of intersection of the two curves must correspond to the steady state solutions. It is clear that the left-hand side is simply a linear function of temperature. Also, it is easy to see that the function on the right-hand side varies between almost zero, since at very low temperatures the value of k_s is very small, and an upper limit of $(-\Delta H)A_f/qC_p\rho$, since at very high temperatures the last term in parenthesis in Eq. 7.1–2 approaches unity. Further study of this function makes it apparent that it has the sigmoidal shape illustrated in Figure 7.1–1.

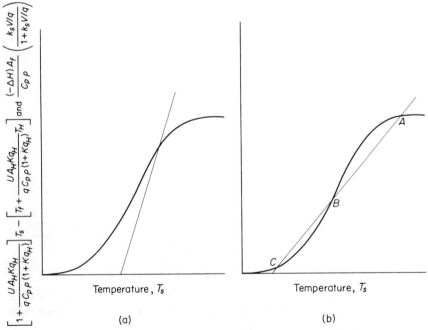

Figure 7.1-1. Steady state solutions of energy equation: (a) single solution; (b) three steady state solutions.

For the system parameters given in Table 7.1–1, we find that there is only a single intersection of the straight line and the sigmoidal curve; hence there is only one possible steady state solution. However, it does not take much imagination to recognize that for other parameters it might be possible to obtain three different steady state temperatures that satisfy Eq. 7.1–2 (see Figure 7.1–1b). Of course, these might not correspond to optimum designs. Alternately, it might be possible to find parameters such that there are multiple solutions of the optimum steady state design problem described in Example 2.1–2. In this last case, the best design would have to be determined by a direct comparison of the solutions.

It is interesting to note that whenever there are three solutions of Eq. 7.1–2, it is possible to show that the middle intersection always corresponds to an unstable equilibrium point.[1] From Figure 7.1–1b we see that the slope of the left-hand side of Eq. 7.1–2 at point B is less than the slope of the right-hand side, which implies that the net rate of heat removal by convective flow is less than the rate of heat generated by the chemical reaction. Thus if the temperature of a reactor operating at this steady state condition increases slightly for some reason, the excess of the heat generation as compared to the heat removal will cause the temperature to continue to increase. Similarly, if the temperature drops slightly below the steady state value, the net rate of heat removal will be greater than the heat generated by the reaction and the temperature will continue to drop.

In mathematical terms we can write this slope condition as

$$1 + \frac{UA_H Kq_H}{qC_p\rho(1 + Kq_H)} < \frac{(-\Delta H)A_f}{qC_p\rho}\left[\frac{(k_s V/q)(E/RT_s^2)}{(1 + k_s V/q)^2}\right]$$

or after substituting Eq. 7.1–1 and rearranging somewhat,

$$\left[1 + \frac{UA_H Kq_H}{qC_p\rho(1 + Kq_H)}\right]\left(1 + \frac{k_s V}{q}\right) < \frac{(-\Delta H)}{C_p\rho}\frac{E}{RT_s^2}\frac{k_s VA_s}{q} \qquad 7.1\text{–}3$$

If we let

$$\alpha = \frac{k_s V}{q}, \quad \beta = 1 + \frac{UA_H Kq_H}{qC_p\rho(1 + Kq_H)}, \quad \gamma = \frac{(-\Delta H)}{C_p\rho}\frac{E}{RT_s^2}\frac{k_s VA_s}{q} \qquad 7.1\text{–}4$$

the result can be put in the form

$$(1 + \alpha)(1 + \beta) > \gamma \qquad 7.1\text{–}5$$

It is tempting to apply this same reasoning concerning the slopes to show that the extreme equilibrium points, points A and C on Figure 7.1–1b, are stable. In other words, we see that the net rate of heat removed after a slight increase in temperature exceeds the rate of heat generation so that the reactor should tend to cool down. Moreover, after a slight drop in temperature the heat generation is larger than the heat removal so that the temperature should increase. Unfortunately, this type of reasoning also indicates that anytime

[1] C. van Heerden, *Ind. Eng. Chem.*, **45**, 1242 (1953).

there is only one solution of Eq. 7.1–2 (see Figure 7.1–1a), the system will be stable, and yet we know that the parameters given in Table 7.1–1 correspond to a case where an unstable system is obtained. Hence the stability criterion given by Eq. 7.1–5 is a sufficient, but not necessary, condition to have an unstable system.

To develop more complete stability criteria,[2] we can examine the linearized equations describing the dynamic behavior of the reactor in some sufficiently small neighborhood of any of the possible steady state solutions. In Example 3.3–1 we showed that the linearized equations could be written as

$$\frac{dx_1}{d\tau} = a_{11}x_1 + a_{12}x_2 + b_{11}u_1 \qquad\qquad 3.3\text{--}12$$

$$\frac{dx_2}{d\tau} = a_{21}x_1 + a_{22}x_2 + b_{21}u_1 + b_{22}u_2 \qquad\qquad 3.3\text{--}13$$

where

$$x_1 = \frac{A - A_s}{A_f}, \quad x_2 = \frac{(T - T_s)C_p\rho}{(-\Delta H)A_f}, \quad \tau = \frac{qt}{V}$$

$$a_{11} = -\left(1 + \frac{k_s V}{q}\right) = -(1 + \alpha), \quad a_{12} = -\frac{(-\Delta H)}{C_p\rho}\frac{E}{RT_s^2}\frac{k_s V A_s}{q} = -\gamma$$

$$a_{21} = \frac{k_s V}{q} = \alpha,$$

$$a_{22} = -\left[1 + \frac{UA_H Kq_H}{qC_p\rho(1 + Kq_H)} - \frac{(-\Delta H)}{C_p\rho}\frac{E}{RT_s^2}\frac{k_s V A_s}{q}\right] = -(1 + \beta - \gamma),$$

$$b_{11} = \frac{A_f - A_s}{A_f}, \quad b_{21} = \frac{(T_f - T_s)C_p\rho}{(-\Delta H)A_f}, \quad b_{22} = \frac{UA_H Kq_H(T_H - T_s)}{q(1 + Kq_H)^2(-\Delta H)A_f}$$

$$7.1\text{--}6$$

The characteristic equation is given by Eq. 3.3–62

$$\lambda^2 - (a_{11} + a_{22})\lambda + a_{11}a_{22} - a_{12}a_{21} = 0 \qquad\qquad 3.3\text{--}62$$

so that the characteristic roots are

$$\lambda = \tfrac{1}{2}[(a_{11} + a_{22}) \pm \sqrt{(a_{11} + a_{22})^2 - 4(a_{11}a_{22} - a_{12}a_{21})}] \qquad 3.3\text{--}63$$

It is apparent that we will obtain two real roots having opposite signs—that is, a saddlepoint—whenever

$$a_{11}a_{22} - a_{12}a_{21} < 0 \qquad\qquad 7.1\text{--}7$$

After substituting the expressions given in Eq. 7.1–6, we can write

$$(1 + \alpha)(1 + \beta) < \gamma \qquad\qquad 7.1\text{--}8$$

which is identical to the result found by considering the slopes, see Eq. 7.1–5.

---

[2] The stability analysis and control study presented below follow the procedure described by R. Aris and N. R. Amundson, *Chem. Eng. Sci.*, **7**, 121, 132 (1958).

When we obtain two real roots having opposite signs, we say that the steady state solution is a saddlepoint because the linearized system response has the form $y = c_1 e^{\lambda \tau} + c_2 e^{-\lambda \tau}$, which obviously is unstable; that is, the system tends to grow without a bound.

If the sign of the preceding inequality is reversed, the slope of the heat-generated curve will be less than the slope of the heat-removal curve and from Eq. 3.3–63 both of the characteristic roots will have the same sign. When both roots are real the local response will be of the form $y = c_1 e^{\lambda_1 \tau} + c_2 e^{\lambda_2 \tau}$ and we say that the equilibrium point is a node, whereas if we obtain complex conjugate roots the response will be $y = e^{\alpha \tau} (c_1 \sin \omega \tau + c_2 \cos \omega \tau)$ and we call the equilibrium point a focus. Of course, the system will still be unstable if these roots are positive, or have positive real parts, so that the conditions for a locally stable equilibrium point are

$$a_{11} + a_{22} < 0 \quad \text{or} \quad 1 + \alpha + 1 + \beta - \gamma > 0 \qquad \text{7.1–9}$$

$$a_{11} a_{22} - a_{12} a_{21} > 0 \quad \text{or} \quad (1 + \alpha)(1 + \beta) - \gamma > 0 \qquad \text{7.1–10}$$

For the optimum design given in Table 7.1–1, the first criterion is violated even though the second is satisfied.

Proportional Control

It is somewhat disheartening to go to all the trouble of establishing the optimum steady state design only to find that the system is unstable and will never operate at steady state conditions. This difficulty emphasizes the importance of a dynamic analysis and the need to develop techniques for the dynamic control of chemical plants. In order to illustrate how a control system could be used to make this unstable process become stable, let us suppose that we can instantaneously measure the temperature in the reactor, that we have a device which compares this measured value with the desired optimum steady state value so that we can obtain an error signal, and, finally, that we have a unit which changes the flow rate of heating fluid proportional to this error. We will connect the equipment so that whenever the actual reactor temperature exceeds the desired value, we decrease the flow rate of heating fluid, which tends to make the reactor temperature decrease. Thus we have

$$q_H - q_{Hs} = -K_1(T - T_s) \qquad \text{7.1–11}$$

or in terms of the dimensionless deviation variables,

$$u_2 = -K_c x_2 \qquad \text{7.1–12}$$

where all the proportionality factors have been included in K_c, which we will refer to as the controller gain. For a case where the feed rate is maintained constant, so that $u_1 = 0$, Eqs. 3.3–12 and 3.3–13 together with the control equation, Eq. 7.1–12, give the result

$$\frac{dx_1}{d\tau} = a_{11}x_1 + a_{12}x_2 \qquad\qquad 7.1\text{–}13$$

$$\frac{dx_2}{d\tau} = a_{21}x_1 + (a_{22} - b_{22}K_c)x_2 \qquad\qquad 7.1\text{–}14$$

Now the characteristic equation becomes

$$\lambda^2 - (a_{11} + a_{22} - b_{22}K_c)\lambda + a_{11}a_{22} - a_{12}a_{21} - a_{11}b_{22}K_c = 0 \qquad 7.1\text{–}15$$

and the characteristic roots are

$$\lambda = \tfrac{1}{2}[(a_{11} + a_{22} - b_{22}K_c)$$
$$\pm \sqrt{(a_{11} + a_{22} - b_{22}K_c)^2 - 4(a_{11}a_{22} - a_{12}a_{21} - a_{11}b_{22}K_c)}]$$
$$7.1\text{–}16$$

For the controlled system to be stable, we must have two roots that have the same sign

$$4(a_{11}a_{22} - a_{12}a_{21} - a_{11}b_{22}K_c) > 0 \qquad\qquad 7.1\text{–}17$$

and both must be negative or have a negative real part

$$a_{11} + a_{22} - b_{22}K_c < 0 \qquad\qquad 7.1\text{–}18$$

By substituting the expressions given in Eq. 7.1–6, the stability criteria can be written

$$(1 + \alpha)(1 + \beta) - \gamma + (1 + \alpha)b_{22}K_c > 0 \qquad\qquad 7.1\text{–}19$$

$$(1 + \alpha + 1 + \beta - \gamma + b_{22}K_c) > 0 \qquad\qquad 7.1\text{–}20$$

Clearly, if we make the controller gain large enough, we can always satisfy both criteria and make the system stable. This implies that even if the optimum steady state design corresponds to the intermediate steady state solution for a case where three steady states are obtained (i.e., a saddlepoint), our controller will stabilize the process.

It should be noted that the sign of the control action term is very important. For example, if we had chosen to increase the heating flow rate when the actual reactor temperature exceeded the optimum design value, our control equation would have been

$$u_2 = K_c x_2$$

and the stability criteria (the conditions we must satisfy to ensure that the characteristic roots have the same sign and are negative, or have negative real parts, so that any deviation from steady state conditions will disappear as time progresses) would have been

$$(1 + \alpha)(1 + \beta) - \gamma - (1 + \alpha)b_{22}K_c > 0 \qquad\qquad 7.1\text{–}21$$

$$(1 + \alpha + 1 + \beta - \gamma - b_{22}K_c) > 0 \qquad\qquad 7.1\text{–}22$$

With this arrangement, high controller gains will make the system unstable.

This case is called *positive feedback control* because the controller tends to amplify deviations from steady state, whereas the first case is called *negative feedback control* because the controller tends to damp out the deviations.

An alternative approach for controlling the reactor would be to make the change in the heating flow rate proportional to the measured deviations in the reactant composition

$$u_2 = K_c x_1 \qquad 7.1\text{--}23$$

We use a positive sign in this expression because increasing q_H will cause T to increase, which will make A decrease. Thus if our measured value of A goes up, we raise q_H to compensate for it. For this case the system equations become

$$\frac{dx_1}{d\tau} = a_{11}x_1 + a_{12}x_2 \qquad 7.1\text{--}24$$

$$\frac{dx_2}{d\tau} = (a_{21} + b_{22}K_c)x_1 + a_{22}x_2 \qquad 7.1\text{--}25$$

The characteristic equation is

$$\lambda^2 - (a_{11} + a_{22})\lambda + a_{11}a_{22} - a_{12}a_{21} - a_{12}b_{22}K_c = 0 \qquad 7.1\text{--}26$$

so that the characteristic roots are

$$\lambda = \tfrac{1}{2}[(a_{11} + a_{22}) \pm \sqrt{(a_{11} + a_{22})^2 - 4(a_{11}a_{22} - a_{12}a_{21} - a_{12}b_{22}K_c)}]$$

$$7.1\text{--}27$$

and the stability criteria become

$$a_{11}a_{22} - a_{12}a_{21} - a_{12}b_{22}K_c > 0 \qquad a_{11} + a_{22} < 0$$

or

$$(1 + \alpha)(1 + \beta) - \gamma + \gamma b_{22}K_c > 0 \qquad 7.1\text{--}28$$

$$1 + \alpha + 1 + \beta - \gamma > 0 \qquad 7.1\text{--}29$$

Thus we can always stabilize a saddlepoint (we can ensure that the roots will have the same sign) with this kind of a controller by making K_c sufficiently large, but we can never make the system described in Table 7.1–1 (where the real part of the root is positive) stable.

This simple illustration shows that it can make a great deal of difference which variable we measure and attempt to force to return to the desired steady state value. It should not be surprising, therefore, that the variable we choose to manipulate can have the same kind of effect on the system stability. For example, in order to assess whether or not the feed rate could be used as a control variable to compensate for either composition or temperature fluctuations, we would have to repeat the kind of analysis just described. However, since the approach is the same, we leave this as an exercise for the reader and proceed to other kinds of control modes.

Derivative Control

Instead of varying the heating flow rate proportional to the error between the actual reactor temperature (or composition) and the desired steady state value, we might manipulate this control variable proportional to the rate of change of the error—that is, its derivative. With proportional control we make large changes in the control variable when the error is large, whereas with derivative control we make large changes when the error is changing rapidly. If we let

$$q_H - q_{Hs} = -K_1 \frac{d(T - T_s)}{d\tau} \qquad 7.1\text{-}30$$

then in terms of the dimensionless deviation variables we have

$$u_2 = -K_D \frac{dx_2}{d\tau} \qquad 7.1\text{-}31$$

Taking the Laplace transform of this equation

$$\tilde{u}_2 = -K_D s\tilde{x}_2 - K_D x_{20} \qquad 7.1\text{-}32$$

and the linearized system equations

$$s\tilde{x}_1 = a_{11}\tilde{x}_1 + a_{12}\tilde{x}_2 + x_{10} \qquad 7.1\text{-}33$$

$$s\tilde{x}_2 = a_{21}\tilde{x}_1 + a_{22}\tilde{x}_2 + b_{22}\tilde{u}_2 + x_{20} \qquad 7.1\text{-}34$$

eliminating $\tilde{u}_2$, and solving for $\tilde{x}_2$, we obtain

$$\tilde{x}_2 = \frac{a_{21}x_{10} + (s - a_{11})x_{20}(1 - b_{22}K_D)}{(1 + b_{22}K_D)s^2 - (a_{11} + a_{22} + a_{11}b_{22}K_D)s + a_{11}a_{22} - a_{12}a_{21}} \qquad 7.1\text{-}35$$

The system stability depends on the roots of the denominator of this expression, and, as shown previously, these are identical to the characteristic roots. Thus

$$\lambda = \frac{1}{2(1 + b_{22}K_D)}[(a_{11} + a_{22} + a_{11}b_{22}K_D)$$
$$\pm \sqrt{(a_{11} + a_{22} + a_{11}b_{22}K_D)^2 - 4(a_{11}a_{22} - a_{12}a_{21})(1 + b_{22}K_D)}] \qquad 7.1\text{-}36$$

and the stability criteria are

$$(a_{11}a_{22} - a_{12}a_{21})(1 + b_{22}K_D) > 0$$

$$\frac{a_{11} + a_{22} + a_{11}b_{22}K_D}{1 + b_{22}K_D} < 0$$

or

$$[(1 + \alpha)(1 + \beta) - \gamma](1 + b_{22}K_D) > 0 \qquad 7.1\text{-}37$$

$$1 + \alpha + 1 + \beta - \gamma + (1 + \alpha)b_{22}K_D > 0 \qquad 7.1\text{-}38$$

By making K_D sufficiently large, we can force the system described in Table 7.1-1 to be stable, but we cannot use this controller to stabilize a saddlepoint.

A similar analysis could be used to study the effect of derivative control on system stability for other measured and manipulated variables.

Integral Control

We might also consider the possibility of changing the heating flow rate proportional to the integral of the error

$$q_H - q_{Hs} = -K_1 \int (T - T_s) \, d\tau \qquad 7.1\text{-}39$$

or

$$u_2 = -K_I \int x_2 \, d\tau \qquad 7.1\text{-}40$$

The Laplace transform of the control law is

$$\tilde{u}_2 = -\frac{K_I \tilde{x}_2}{s} \qquad 7.1\text{-}41$$

and the characteristic equation becomes

$$s^3 + (1 + \alpha + 1 + \beta - \gamma)s^2 + [(1 + \alpha)(1 + \beta) - \gamma + b_{22}K_I]s$$
$$+ (1 + \alpha)b_{22}K_I = 0 \qquad 7.1\text{-}42$$

The stability of the system is determined by examining the roots of this cubic equation. Although a procedure for accomplishing this task will be described in Section 7.4, the results are given below:

$$(1 + \alpha)(1 + \beta) - \gamma + b_{22}K_I > 0 \qquad 7.1\text{-}43$$

$$1 + \alpha + 1 + \beta - \gamma > 0 \qquad 7.1\text{-}44$$

$$(1 + \alpha)b_{22}K_I > 0 \qquad 7.1\text{-}45$$

$$\frac{(1 + \alpha + 1 + \beta - \gamma)[(1 + \alpha)(1 + \beta) - \gamma] + (1 + \beta - \gamma)b_{22}K_I}{1 + \alpha + 1 + \beta - \gamma} > 0$$

$$7.1\text{-}46$$

Thus we find that an integral controller will not stabilize the system described in Table 7.1–1; that is, Eq. 7.1–44 does not depend on the controller gain. Also, it will not always be possible to satisfy the stability criterion given by Eq. 7.1–46.

We could attempt to use other measured variables and control variables to stabilize our reactor problem, or we could study the effect of a linear combination of three control modes on the process

$$-u_2 = K_c\left(x_2 + K_D \frac{dx_2}{dt} + K_I \int x_2 \, dt\right) \qquad 7.1\text{-}47$$

or

$$-\tilde{u}_2 = K_c\left(\tilde{x}_2 + K_D s\tilde{x}_2 + \frac{K_I \tilde{x}_2}{s}\right) \qquad 7.1\text{-}48$$

However, the approach is identical to that described above; therefore, instead of pursuing this matter, we will attempt to determine the quantitative effect that a controller has on some simple systems.

**SECTION 7.2 EFFECT OF FEEDBACK CONTROL ON
SIMPLE SYSTEMS**

We have seen that if we manipulate some of the input variables—that is, the control variables—in such a way as to correct the deviations of the output variables from their desired values, we can often make an unstable process become stable. This step can be accomplished because the controller changes the characteristic roots of the system equations. Consequently, we might expect to be able to use this same idea to correct for the effects that disturbances have on stable plants. In other words, if we do not particularly like the dynamic characteristics of some process, we can add a control system and attempt to obtain a dynamic response that is better in some way.

First-Order Systems

Previously we showed (see Section 4.3) that the response of a first-order system could be written as

$$\frac{dy}{dt} = \lambda_1 y + u_1 + v_1 \qquad\qquad 7.2\text{-}1$$

or after dividing by λ_1, redefining some of the variables, and taking the Laplace transform,

$$(\tau s + 1)\tilde{y} = \tilde{u} + \tilde{v} \qquad\qquad 7.2\text{-}2$$

where $\tilde{y}$ might correspond to the composition in an isothermal stirred-tank reactor, $\tilde{u}$ might be the flow rate to the reactor, and $\tilde{v}$ the feed composition.

Now if we attempt to adjust the feed rate proportional to the deviation of y from some desired value y_0 (where $y_0 = 0$ if our variables represent deviations from an optimum steady state design), the system equations become

$$(\tau s + 1)\tilde{y} = K_c(\tilde{y}_0 - \tilde{y}) + \tilde{v} \qquad\qquad 7.2\text{-}3$$

or

$$\tilde{y} = \frac{K_c}{\tau s + 1 + K_c}\tilde{y}_0 + \frac{1}{\tau s + 1 + K_c}\tilde{v} \qquad\qquad 7.2\text{-}4$$

This result can be used to define transfer functions for the change in the output with respect to changes in the *set point* of the controller (i.e., the desired output $\tilde{y}_0$) or with respect to a disturbance entering the system, $\tilde{v}$.

$$\frac{\tilde{y}}{\tilde{y}_0} = \frac{K_c}{\tau s + 1 + K_c} \qquad\qquad \frac{\tilde{y}}{\tilde{v}} = \frac{1}{\tau s + 1 + K_c} \qquad\qquad 7.2\text{-}5$$

In many cases it is helpful to draw a block diagram of the system. The output from a block is merely the product of the function written inside the block and the input signal, whereas the output from a circle is the sum of the input signals. The diagram (see Figure 7.2–1) clearly indicates the feed-

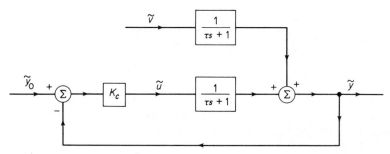

Figure 7.2-1. Feedback control of a first-order system.

back nature of the control; that is, the measured value of the output is compared with the desired value to generate an error signal, which is used to adjust one of the plant inputs in an attempt to compensate or correct the output deviation. Also, we have a negative feedback control system, according to the sign below the summation point on the left-hand side of the diagram, because whenever the output exceeds the desired value we decrease the amount of control. Although there is a whole algebra for treating block diagrams, the important thing to note at this point is that the transfer functions can be written down immediately once the diagram is available. For example, if we combine the transfer function for the controller and the plant, introduce an error signal, and generalize the results somewhat [by replacing $1/(\tau s + 1)$ by G_v and $K_c/(\tau s + 1)$ by G] (see Figure 7.2–2), we can show that the transfer functions are

$$\frac{\tilde{y}}{\tilde{y}_0} = \frac{G}{1 + G} \qquad \frac{\tilde{y}}{\tilde{v}} = \frac{G_v}{1 + G} \qquad\qquad 7.2\text{–}6$$

Each of these transfer functions can be obtained by writing the transfer func-

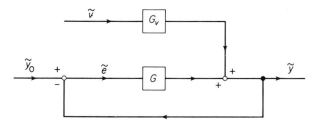

Figure 7.2-2. Feedback control system.

tion between the output and the input signal of interest, divided by unity plus the product of the transfer functions in the feedback loop. The validity of this result can be demonstrated by examining the equations implied by the diagram

$$\tilde{y} = G_v \tilde{v} + G\tilde{e} \qquad\qquad 7.2\text{–}7$$

$$\tilde{y} = G_v \tilde{v} + G(\tilde{y}_0 - \tilde{y}) \qquad\qquad 7.2\text{–}8$$

or

$$(1 + G)\tilde{y} = G_v\tilde{v} + G\tilde{y}_0 \qquad\qquad 7.2\text{--}9$$

so that

$$\frac{\tilde{y}}{\tilde{y}_0} = \frac{G}{1 + G} \qquad \frac{\tilde{y}}{\tilde{v}} = \frac{G_v}{1 + G} \qquad\qquad 7.2\text{--}10$$

In order to find what effect the control system exerts on the plant, we consider a step change in the disturbance variable, $\tilde{v} = A/s$. For a first-order system, it is a simple matter to calculate the output

$$\tilde{y} = \frac{1}{(1 + K_c)\{[\tau/(1 + K_c)]s + 1\}}\frac{A}{s} \qquad\qquad 7.2\text{--}11$$

so that

$$y = \frac{A}{1 + K_c}(1 - e^{-t/\tau_c}) \qquad\qquad 7.2\text{--}12$$

where

$$\tau_c = \frac{\tau}{1 + K_c} \qquad\qquad 7.2\text{--}13$$

and is the effective time constant of the controlled process. This result is plotted in Figure 7.2–3.

We see from the graph, or the equation for the response, Eq. 7.2–12, that

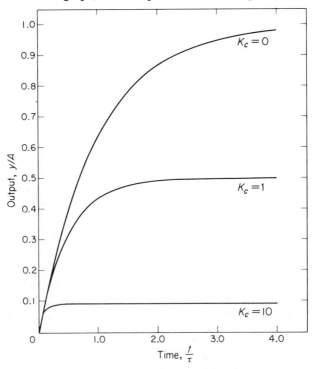

Figure 7.2-3. Step response of a first-order system.

at very long times the output approaches $y = A/(1 + K_c)$. (Of course, the same result can be obtained by applying the final value theorem to Eq. 7.2–11.) This means that if the controller gain is very high, the output will deviate only slightly from the desired value, y_0, instead of approaching a new steady state at $y = A$. In other words, the steady state error, or *offset*, decreases as K_c increases. Similarly, the effective time constant, Eq. 7.2–13, of the system is decreased, and therefore the system has a faster response. Even though this result seems to imply that we should make K_c infinitely large, we find that normally we run into difficulty at some finite value of K_c because the dynamics of the control system (the control valve, measuring instruments, etc.), which have been neglected in our discussion, become important at high controller gains. This point will be discussed in more detail later.

If we consider the response to an impulse disturbance, $\tilde{v} = A$, we find that

$$y = \frac{A}{\tau} e^{-t/\tau_c} \qquad \text{7.2–14}$$

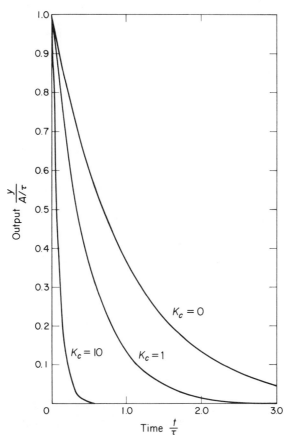

Figure 7.2-4. Impulse response of a first-order system.

so that the output deviation tends to disappear more rapidly than in the un-controlled case (see Figure 7.2–4). Similarly, the frequency response becomes

$$y = \frac{A}{(1 + K_c)\sqrt{1 + \tau_c^2 \omega^2}} \sin(\omega t + \phi) \qquad 7.2\text{–}15$$

where

$$\tan \phi = -\tau_c \omega \qquad 7.2\text{–}16$$

Thus, increasing the controller gain tends to decrease the amplitude of the output fluctuations and it decreases the phase lag (see the Bode plot given as Figure 7.2–5). Since an arbitrary periodic disturbance can be represented in terms of a Fourier series, we find that the controller will damp this disturbance. Also, we will show later that the tendency of the controller to decrease the phase lag is advantageous.

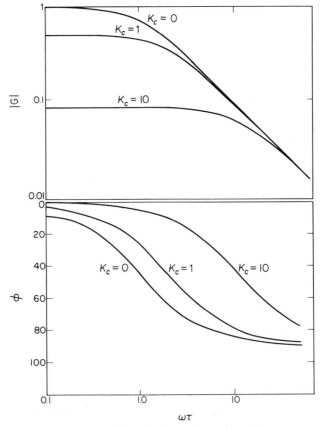

Figure 7.2-5. Bode plot for a first-order system.

These simple studies indicate that the proportional feedback control system improves the dynamic characteristics of the plant because it tends to damp out any disturbances that tend to force the process away from its optimum steady state design value. Again, this phenomenon occurs because the controller changes the effective time constant, or characteristic root, of the plant.

Second-Order Systems

The approach used to determine the effect of a feedback control system on a second-order plant is essentially the same as that described above. Although a number of block diagrams describe various kinds of second-order systems, we consider the particular case illustrated in Figure 7.2–6 for an overdamped plant. From the diagram we see that

$$\tilde{y} = \left(\frac{K_1}{\tau_1 s + 1}\right)\left(\frac{K_2}{\tau_2 s + 1}\right)\tilde{u} + \left(\frac{K_v}{\tau_1 s + 1}\right)\tilde{v} \qquad 7.2\text{--}17$$

and that

$$\tilde{u} = K_c(\tilde{y}_0 - \tilde{y}) \qquad 7.2\text{--}18$$

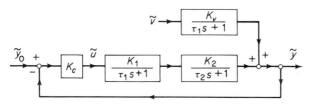

Figure 7.2-6. Feedback control of a second-order overdamped system.

Eliminating the control variable between these equations gives

$$\tilde{y} = \left(\frac{K_1}{\tau_1 s + 1}\right)\left(\frac{K_2}{\tau_2 s + 1}\right)K_c(\tilde{y}_0 - \tilde{y}) + \left(\frac{K_v}{\tau_1 s + 1}\right)\tilde{v}$$

or after some rearrangement

$$\tilde{y} = \frac{K_c K_1 K_2/(\tau_1 s + 1)(\tau_2 s + 1)}{\{1 + [K_c K_1 K_2/(\tau_1 s + 1)(\tau_2 s + 1)]\}}\tilde{y}_0$$

$$+ \frac{K_v/(\tau_1 s + 1)}{\{1 + [K_c K_1 K_2/(\tau_1 s + 1)(\tau_2 s + 1)]\}}\tilde{v} \qquad 7.2\text{--}19$$

Referring back to Eq. 7.2–6, we see that we could have obtained this same expression directly by applying our rule about block diagrams. For the case where we consider only load changes, so that $\tilde{y}_0 = 0$, our result becomes

$$\tilde{y} = \frac{K_v(\tau_2 s + 1)}{\tau_1 \tau_2 s^2 + (\tau_1 + \tau_2)s + 1 + K}$$

$$\tilde{y} = \left(\frac{K_v}{1 + K}\right)\left[\frac{\tau_2 s + 1}{[\tau_1 \tau_2/(1 + K)]s^2 + [(\tau_1 + \tau_2)/(1 + K)]s + 1}\right]\tilde{v} \qquad 7.2\text{--}20$$

which can be written in the standard form for a second-order system

$$\tilde{y} = \left(\frac{K_v}{1+K}\right)\left(\frac{\tau_2 s + 1}{\tau s^2 + 2\gamma\tau s + 1}\right)\tilde{v} \qquad 7.2\text{-}21$$

where

$$\tau = \sqrt{\frac{\tau_1 \tau_2}{1+K}}, \quad \gamma = \left(\frac{\tau_1 + \tau_2}{2\sqrt{\tau_1 \tau_2}}\right)\left(\frac{1}{1+K}\right)^{1/2}, \quad K = K_c K_1 K_2 \qquad 7.2\text{-}22$$

Now it is clear that the damping coefficient of the controlled plant, γ, decreases as the controller gain K_c (and therefore K) is increased. Hence the originally overdamped plant can be made to have an underdamped response by using a sufficiently high controller gain. Assuming this to be the case—that is, $\gamma < 1$ —the step response of the plant, $\tilde{v} = A/s$, becomes

$$y(t) = \frac{AK_v}{1+K}\left[1 + \frac{[1 - (2\gamma\tau_2/\tau) + (\tau_2^2/\tau^2)]^{1/2}}{\sqrt{1-\gamma^2}}e^{-\gamma t/\tau}\sin\left(\sqrt{1-\gamma^2}\,\frac{t}{\tau} + \phi\right)\right]$$

$$7.2\text{-}23$$

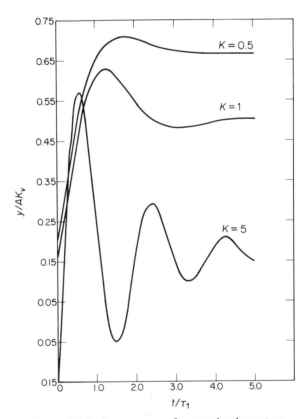

Figure 7.2-7. Step response of a second-order system.

where

$$\phi = \tan^{-1}\left(\frac{\sqrt{1-\gamma^2}\,\tau_2/\tau}{1-(\gamma\tau_2/\tau)}\right) - \tan^{-1}\left(\frac{\sqrt{1-\gamma^2}}{-\gamma}\right) \qquad 7.2\text{-}24$$

This result makes it clear that large values of the controller gain will give a small offset, $AK_v/(1+K)$. However, at the same time the damping coefficient, γ, decreases, the natural frequency of the system, $\omega_n = 1/\tau$, increases, so that the system responds faster but the response is more oscillatory. In fact, as K_c approaches infinity, the offset approaches zero, but the system becomes equivalent to a harmonic oscillator with a very high frequency. Some typical response curves for various values of K_c are shown in Figure 7.2–7.

A similar approach can be used to determine the impulse and frequency response. From our previous discussion in Chapter 4 we know that we want to achieve an underdamped system without too much oscillation. A technique for establishing the value of the controller gain that corresponds to the most desirable response will be presented in Section 7.5.

Third-Order Systems

The procedure just discussed can easily be extended to higher-order systems. Little would be gained by this effort unless some new kind of behavior is encountered. However, we do indeed find that third-order systems with a proportional controller in the feedback loop can exhibit very different characteristics than a first- or second-order plant, and therefore we must consider this case. Fortunately, it is not necessary to go beyond this point.

As a specific example, we consider the plant shown in Figure 7.2–8. The transfer function relating the output to the disturbances is

$$\frac{\tilde{y}}{\tilde{v}} = \frac{G_n}{1+G} = \frac{K_v/(\tau_1 s + 1)}{\{1 + [K_c K_1 K_2 K_3/(\tau_1 s + 1)(\tau_2 s + 1)(\tau_3 s + 1)]\}}$$

$$= \frac{K_v(\tau_2 s + 1)(\tau_3 s + 1)}{\tau_1\tau_2\tau_3 s^3 + (\tau_1\tau_2 + \tau_1\tau_3 + \tau_2\tau_3)s^2 + (\tau_1 + \tau_2 + \tau_3)s + 1 + K}$$

$$7.2\text{-}25$$

where

$$K = K_c K_1 K_2 K_3 \qquad 7.2\text{-}26$$

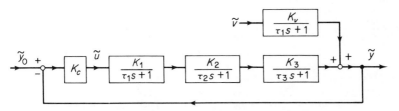

Figure 7.2-8. Feedback control of a third-order system.

In order to make further progress, it is necessary to find the roots of the third-order polynomial in the denominator. It is always possible to write this as the product of a real factor and a quadratic factor; then, depending on the system parameters, the quadratic can be written as the product of two real roots or a pair of complex conjugate roots. Thus we let

$$\tau_1\tau_2\tau_3 s^3 + (\tau_1\tau_2 + \tau_1\tau_3 + \tau_2\tau_3)s^2 + (\tau_1 + \tau_2 + \tau_3)s + 1 + K$$
$$= (\tau_a s + 1)(\tau_b^2 s^2 + 2\gamma\tau_b s + 1)(1 + K) \qquad 7.2\text{-}27$$

where the constants τ_a, τ_b, and γ can be obtained by expanding the right-hand side of the preceding expression and comparing coefficients.

At very low values of the controller gain, K_c, and therefore low values of K, we expect to obtain two real roots for the quadratic expression on the right-hand side of this equation (see Table 7.1–1 for some typical values). This means that there would be three real roots, which would be similar to the system shown in Figure 7.2–8 without the feedback loop. We find that as we increase K_c, the value of τ_a decreases, τ_b decreases, and γ decreases. Of course, as soon as γ becomes less than unity, the controlled plant resembles a first-order system and an underdamped second-order system in series. This is the kind of behavior we might expect from our previous discussion of the effect of a proportional feedback controller on a second-order overdamped plant.

However, for the third-order plant, we find that as we continue to increase K_c, or K, eventually we come to a point where γ becomes equal to zero, and after that it takes on negative values. In other words, the roots of the quadratic expression change from complex conjugates with negative real parts, to pure imaginary, and then to complex conjugates with positive real parts. Referring back to Section 4.5 and solving for the impulse response, say, for a case where $-1 < \gamma < 0$, we find that the output is

$$y(t) = \frac{A}{\tau_b} \frac{1}{\sqrt{1 - \gamma^2}} e^{\gamma t/\tau_b} \sin \frac{\sqrt{1 - \gamma^2}\, t}{\tau_b} \qquad 7.2\text{-}28$$

Because of the presence of the positive exponential term, the output will tend to move away from the desired steady state operating point. Thus for a pro-

TABLE 7.2–1

VARIATION IN SYSTEM TIME CONSTANTS

If $\tau_1 = 1$, $\tau_2 = 2$, and $\tau_3 = 5$ in Eq. 7.2-27, we find that

K	τ_a	τ_b	γ
0	1.000	3.162	1.107
1.0	0.863	2.407	0.652
5.0	0.697	1.546	0.206
10.0	0.616	1.215	0.046
12.6	0.588	1.118	0
15.0	0.568	1.049	−0.032

portional feedback control of a third-order plant, it is possible to increase the controller gain to a large enough value that the original desired operating point becomes locally unstable. The transition between stable and unstable operation occurs when the damping coefficient in Eq. 7.2-27 is equal to zero. By expanding the right-hand side of this equation and comparing coefficients, it is possible to show that $\gamma = 0$ when

$$K_{\max} = (K_1 K_2 K_3) K_{c\,\max} = \left(1 + \frac{\tau_2}{\tau_1} + \frac{\tau_3}{\tau_1}\right)\left(1 + \frac{1}{\tau_2/\tau_1} + \frac{1}{\tau_3/\tau_1}\right) - 1$$

7.2-29

Also, the frequency of the oscillations generated by the system when $\gamma = 0$ is

$$\omega_c = \sqrt{\frac{1 + K_{\max}}{\tau_1(\tau_2 + \tau_3) + \tau_2\tau_3}}$$

7.2-30

From Eq. 7.2-29 it is apparent that the maximum allowable controller gain for stable operation of the plant will be very large if any one of the three time constants is either very large or very small compared with the others. This result provides the reason why we cannot actually allow the controller gain for the first- and second-order systems we considered earlier to approach infinity. For any real process, the measuring instrument, signal transmitter, control valve, etc., will have certain time constants, so that every system is at least third-order. Whenever the time constants of this auxiliary equipment are very small in comparison with those of the main process, we can neglect them if we are merely studying the dynamic response. However, they are not negligible in the design of a closed loop control system and will impose an upper limit, although possibly a very large one, on the maximum controller gain.

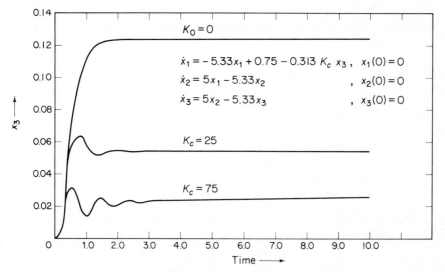

Figure 7.2-9. Step response of a third-order system.

An illustration of the step response of a third-order system is shown in Figure 7.2–9 for various controller gains. The offset decreases, but the system becomes more oscillatory, as the controller gain is increased until it eventually becomes unstable. Thus, the controller gives us an improved response for a certain region of controller gains, but then becomes a definite liability rather than an asset.

It would be possible to develop the system transfer functions and analytical expressions for the system response for a great number of other examples of the proportional control of linear systems. However, the procedure required for this type of analysis should be clear from the discussion above. In general, we find that: the controller changes the characteristic roots of the plant, as the controller gain is increased the offset decreases and the plant resembles an underdamped system, and that excessive controller settings will make the desired steady state operating condition become unstable. Quantitative procedures for the design of proportional, and other, control systems will be described in detail later in this chapter.

Integral Control

Integral control action means that the value of the control variable is set equal to a constant—the integral gain—multiplied by the integral of the error signal[1]

$$u = K_c K_i \int (y_0 - y)\, dt = \frac{K_c}{T_i} \int (y_0 - y)\, dt \qquad 7.2\text{-}31$$

By differentiating this expression, we find that the time derivative of the control variable is proportional to the error; consequently, integral control is sometimes called *proportional speed floating control*. As a particular example

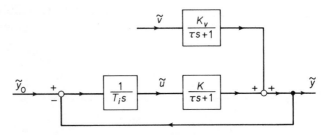

Figure 7.2-10. Integral control of a first-order plant.

[1] Because an integral controller generally contains a proportional mode, the control law is written as

$$u = K_c \left[(y_0 - y) + \frac{1}{T_i} \int (y_0 - y)\, dt \right]$$

so that the product K_c/T_i appears.

of the effect of an integral controller on the system response, we consider the first-order process shown in Figure 7.2–10. The system transfer function is

$$\frac{\tilde{y}}{\tilde{v}} = \frac{G_v}{1 + G} = \frac{K_v/(\tau s + 1)}{1 + (1/T_i s)[K/(\tau s + 1)]}$$

$$= \left(\frac{K_v T_i}{K}\right) \frac{s}{[(T_i \tau/K)s^2 + (T_i/K)s + 1]} \qquad 7.2\text{-}32$$

The denominator of this transfer function is a second-order polynomial; therefore the dynamic response will depend on two characteristic roots. In other words, one effect of adding an integral controller to a process is to increase the order of the system.

Once the input disturbance $v(t)$ has been specified, the exact response can be obtained via the methods described in Chapter 4. For example, for a step response, $\tilde{v} = A/s$, we find that

$$y = \frac{A K_v T_i}{K}\left[\frac{\omega}{\sqrt{1 - \gamma^2}} e^{-\gamma \omega t} \sin(\omega \sqrt{1 - \gamma^2}\, t)\right] \qquad 7.2\text{-}33$$

where

$$\omega = \sqrt{K/T_i \tau} \quad \text{and} \quad \gamma = \tfrac{1}{2}\sqrt{T_i/\tau K} \qquad 7.2\text{-}34$$

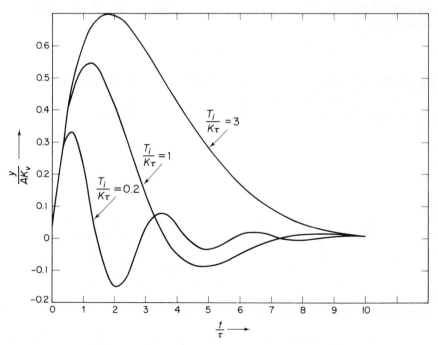

Figure 7.2-11. Step response of a first-order plant with integral control.

This result is similar to that of a second-order system except that there is no offset. As we decrease the value of T_i (often called the *integral time;* $1/T_i$ is called the *reset rate*), which is equivalent to increasing the gain of the integral controller (see Eq. 7.2–31), we note that the damping coefficient decreases and the frequency of the oscillations increases. Some illustrations of the plant response are shown in Figure 7.2–11.

The fact that integral control action removes offset, even for step inputs, is valid in general, as can be shown either by solving for the system output or by applying the final value theorem to the system transfer function (see Section 4.4). For this reason integral control is referred to as *reset action;* that is, it is not necessary to reset the set point to remove offset after a step input, as it is with proportional control. However, the integral mode increases the order of the plant, which tends to make it more unstable (i.e., a second-order plant with integral control is equivalent to a third-order system, which becomes unstable at some finite integral time). It is common practice to combine integral control with a proportional mode.

Derivative Control

Although derivative action is seldom used without a proportional mode, for simplicity we consider the plant shown in Figure 7.2–12. We define derivative control as

$$u = K_D \frac{d(y_0 - y)}{dt} \qquad 7.2\text{–}35$$

and from the block diagram we find that the system transfer function is

$$\frac{\tilde{y}}{\tilde{v}} = \frac{G_v}{1 + G} = \frac{K_v/(\tau s + 1)}{1 + K_D s[K/(\tau s + 1)]} = \frac{K_v}{(K_D K + \tau)s + 1} \qquad 7.2\text{–}36$$

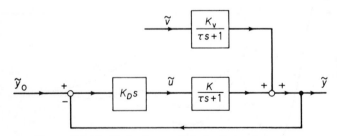

Figure 7.2-12. Derivative control of a first-order system.

Then the response to a step input, $\tilde{v} = A/s$, is

$$y(t) = AK_v\left[1 - \exp\left(-\frac{t}{K_D K + \tau}\right)\right] \qquad 7.2\text{–}37$$

From this result it is clear that in contrast to proportional control, derivative action does not affect the final steady state the system approaches; that is, it does not reduce the offset. Also, as the controller gain K_D is increased, the effective time constant for the plant, $K_D K + \tau$, increases. This means that the response of the controlled system is slower than that of the original process.

At first the fact that derivative control slows down the system dynamics seems to imply that it is not advantageous. However, this kind of behavior means that the system becomes more stable. By examining more complicated plants, it is possible to show that the addition of a derivative mode always introduces a stabilizing effect, thereby making it possible to use higher proportional gains and/or lower reset times (with a corresponding decrease in offset) before the system becomes unstable.

Simulation

In the preceding discussion we considered the effect of various kinds of controllers on very simple linear systems. We expect that a linear analysis of this type should be valid in some sufficiently small neighborhood of a steady state operating point and that it should be possible to estimate the size of this region by applying the techniques described in Chapter 6. Of course, even the perturbation approach only gives results that are meaningful in a local neighborhood of the desired steady state, and in some cases this information will prove inadequate for a complete understanding of the problem. As an example of this type, we again return to our nonisothermal, stirred-tank reactor problem and review some results published by Aris and Amundson.[2] The system parameters are given in Table 7.2–2. Despite the fact that they do not correspond to a case where there is an optimum, economic reactor design, because the reactor contains a cooling coil instead of a heating unit, they do provide an interesting illustration of the effect of a proportional feedback controller on the number and stability of the steady state solutions of a set of nonlinear equations.

TABLE 7.2–2

SYSTEM PARAMETERS*

$$\frac{C_p \rho}{(-\Delta H)A_f} = \frac{1}{200\,°K}\,, \quad \frac{C_p \rho T_f}{(-\Delta H)A_f} = 1.75, \quad \frac{C_p \rho T_c}{(-\Delta H)A_f} = 1.75$$

$$\frac{UA_c Kq_c}{q(-\Delta H)A_f(1 + Kq_c)} = 1.0, \quad \frac{k_0 V}{q} = e^{25}, \quad \frac{EC_p \rho}{R(-\Delta H)A_f} = 50$$

$$V = 50 \text{ liters}, \ q = 50 \text{ liters/min}, \ E = 20,000 \text{ cal/mole}.$$

* Reproduced from R. Aris and N. R. Amundson, *Chem. Eng. Sci.* **7**, 132 (1958), by permission.

As a first step, we substitute the system parameters into Eq. 7.1–2 and plot the right- and left-hand sides of this equation against temperature. The results,

[2] See footnote 2, p. 7.

shown in Figure 7.2–13, indicate that there are three possible steady state solutions, and we assume that the intermediate solution is the desired one. Evaluating the characteristic roots of the linearized equations at each of the steady state solutions, Eq. 3.3–63, we find that the two extreme points are stable (two negative real roots are obtained for each point) and that the desired operating condition is unstable (a pair of real roots having opposite signs is obtained). Now we can use the analytical solutions of the linearized system equations,

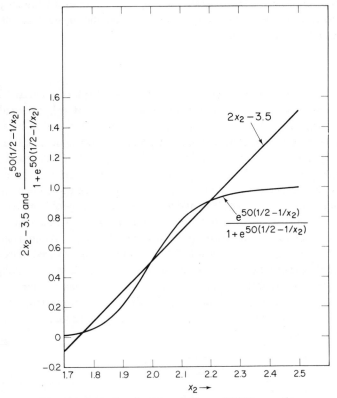

Figure 7.2-13. Steady state solutions of CSTR equations.

with the roots calculated above, to establish the dynamic behavior in the immediate vicinity of the steady state solutions. However, this information still does not give us a clear picture of the response characteristics of the reactor. Therefore Aris and Amundson solved the nonlinear dynamic equations on an analog computer. This kind of an approach, where numerical solutions of the system equations are generated with either an analong or a digital computer, is often called *simulation* because the computer outputs should resemble the output of the real physical system, at least within the limitations of the assumptions used in the model.

Aris and Amundson's results are shown in Figure 7.2–14 for a particular initial concentration and temperature. Instead of generating a large number of curves of this type for various initial conditions and then attempting to understand their significance, it is simpler to plot the corresponding values of composition and temperature for each solution as a single trajectory on a *phase plane* (see Figure 7.2–15). On this graph, the trajectory DA is equivalent to the two graphs in Figure 7.2–14. It would be possible to mark time increments as a parameter on the trajectory, but doing so tends to make the graph rather cluttered and thus obscures the behavior of the system.

Figure 7.2–15 provides a clear picture of the dynamic characteristics of the reactor. Starting with any combination of compositions and temperatures to

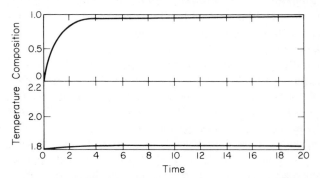

Figure 7.2-14. Reactor response. [Reproduced from R. Aris and N. R. Amundson, *Chem. Eng. Sci.,* **7,** 132 (1958), by permission of Pergamon Press.]

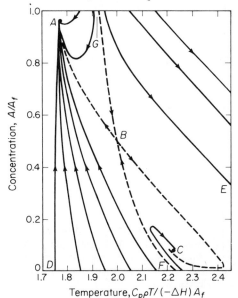

Figure 7.2-15. Reactor phase plane. [Reproduced from R. Aris. and N. R. Amundson, *Chem. Eng. Sci.* **7,** 121 (1958), by permission of Pergamon Press.]

the left of the curve *FG* (called a separatrix because it separates the phase plane into two distinct regions), we always approach the low conversion steady state, *A*. Alternately, any starting conditions to the right of *FG* lead us to the high conversion steady state, *C*. However, if we attempt to start up the reactor with pure feed and a temperature greater than that at *G*, we find that the temperature rises to very high values (out of the range of our graph) before the reactor returns to the steady state at *C*. Thus this kind of a start-up procedure might be disadvantageous, depending on the melting point of the reactor, the possibility of side reactions at high temperatures, or the rate of temperature rise (since severe thermal shocks might lead to a reactor failure).

The only possibility of arriving at the steady state *B* is if we carefully adjusted the initial composition and temperature to lie along one of the paths *FB* or *GB*. In terms of our analytical solution for this case (where there are two real roots having opposite signs), we would have to adjust the initial conditions so that the coefficient of the positive exponential factor was equal to zero. This procedure is impractical, however, because any small error will take us off the path *FB* or *GB*, and eventually we would go either to the steady state *A* or to *C*. Similarly, if we are operating at point *C* and a disturbance enters the reactor which takes us to the region to the left of the separatrix *FG*, the system will proceed to the steady state *A* rather than return to point *C*. Thus we see that even though points *A* and *C* are locally stable—that is, after a small disturbance the reactor will return to steady state—they are not globally stable.

If we again assume that point *B* is the desired steady state operating condition, we need to find some way to make this point stable. From our linear

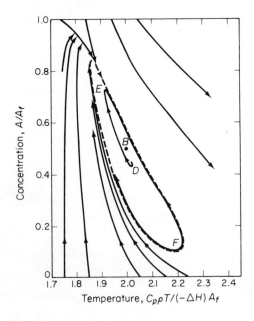

Figure 7.2-16. Phase plane for $K_c = 6$. [Reproduced from R. Aris and N. R. Amundson, *Chem. Eng. Sci.*, **7**, 132 (1958), by permission of Pergamon Press.]

analysis in Section 7.1 we know that installing a feedback controller to manipulate the coolant flow rate proportional to the deviation of the reactor temperature from its desired steady state value will at least make it locally stable. But in order to ascertain the global behavior, we must solve the nonlinear dynamic equations simultaneously with the expression describing the control action, Eq. 7.1–11, for the case where we are cooling rather than heating.

Aris and Amundson present a detailed description of the behavior of the system as the controller gain is increased. Their results indicate that at $K_c = 2$ the steady state solution at C is unstable, so that all trajectories terminate at the remaining stable, steady state point at A. A similar type of behavior is observed until we increase the gain to a value $K_c = 5.9$. At this condition both the equilibrium points at A and C disappear, and the characteristic roots corresponding to point B are complex conjugates with positive real parts. Since point B is unstable, and yet the trajectories have to go somewhere, we find that the reactor outputs become periodic (see Figures 7.2–16 and 7.2–17). In other

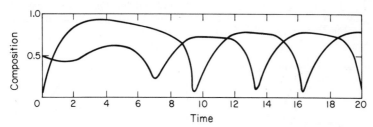

Figure 7.2-17. Reactor response for $K_c = 7$. [Reproduced from R. Aris and N. R. Amundson, *Chem. Eng. Sci.*, **7**, 132 (1958), by permission of Pergamon Press.]

words, the reactor generates oscillations, even though the inputs are maintained constant. We call this kind of a system a chemical oscillator and will discuss its behavior in considerable detail in Chapter 10.

As the controller gain is increased further, the size of the oscillations becomes smaller, until at about $K_c = 9$ the steady state solution at point B becomes stable; that is, the characteristic roots are complex conjugates with negative real parts. Since B is the only equilibrium point remaining, the reactor becomes globally stable (see Figures 7.2–18 and 7.2–19). Even though setting $K_c = 10$ means that we will be able to operate at the desired design condition, point B, Figures 7.2–18 and 7.2–20 illustrate that the dynamic response of the reactor is too oscillatory, so that it will take too long for disturbances to die out. By increasing the gain to $K_c = 30$ (see Figures 7.2–19 and 7.2–20), we obtain a better dynamic performance.

As we look back over the example just discussed, a number of important points become clear. First, we recognize that an optimum, economic, steady state design will not necessarily correspond to a stable operating condition.

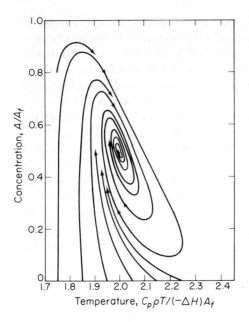

Figure 7.2-18. Phase plane for $K_c = 10$. [Reproduced from R. Aris and N. R. Amundson, *Chem. Eng. Sci.*, **7**, 132 (1958), by permission of Pergamon Press.]

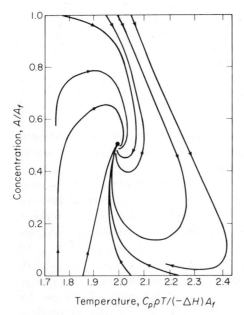

Figure 7.2-19. Phase plane for $K_c = 30$. [Reproduced from R. Aris and N. R. Amundson, *Chem. Eng. Sci.*, **7**, 132 (1958), by permission of Pergamon Press.]

Second, even if a linear analysis indicates that a desired steady state operating point is stable, it is not necessarily stable for large disturbances—for example, point C with $K_c = 0$ in Figure 7.2–15. Third, a feedback control system can be used to change both the number and the stability characteristics of the steady state operating points. Fourth, a linearized dynamic model of a plant, with or

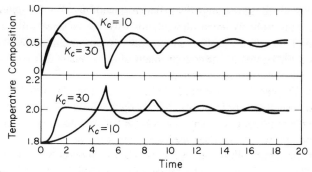

Figure 7.2-20. Reactor response. [Reproduced from R. Aris and N. R. Amundson, *Chem. Eng. Sci.*, **7**, 132, (1958), by permission of Pergamon Press.]

without a controller, will provide information about the dynamic response in some small neighborhood of a steady state, but it does not always give an adequate understanding of the system behavior (e.g., when a limit cycle is generated around a single, unstable equilibrium point). Fifth, sometimes it is necessary to generate an extensive number of numerical solutions of the system equations in order to gain a clear picture of the dynamic characteristics of even a simple nonlinear plant.

Despite the obvious difficulties that may be encountered in the design of a control system, we must remember that empirical evidence indicates that most chemical processes are stable, overdamped systems. Then, intuitively, we expect that the linearized dynamic model will provide an adequate description of the plant over an operating range wide enough for engineering purposes. In any event, the linear analysis will allow a first estimate of the design of a controller, which we can check later by using simulation techniques. Before proceeding with a discussion of design methods, it seems worthwhile to define carefully the control problem we are considering and to relate it to our earlier discussion of optimum steady state control.

The Problem of Designing a Regulating Control System

Whenever an optimum economic design turns out to be unstable, but we can make the system stable by adding a controller, it is clear that this procedure is advantageous. Similarly, for cases where the design condition is locally stable, but not globally stable, the installation of a controller might make the difference between satisfactory operation and disaster. Consequently, it is simple to see the virtues of control systems from this stability point of view. However, the value of using feedback control systems to improve the performance of stable linear plants, such as those discussed in the early part of this section, requires a more careful evaluation. Although it is true that the controller changes the characteristic roots, generally speeds up the response of the process, and decreases the effect that a step disturbance has on driving the plant to a new steady state condition, we must understand why we consider this behavior an asset. In other words, we need to state explicitly the implicit assumptions used to formulate this kind of control problem and to relate this problem to our previous discussion of process control.

Back in Chapter 2 we described the procedure for obtaining the optimum steady state design of simple plants. We recognized that some of the inputs, the disturbances, would change with time and that it was necessary to adjust some of the other inputs, the control variables, in order to achieve the maximum profit from the plant. Thus the optimum operating conditions varied with time as the disturbances changed. This procedure was called optimum steady state control. It was limited to situations where the dynamic features of the plant were negligible and where all the disturbances could be measured (or at least calculated, using the steady state equations describing the system and measured values of some of the process outputs). In cases where the dynamic characteristics of the plant are significant, which we can estimate by means of the techniques described in Chapters 4 through 6, together with some experimental tests, we must develop a new approach.

Perhaps the most obvious way of extending the optimum steady state control method to dynamic systems is simply to say that we should replace the steady state system equations by their dynamic counterparts, replace the instantaneous profit or total cost expression by the time average value of this equation over some long operating interval, assume that we can instantaneously measure all the disturbances entering the plant, and then look for the way in which we should adjust the control variables so that we maximize the time average profit (or minimize the cost). After reflecting on this approach for a few minutes, we realize that actually it would be better if we knew the behavior of the disturbances over the entire operating interval at the outset of the problem. The reason is that, in some cases, it seems plausible that we should start to manipulate the control variable even before the disturbance enters the plant—for example, if we are attempting to correct for a single step input. Of course, this means that we must know the future behavior of all disturbances. Methods for making quantitative predictions of the future must be considered as beyond the scope of this text. In fact, normally it is not practical even to attempt to measure all the disturbances, such as the feed composition to a catalytic cracking unit which probably contains thousands of chemical compounds. In addition, our control problem formulation requires that we determine the manner in which we should manipulate the control variables over the whole operating interval of interest. In other words, we are looking for functions of time, rather than particular values of variables, that maximize the time average profit. This kind of problem requires application of the calculus of variations, instead of classical calculus, and up until a few years ago this approach was beyond the mathematical capability of most engineers.

Because of the difficulties associated with the optimal dynamic control problem discussed above and because of the desire, and in some cases the need, for developing automatic control systems that could be used to improve the performance of a process, it was necessary to formulate a simpler control problem. By recognizing that the disturbances are expected to fluctuate around some time average value, which is the value used to determine the optimum steady state design, we might expect that on a time average basis it would be advantageous always to operate as close as possible to the optimum steady state design condition. Although this assumption is not always correct (as we shall see in Chapter 10) and is different from the approach taken in the optimum steady state control problem, it provides the basis for classical control theory, which is often called *servomechanism theory*.

If we implement this approach in a feedback manner, essentially we can ignore the disturbances. Thus whenever the output deviates from its optimum steady state value, because of a disturbance or any other reason, we change the control variables in such a way as to try to make it return to the desired value in some appropriate manner. A control system of this type, one designed to maintain the output constant, is called a *regulating control system*. It is the only kind of control system that will be considered in this book. In the present chapter we study single-input, single-output plants; in Chapter 8 we attempt to extend the results to multivariable controllers; in Chapter 9 we use the calculus of variations to develop optimal regulating controllers; and then in Chapter 10 we look at some cases where the basic assumption that a regulating controller is the best is not valid. Despite this emphasis on regulator design, the reader should be aware that another class of control problem exists.

Servomechanism Control Problems

There are a great number of physical systems where no attempt is made to maintain the system outputs constant. For example, if we are attempting to aim an antiaircraft gun at a plane, we want the position of the gun to follow the path of the plane. Similarly, if we are designing a machine to cut out metal parts, we want the cutter to follow a given pattern. In other words, we want the output of the system to follow a known path rather than remain constant. The control systems discussed earlier in this section can be used to improve the way the system follows this desired path by feeding the known information into the set-point position of the controller. The controller then compares the actual output with the desired output and makes corrections whenever a deviation is observed. This kind of controller is usually called a *servomechanism system*. Thus the only difference between regulating and servomechanism control systems is that in the first case we maintain the set point constant and try to compensate for disturbances, which are often called *load changes*, whereas in the latter case we concentrate on set-point variations.

Servomechanism problems are important in chemical engineering for the design of control systems for batch processes. Any batch system always operates in a dynamic manner, and there is no optimum steady state design. However, for known inputs, there is an economic, optimum operating period, which includes the actual operating interval, dumping time, and refilling time. Again, some of the inputs, such as feed composition, will change with time, thereby affecting the conditions during and at the end of the operating period. Therefore we can change some of the other inputs, the control variables, in an attempt always to obtain the original design conditions in the batch system as they change with time; that is, we try to force the process to follow the known design path without considering what disturbances might cause it to deviate from that path. Since only minor modifications of the theory are required in order to design servomechanism control systems, they will not be considered further.

SECTION 7.3 ELEMENTS IN THE CONTROL LOOP

In our previous discussion of control systems, we assumed that somehow we could measure one or more system outputs of interest, employ some kind of a device to compare these measured values with the desired outputs, and, finally, use this information about the error to adjust one or more of the system inputs, the control variables, in order to compensate for the error. It hardly seems advisable to develop any kind of procedure for the design of control systems based on such a vague set of assumptions. Consequently, before proceeding with a discussion of design methods, we will describe some typical pieces of hardware that enable us to accomplish the tasks under consideration. However, since the amount of equipment available is vast and since new kinds of devices are rapidly being developed, no attempt will be made to treat the topic in detail. All we hope to accomplish is to show that it is possible to satisfy our assumptions for one simple system and to indicate the kinds of limitations that might be introduced when we attempt to implement the theory. A host of additional information is available, and the interested reader is encouraged to consult the literature.

System Under Consideration

A sizable portion of our discussion of control systems has been directed to the control of a nonisothermal stirred-tank reactor where we decided to measure the reactor temperature, compare the temperature with the desired optimum steady state design value, and use this information to adjust the flow rate of heating (or cooling) fluid. Hence the actual means for implementing this procedure will be considered. Of course, as pointed out in Section 7.1, we could adjust the heating-fluid flow rate based on the deviation between the measured composition and the desired value, or we could adjust the feed rate to the reactor as some function of either the composition or temperature error. We expect that it will be easier to measure temperature than composition and that manipulating the heating-fluid flow rate will provide adequate control without having drastic effects on the production rate, so that our proposed system seems to be the most reasonable to consider first. However, a case-study approach might be warranted, and the final decision should be based on the quality of control obtained for a particular cost of hardware.

Measuring and Transmitting Device

Every engineer is familiar with a number of possible ways of measuring temperature. However, for our purpose, we want to be able to compare the measured value with some specified value automatically, and to generate a signal that depends on this difference. This means that an ordinary thermometer will not be adequate unless we can find some way of modifying its design so that it will transmit a signal that is related to the normal expansion of the liquid in the bulb. One method for including a transmitting device is to place the liquid under a high pressure, in the range from 400 to 1200 psi, and have the top of the bulb connected by a piece of fine, armored tubing to a bellows (see Figure 7.3–1). Changes in liquid level are thus converted to pressure

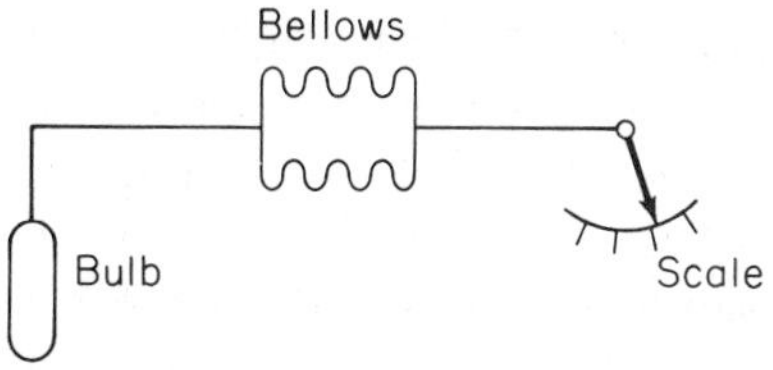

Figure 7.3-1. A temperature recording device.

variations, which are hydraulically transmitted to a receiving device. For a temperature recorder, the transmission line may be up to 200 ft long, and the receiver is usually a Bourdon tube or helix that converts a pressure change into a mechanical displacement. Typical bulb sizes are in the range of $\frac{1}{2}$ in. in diameter and 4 or 5 in. long.

It is also possible to use bulbs filled with gases, such as nitrogen. However, larger bulbs are normally required with this approach; as a result, the dynamic response of the device is slower. Another alternative is to use a bulb half-filled with a pure liquid in equilibrium with its own vapor, which fills the remainder of the bulb. The pressure in the vapor space, and transmission line, again provides an accurate measure of the bulb temperature. This type of instrument is less sensitive to ambient temperature changes and has a faster response than either the liquid- or gas-filled bulbs, but the pressure output is a nonlinear function of the measured temperature.

Although the temperature-measuring instruments described produce a pressure signal, which is required for a pneumatic controller, it is also possible to obtain electrical signals. The most common device of this type, and the most widely used thermometer in industry, is the thermocouple. It consists of two wires of different metals joined together at the ends; the junction is placed at the position where the measurement is to be made. When the opposite ends are connected, and providing that the temperature at this point is different from the measuring location, a current proportional to the net heat exchange at the two junctions will flow in the circuit. Since the signal transmission in this case is electronic, the transmission dynamics are negligible. This is in contrast to the previous case of expansion thermometers, where it is often necessary to use boosters or amplifying pilots. Obviously thermocouples are more compatible for use with an electronic controller rather than a pneumatic instrument. In fact, in the latter case, we would have to include another device, called a transducer, to convert the electrical signal into a pressure signal.

Controllers

Up until the last few years, almost all the controllers used in industry were pneumatic. They were inexpensive, rugged, maintenance free, safe to operate in explosive atmospheres, and relatively simple to connect to other devices. However, as high-speed digital computers became readily available and interest developed in using them as on-line computing and controlling devices, more effort was directed toward the improvement of electronic controllers and electronically activated control valves. At the present time, the two types of systems are economically competitive, and the ease of transmitting and manipulating electrical signals is roughly balanced by the difficulty of designing an electronic valve actuator. Since our main interest here is to show one possible technique for implementing our control philosophy with a typical set of hardware components, we will limit our attention to pneumatic systems.

A schematic diagram of a proportional controller is given in Figure 7.3–2. We have replaced the recording device on the spring-loaded receiving bellows by some hinged levers, which are connected to an adjustable set-point dial.

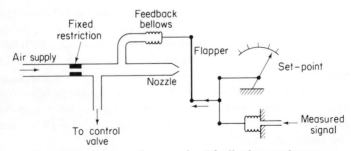

Figure 7.3-2. Pneumatic proportional feedback control system.

Thus whenever either the pressure corresponding to the measured signal or the set-point adjustment is changed, the bottom of the flapper will be moved, to the left or to the right, in a manner proportional to the error between the desired and measured signals. The proportionality constant depends on the bellows characteristics, the lever advantages, and so forth. Clean air is supplied at 20 psig, flows through a fixed restriction, then through a nozzle, and against the face of the flapper. As the error signal moves the flapper to the left (right), the pressure in the nozzle increases (decreases) because of the presence of the fixed restriction in the nozzle and the variable area for flow out through the tip. Generally the design is such that the flapper only moves one- or two-thousandths of an inch, and this corresponds to a change of from 3 to 15 psig in the nozzle. This means that the error signal is amplified by a factor of about 10,000 psi/in. Such a tremendously high amplification factor makes the nozzle-flapper combination very sensitive; therefore a feedback bellows is installed to obtain better operation. With the feedback bellows, an increase (decrease) of the pressure in the nozzle tends to move the top of the flapper to the right (left), which partially compensates for the original change caused by the error signal.

In order to show that this configuration does lead to a nozzle pressure that is proportional to the error signal, so that the system acts as a proportional controller, we consider the sketch given in Figure 7.3-3. We assume that the

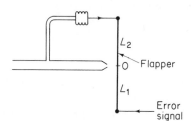

Figure 7.3-3. Proportional controller.

controller dynamics are negligible in comparison with those of the process, which means that a steady state analysis will give us a valid description of the system performance. For a case where the error signal causes the lower end of the flapper to move to the left a distance $-k_1 e$, which causes the pressure in the nozzle to increase by an amount ΔP, the feedback bellows will force the top of the flapper to the right a distance $\Delta P A_b/k_2$, where A_b is the cross-sectional area of the end of the bellows and k_2 is the spring constant of the bellows. Thus the net displacement to the left of point 0 on the flapper, $-\Delta L$, is

$$\Delta L = -\left(\frac{L_2}{L_1 + L_2}\right)k_1 e + \left(\frac{L_1}{L_1 + L_2}\right)\frac{\Delta P A_b}{k_2} \qquad 7.3\text{-}1$$

Now if we fix the design so that only small displacements of the flapper are allowed, we can assume that the relationship between the increase in the nozzle pressure and the flapper displacement is linear

$$\Delta P = -k_3 \, \Delta L \qquad\qquad 7.3\text{-}2$$

where k_3 is the proportionality constant and the negative sign indicates that motion to the left corresponds to a pressure increase. Using this expression to eliminate ΔL from Eq. 7.3-1 and rearranging the result, we obtain

$$\Delta P = \left\{ \frac{[L_2/(L_1 + L_2)]k_1 k_3}{1 + [L_1/(L_1 + L_2)] \, A_b k_3/k_2} \right\} e \qquad\qquad 7.3\text{-}3$$

which shows that the pressure change is proportional to the error signal.

In addition, the preceding relationship makes it possible to demonstrate the advantages of adding the feedback bellows to the controller. We expect that the nozzle-flapper combination will have a very high amplification factor, which means that the value of k_3 in Eq. 7.3-2 is very large. With large values of k_3 and the proper design of the lever arms and feedback bellows, we can ensure that

$$\left(\frac{L_1}{L_1 + L_2} \right) \frac{A_b k_3}{k_2} \gg 1 \qquad\qquad 7.3\text{-}4$$

However, when this is true, Eq. 7.3-3 reduces to the simpler form

$$\Delta P = \left(\frac{L_2}{L_1} \right) \left(\frac{k_1 k_2}{A_b} \right) e \qquad\qquad 7.3\text{-}5$$

and clearly this result is independent of the amplification factor. Thus we see that the feedback bellows arrangement decreases the sensitivity of the system and should give a better performance.

Control Valve

As illustrated in Figure 7.3-2, the output of a pneumatic, proportional controller is a pressure signal in the range of 3 to 15 psig. Now we want to find a way to use this signal to change the flow rate of one of the system inputs. A device that can be used for this purpose is shown in Figure 7.3-4.[1] Here an

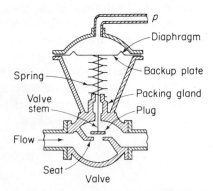

Figure 7.3-4. Diaphragm valve. (Reproduced from E. F. Johnson, *Automatic Process Control*, p. 150, McGraw-Hill, N. Y., 1967, by permission of McGraw-Hill Book Co.)

[1] The two schematic drawings of control valves were taken from E. F. Johnson, *Automatic Process Control*, pp. 150-151, McGraw-Hill, N.Y., 1967.

increase in the pressure signal from the controller causes the diaphragm to move downward. Normally the diaphragm is made of rubber fabric and is supported by a spring-loaded backup plate. The valve stem is attached to the backup plate, so that the downward motion causes the flow rate to decrease. For common designs, the valve stem position changes proportional to the pressure imposed on the diaphragm, and the maximum travel of the stem is 2 to 3 in.

In certain situations where the system is very sensitive to flow rate changes or the control is critical, we might not want to rely on the proportionality between the pressure on the diaphragm and valve position. For these cases we can again install an auxiliary feedback system, similar to that used in the pro-

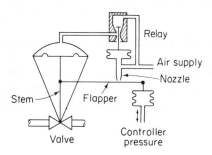

Figure **7.3-5.** Valve positioner. (Reproduced from E. F. Johnson, *Automatic Process Control*, p. 151, McGraw-Hill, N. Y. 1967, by permission of McGraw-Hill Book Co.)

portional controller. This kind of a device is called a valve positioner. One possible arrangement is shown in Figure 7.3–5. By using an air supply at 100 psig for the positioner, we can impose greater forces on the diaphragm and, correspondingly, the valve stem.

Limitations of the Control Hardware

In the foregoing discussion we have attempted to show that it is possible to measure the temperatute of a system automatically and to change the position of a valve stem proportional to the difference between this measured value and some desired value. In addition to the method we proposed, the same objective could be accomplished in many other ways. Similarly, equipment is available for measuring other process variables and for creating control signals that are proportional to the derivative and/or the integral of an error signal. This field of instrumentation is vast, and space considerations preclude more than a qualitative discussion. However, there are a few other important points that the reader should recognize, so that the theoretical design of a control system will not be in conflict with practical hardware limitations.

Perhaps the first factor we should note is that each of the elements in the control loop is a dynamic system. For example, the dynamic response of a thermometer to a step change in temperature is never instantaneous; it takes a finite time for the pressure signal generated by an expansion thermometer to be transmitted to the receiver in the controller; the receiving bellows has dynamic characteristics; the levers and flapper associated with the generation of the error signal require a finite time to change position; the transmission of the pressure signal at the tip of the nozzle to the feedback bellows or the top of the valve diaphragm is not instantaneous; and the motion of the valve stem depends on its inertia. None of these factors was considered in our previous analyses of control systems. In effect, we implicitly assumed that the dynamic characteristics of every element in the control loop were negligible in comparison with the process dynamics. There are a great number of cases where this assumption is valid. However, in our discussion of third-order systems in Section 7.2, we found that the measurement and controller dynamics can never be completely neglected if the plant under study is a first- or second-order system. Otherwise the analysis is completely misleading because it predicts that the controller gain should be made very large in order to obtain the best performance, whereas actually the closed-loop system will become unstable at some finite gain. The dynamic behavior of the various elements in the loop can be determined via the theoretical and experimental techniques presented in Chapters 3 through 6. Then the block diagram for the control system should appear as shown in Figure 7.3–6, where G_p is the transfer function of the plant, G_m

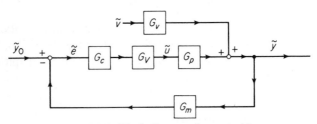

Figure 7.3-6. Block diagram of a control loop.

is the transfer function of the measuring instrument and transmission line, G_c is the transfer function of the controller and transmission line, and G_V is the transfer function of the valve.

A second factor that must be considered is that in order to obtain a proportional control law, each element in the control loop must be described by a linear, input–output relationship. This condition is normally satisfied for a pneumatic controller because of the presence of the feedback bellows. However, a wide variety of measuring instruments, such as a vapor-pressure thermometer, have nonlinear input–output characteristics. Similarly, for most control valves, the flow rate is a nonlinear function of the valve stem position, although it is possible to purchase "linear" valves at somewhat higher prices. The common technique for treating nonlinear elements in the control loop is the same as that for handling nonlinear plants; that is, the system equations are linearized around some steady state operating point of interest, a first estimate of the dynamic performance is based on the linearized analysis, and, eventually, the adequacy of the results obtained in this way are verified experimentally. A related but much more profound difficulty is encountered when a control valve exhibits hysteresis. This phenemenon occurs when a plot of the valve stem position versus the flow rate through the valve gives different curves for when the valve is being opened and when it is being closed (see Figure 7.3–7). This difference is due to the mechanical construction of the valve and the nature of many springs. An interesting study of the importance of valve hysteresis was published by Ramirez and Turner,[2] who found that a linearized analysis of a reactor

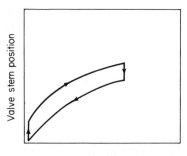

Flow rate   **Figure 7.3-7.** Valve hysteresis.

problem indicated that the system should be stable but that a simulation of the nonlinear equations, which included valve hysteresis, demonstrated that the behavior would be unstable. It is possible to use a describing function technique to account approximately for the dynamic behavior of a system with hysteresis, as will be shown later in this chapter.

Another major consideration, which we tended to overlook, is that each component of hardware in the control loop has a limited operating range. For example, liquid-expansion thermometers normally operate between -100 and $1200\,°\text{F}$, whereas gas-expansion thermometers and vapor-pressure thermometers span -100 to $700\,°\text{F}$. Similarly, the flow characteristics of a valve depend on its size, as well as the type of valve under consideration. The fact that different equipment must be used for different operating regions means that it is often necessary to make a rough estimate of the design of the control system before it is possible to select the actual equipment that might be used. Then more accurate models for both the steady state and dynamic behavior of this equipment can be developed. This kind of a difficulty, where it is necessary to know something about the answer before it is possible to formulate a quantitative statement of the problem, is commonly encountered in all aspects of engineering.

One additional factor, which is related to those just discussed, is the necessity for properly identifying the units used in the control equations. For example, if we write a proportional control law relating the change in the coolant flow

[2] W. F. Ramirez, and B. A. Turner, *AIChE Journal*, **15**, 853 (1969).

rate to a measured temperature deviation

$$(q_c - q_{cs}) = K_c(T - T_s) \qquad 7.3\text{-}6$$

it is apparent that the controller gain, K_c, must have the following units: cubic feet second^{-1} °F^{-1}. Actually, when the controller gain is written this way, it also includes all the conversion factors for the measuring system, the controller itself, and the control valve. However, by considering each element in the control loop individually and by establishing the system equations for that element, the reader should have little difficulty in determining the appropriate values that enter into K_c. This kind of a procedure is frequently necessary when an attempt is made to set the dial on the controller to a particular value, because the dial settings are normally given in terms of a quantity called the proportional band. The *proportional band* is defined as the percentage of the maximum range of the input variable of interest that will cause the controller output to cover its full range, or

$$\text{Percent proportional band} = 100 \frac{\Delta e}{\Delta e_{\max}} \qquad 7.3\text{-}7$$

where $\Delta e_{\max}$ is the maximum range of the error signal and Δe is the change in the error that will make the controller output change its maximum amount —for example, from 3 to 15 psig for a pneumatic controller. In our previous discussion of proportional controllers, we showed that the controller output, the change in the pressure in the nozzle, was proportional to its input, the error signal, so that the error that produces a full scale change in the output is

$$\Delta P_{\max} = K_c \Delta e \qquad 7.3\text{-}8$$

Combining the two preceding expressions (Eqs. 7.3–7 and 7.3–8), we find that the relationship between the controller gain and the proportional band is

$$\text{Percent proportional band} = \frac{100}{K_c} \frac{\Delta P_{\max}}{\Delta e_{\max}} \qquad 7.3\text{-}9$$

However, the definitions of K_c in Eq. 7.3–6 and Eq. 7.3–8 are different, because the former includes the control valve characteristics, whereas the latter is based only on the measuring system and the controller. We will use the more abstract formulation, such as is given in Eq. 7.3–6, in this text primarily because it conserves space and avoids unnecessary complication. It is important to recognize that this is not the most common approach.

The Effect of Hardware Dynamics

Before dispensing with our somewhat cursory discussion of the practical features of a control system, we should attempt to see if it is possible to make any general statements about the effect of the elements in the control loop on the system performance. If we consider the block diagram given in Figure

7.3–6, we can develop the system equations

$$\tilde{y} = G_v \tilde{v} + G_p \tilde{u}$$
$$= G_v \tilde{v} + G_p G_V G_c \tilde{e}$$
$$= G_v \tilde{v} + G_p G_V G_c (\tilde{y}_0 - G_m \tilde{y})$$

Solving for $\tilde{y}$, we obtain

$$\tilde{y} = \frac{G_v}{1 + G_m G_c G_V G_p} \tilde{v} + \frac{G_p G_V G_c}{1 + G_m G_c G_V G_p} \tilde{y}_0 \qquad 7.3\text{–}10$$

This result is exactly what we would have expected from our discussion in Section 7.2. The transfer function for a closed-loop system can be written from a block diagram by taking the product of the transfer functions between any output and input signals and then dividing by unity plus the product of all the transfer functions in the loop. The fact that the dynamics of the measuring system appear in the feedback portion of the loop does not introduce anything significantly new into the problem.

For a case where each component in the control loop, except for the process, is described by a linear input–output relationship and no dynamic characteristics, the transfer functions G_m, G_c, and G_V will just be constants and the preceding expression will reduce to our previous results. For cases where the dynamics are important, we assume that the transfer functions can be represented by the ratio of two polynomials in the Laplace parameter s and that, except for the controller, the order of the polynomial in the numerator is smaller than that of the denominator. This assumption is consistent with our knowledge of the dynamic behavior of physical systems, which was discussed in Chapters 3 through 5. Whenever the controller contains a derivative mode, its output is proportional to the derivative of the input signal, so that it is an exception to the general case.

From our discussion in the early part of Section 7.2, we know that the response of a closed-loop system generally becomes less satisfactory as the order of the system increases. In particular, stable third-order plants can become unstable in a closed-loop configuration. By examining the form of the transfer functions in Eq. 7.3–10, we see that the effect of using dynamic models for the measuring instrument and valve is to increase the order of the closed-loop system, thereby decreasing the performance. Partial compensation for this effect can be obtained by including a derivative mode in the controller.

On an intuitive basis we would also expect that whenever the dynamics of the measurement instrument or the control valve were significant, we would observe a poorer response of the controlled plant. This factor becomes painfully clear if we consider a somewhat exaggerated case where we want to control the composition of a reactor but do not have any way of continuously measuring the output composition. If we take a grab-sample from the reactor, spend one-half hour carrying it over to a laboratory, carry out a wet-method quantitative analysis that takes 2 hours, and go back to the plant and change the valve setting, we have about a 3–hour dead time in our manual feedback loop. Of course, if a new disturbance enters the plant during this time, the adjustment we make might amplify the effects of this disturbance rather than compensate for it. Thus it might be better not to make any attempt to control the reactor using feedback principles; or we could look for another process variable to measure almost instantaneously and, by maintaining this variable constant, hope that the reactor composition will remain relatively constant. In any event, we should make every attempt to minimize the dynamic characteristics of the elements in the feedback loop.

SECTION 7.4 STABILITY

The stability of a system certainly will be one of the most important considerations in the design of a control system. If an optimum, economic steady state design turns out to be locally unstable, all our design effort will have been wasted unless a controller can be used to make it stable. Alternatively, if we attempt to use a control system to improve the dynamic response characteristics of a stable, optimum, economic steady state design, and during this process make the desired operating condition become unstable, then we will have defeated our purpose. Although we have already discussed system stability several times, it seems worthwhile to review our previous techniques and present some special methods for determining stability. Criteria for the stability of both open-loop and closed-loop systems will be of interest.

Characteristic Roots

The definition of stability we have used in our previous analyses depends on the characteristic roots of the linearized system equations. We know that the linearized equations provide a valid description of any nonlinear process in some small region around a *singular point*—that is, a steady state solution of the dynamic equations—and therefore whenever the characteristic equation has one or more positive real roots, or one or more pairs of complex conjugate roots with positive real parts, the solution will contain terms like

$$c_1 e^{\lambda t} \quad \text{or} \quad e^{\lambda t} (c_1 \sin \omega t + c_2 \cos \omega t) \qquad \text{7.4–1}$$

where λ is the positive root or the positive real part of the root, ω is the imaginary part of the root, and c_1 and c_2 are integration constants. The existence of terms of this kind means that the output will grow without a bound as time progresses; or in actual practice that the system will move away from the singular point until the linearized equations no longer provide a valid approximation of the plant dynamics. At this point we should emphasize that the characteristic roots correctly predict the stability of the system only if no pairs of pure imaginary roots exist. The case of pure imaginary roots corresponds to a transition from stable to unstable operation or vice versa, because the real part of the roots is changing sign. In this case a more detailed analysis, involving quadratic and higher-order terms in the Taylor series expansion of the nonlinear state equations, is required.

The evaluation of the characteristic roots permits an evaluation of the stability of both open and closed-loop system. An example of a plant that was originally unstable but that could be made stable by the addition of a proportional feedback controller with a sufficiently high gain was presented in Section 7.1. Similarly, a case where a stable third-order plant became unstable, when the gain of a proportional feedback controller exceeded a certain value, was discussed in Section 7.2. In this second illustration, it was necessary to find the roots, or at least the behavior of the roots, for a cubic equation, which is much more difficult than using the simple quadratic relationship. Obviously, for more realistic dynamics problems, we will need to find the roots of much higher-order polynomials. Fortunately digital computer programs that can be used for this purpose are readily available, although it must be recognized that for a closed-loop analysis the roots depend on the gains of each control mode and therefore a number of case studies must be made. Consequently, it seems to be advantageous to look for other methods of assessing system stability.

Another limitation encountered in attempts to evaluate the characteristic roots is that in some cases it is difficult to establish the appropriate characteristic equation. For example, the steady state equation, Eq. 3.4–32, for the simple model of a catalytic cracking unit has the same form as the stirred-tank reactor problem, Eq. 7.1–2. Thus we expect to be able to obtain multiple steady state solutions for certain system parameters, some stable and others unstable. In order to test the stability characteristics of any one of these singular points, we must solve Eqs. 3.1–38 through 3.1–43 explicitly for the derivative terms, linearize the right-hand sides of the nonlinear equations around the singular point of interest, arrange the expressions in matrix form, evaluate the determinant

$$\det |A - \lambda I| = 0 \qquad\qquad 3.3\text{–}56$$

to obtain the characteristic equation, and then solve for the characteristic roots. However, expanding the determinant

$$\begin{vmatrix} a_{11} - \lambda & a_{12} & a_{13} & a_{14} & a_{15} & a_{16} \\ a_{21} & a_{22} - \lambda & a_{23} & a_{24} & a_{25} & a_{26} \\ a_{31} & a_{32} & a_{33} - \lambda & a_{34} & a_{35} & a_{36} \\ a_{41} & a_{42} & a_{43} & a_{44} - \lambda & a_{45} & a_{46} \\ a_{51} & a_{52} & a_{53} & a_{54} & a_{55} - \lambda & a_{56} \\ a_{61} & a_{62} & a_{63} & a_{64} & a_{65} & a_{66} - \lambda \end{vmatrix} = 0 \qquad 7.4\text{–}2$$

to obtain a polynomial is not such an easy task, even for this simple system. For a 20th– or 50th–order plant it would be almost overwhelming. For this reason, digital computer programs based on matrix methods are used to determine the characteristic roots without developing the characteristic equation. Of course, the process transfer function, or transfer matrix, is not available if this procedure if followed. Exceptions to these problems arise when the coefficient matrix of the linearized equations is bi-diagonal or tri-diagonal, for then it is relatively simple to find the characteristic roots analytically (using either matrix techniques or the finite-difference methods described in Section 4.6).

In the preceding discussion we have tried to point out that the stability of a steady state operating condition for either an open-loop or a closed-loop system can be determined by evaluating the characteristic roots of the linearized dynamic equations. However, in certain situations this procedure is fairly complicated. As a result, we want to consider other techniques leading to stability criteria. A few alternate procedures are discussed below, and an approximate method (i.e., a root-locus diagram) for plotting the values of the roots as a function of a system parameter, such as controller gain, is presented in Section 7.5.

Final Value Theorem

The final value theorem has been used several times in this text to determine the new steady state condition a system tends to approach after a step input. Thus an unwary investigator might attempt to use this theorem to see if the output becomes unbounded, which would correspond to an unstable plant, or if the new steady state is finite, which would correspond to a stable system. To show that the *final value theorem gives incorrect results*, we consider a very simple, unstable first-order process

$$\tau \frac{dy}{dt} = y_f + y \qquad\qquad 7.4\text{-}3$$

where y represents the deviation from some steady state operation. The Laplace transform of the output for a step input, $\tilde{y}_f = A/s$, is

$$\tilde{y} = \frac{A}{s(\tau s - 1)} \qquad\qquad 7.4\text{-}4$$

The final value theorem predicts that the system will approach a new steady state

$$\lim_{t \to \infty} y(t) = \lim_{s \to 0} (s\tilde{y}) = \lim_{s \to 0} \frac{A}{\tau s - 1} = -A \qquad\qquad 7.4\text{-}5$$

whereas if we invert the Laplace transform, we see that the actual output should be

$$y(t) = A(1 - e^{t/\tau}) \qquad\qquad 7.4\text{-}6$$

which becomes unbounded as time approaches infinity. Hence it must be emphasized that the final value theorem *is not applicable* when the function $s\tilde{f}(s)$ becomes infinite for any positive real value of s—for example, $s = +1/\tau$ for our problem.

Routh's Criterion

One major difficulty encountered in the assessment of the stability of a plant is calculating the roots of higher-order polynomials. Actually, we know that the system will be stable if all the roots are negative or have negative real parts; and if there is some way of establishing this information without finding the magnitude of the roots, we will have solved the stability problem. A method of this type, which is based on the nature of polynomial equations, was developed by Routh. Since the proof is somewhat lengthy, we omit it here[1] and merely describe the procedure. We consider the characteristic equation

$$a_n s^n + a_{n-1} s^{n-1} + a_{n-2} s^{n-2} + \cdots + a_1 s + a_0 = 0 \qquad 7.4\text{-}7$$

and first note that if any of the coefficients a_i are negative, or zero, the system will be unstable. If all the coefficients are positive, we must arrange them in a Routh array as shown below:

$$
\begin{array}{llll}
a_n & a_{n-2} & a_{n-4} & a_{n-6} \quad \cdots \\
a_{n-1} & a_{n-3} & a_{n-5} & a_{n-7} \quad \cdots \\
b_1 & b_2 & b_3 & \cdots \\
c_1 & c_2 & c_3 & \cdots \\
d_1 & d_2 & \cdots \\
e_1 & e_2 & \cdots
\end{array}
\qquad 7.4\text{-}8
$$

where the constants in the third and succeeding rows are given by the expressions

$$b_1 = \frac{a_{n-1}a_{n-2} - a_n a_{n-3}}{a_{n-1}} \qquad b_2 = \frac{a_{n-1}a_{n-4} - a_n a_{n-5}}{a_{n-1}}$$

$$b_3 = \frac{a_{n-1}a_{n-6} - a_n a_{n-7}}{a_{n-1}} \qquad \cdots$$

$$c_1 = \frac{b_1 a_{n-3} - a_{n-1}b_2}{b_1} \qquad c_2 = \frac{b_1 a_{n-5} - a_{n-1}b_3}{b_1} \cdots$$

$$d_1 = \frac{c_1 b_2 - b_1 c_2}{c_1} \qquad d_2 = \frac{c_1 b_3 - b_1 c_3}{c_1} \cdots$$

$$7.4\text{-}9$$

In other words, we obtain these elements in a systematic way from determinants of the elements in the two preceding rows. The number of terms in a row

[1] Details are available in E.J. Routh, *Dynamics of a System of Rigid Bodies: Part II, Advanced*, Macmillan, London, 1965.

decreases as the array is developed, and we continue the procedure until only zeros are obtained as elements. Normally it is necessary to determine the elements in $(n + 1)$ rows before the stopping condition is satisfied.

Now, the number of sign changes in the left-hand column of the Routh array, $a_n, a_{n-1}, b_1, c_1, \ldots$ will be the number of roots having positive real parts. Thus in order for a system to be stable, all the elements in this first column must be positive (and nonzero). It should be noted that if all the elements in any row are divided by a positive constant, it does not change the results of the test. Also, if all the elements in one row turn out to be zeros, while the elements in the preceding row are nonzero, there is a pair of pure imaginary roots, and the test does not give a definite answer to the stability question. For this case, we calculate the elements in the row after the zeros by replacing the zeros by an arbitrarily small number ε and following our normal procedure.

Although it is always much easier to calculate the Routh array, rather than the characteristic roots, in order to determine the stability of either an open or a closed-loop system, we pay the price of not having a quantitative estimate of how fast the system tends toward (or away from) a steady state operating condition. However, there are many cases where the qualitative stability information is adequate for our purposes.

Example 7.4-1 Stability criteria for a CSTR with integral control

The characteristic equation for a nonisothermal, stirred-tank reactor with a feedback control system that manipulates the flow rate of heating fluid proportional to the integral of the measured temperature deviations was given in Section 7.1 as

$$s^3 + (1 + \alpha + 1 + \beta + \gamma)s^2 + [(1 + \alpha)(1 + \beta) - \gamma + b_{22}K_I]s$$
$$+ (1 + \alpha)b_{22}K_I = 0 \qquad \text{7.1-44}$$

Find the stability criteria for the reactor.

Solution

We first require that every coefficient in the equation be positive

$$1 + \alpha + 1 + \beta - \gamma > 0$$
$$(1 + \alpha)(1 + \beta) - \gamma + b_{22}K_I > 0 \qquad \text{7.4-10}$$
$$(1 + \alpha)b_{22}K_I > 0$$

where the constants are defined by Eq. 7.1-6. If these conditions are satisfied for a particular set of system parameters, then we form the Routh array

$$
\begin{array}{ccc}
1 & (1 + \alpha)(1 + \beta) - \gamma + b_{22}K_I & 0 \\
1 + \alpha + 1 + \beta - \gamma & (1 + \alpha)b_{22}K_I & 0 \\
b_1 & 0 & \\
c_1 & &
\end{array}
\qquad \text{7.4-11}
$$

where

$$b_1 = \frac{(1 + \alpha + 1 + \beta - \gamma)[(1 + \alpha)(1 + \beta) - \gamma + b_{22}K_I] - (1 + \alpha)b_{22}K_I}{1 + \alpha + 1 + \beta - \gamma}$$

$$= \frac{(1 + \alpha + 1 + \beta - \gamma)[(1 + \alpha)(1 + \beta) - \gamma] + (1 + \beta - \gamma)b_{22}K_I}{1 + \alpha + 1 + \beta - \gamma}$$

$$c_1 = \frac{b_1(1 + \alpha)b_{22}K_I}{b_1} = (1 + \alpha)b_{22}K_I \qquad\qquad 7.4\text{--}12$$

For stability, all the elements in the left-hand column must be equal to zero. This requirement for the second and fourth elements gives the same information as Eq. 7.4–10. However, we obtain the new condition that

$$c_1 = \frac{(1 + \alpha + 1 + \beta - \gamma)[(1 + \alpha)(1 + \beta) - \gamma] + (1 + \beta - \gamma)b_{22}K_I}{1 + \alpha + 1 + \beta - \gamma}$$

which is identical to Eq. 7.1–46.

Example 7.4–2 Range of controller gains for stable operation

Coughanowr and Koppel[2] discuss the control of a two-tank level system with a three-mode controller. The time constants of the tanks are taken as 20 min and 10 min, and they assume that the measuring system can be represented by a first-order lag with a time constant of 30 sec. When the controller transfer function is written as

$$\frac{\tilde{u}}{\tilde{y}_0 - \tilde{y}} = K\left(1 + T_D s + \frac{1}{T_i s}\right) \qquad\qquad 7.4\text{--}13$$

they give the integral time T_i as 3 min and the derivative time as 40 sec. Then they use the Nyquist stability criterion to determine the range of the gain K (which includes the gain of the proportional control mode and the conversion factors for the various elements in the control loop) corresponding to stable operation. Instead of following their procedure, use the Routh criteria to determine the stability of the closed-loop system.

Solution

From the block diagram of the process (see Figure 7.4–1), we find that the transfer function relating the output to the disturbances—that is, we are assuming that $\tilde{y}_0 = 0$—is

$$\frac{\tilde{y}}{\tilde{v}} = \frac{G_v}{1 + K[1 + T_D s + (1/T_i s)][1/(20s + 1)(10s + 1)(0.5s + 1)]}$$

$$= \frac{s(20s + 1)(10s + 1)(0.5s + 1)G_v}{s(20s + 1)(10s + 1)(0.5s + 1) + K[s + T_D s^2 + (1/T_i)]} \qquad 7.4\text{--}14$$

[2] D. R. Coughanowr and L. B. Koppel, *Process Systems Analysis and Control*, p. 271, McGraw-Hill, N.Y., 1965.

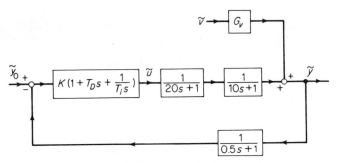

Figure 7.4-1. Block diagram of a feedback control system.

The system stability only depends on the characteristic roots of the denominator. Substituting the given values of $T_D = 2/3$ min. and $T_i = 3$ min, expanding the denominator, and setting it equal to zero, we obtain the characteristic equation

$$300s^4 + 645s^3 + (91.5 + 2K)s^2 + 3(1 + K)s + K = 0 \qquad 7.4\text{-}15$$

For positive values of K, all the coefficients in this equation will be positive. Thus we must form the Routh array

$$
\begin{array}{cccc}
300 & 91.5 + 2K & K & 0 \\
645 & 3(1 + K) & 0 & 0 \\
b_1 & b_2 & 0 & \\
c_1 & 0 & & \\
d_1 & 0 & & \\
0 & & &
\end{array}
$$

where we have inserted zeros where appropriate. The constants are calculated as shown below:

$$b_1 = \frac{645(91.5 + 2K) - 300(3)(1 + K)}{645} = 90.1 + 0.605K$$

$$b_2 = \frac{645K - 0}{645} = K$$

$$c_1 = \frac{b_1 3(1 + K) - 645b_2}{b_1} = \frac{(90.1 + 0.605K)3(1 + K) - 645K}{90.1 + 0.605K}$$

$$= \frac{1.815K^2 - 372.9K + 270.3}{90.1 + 0.605K}$$

$$d_1 = \frac{c_1 b_2 - 0}{c_1} = K \qquad 7.4\text{-}16$$

The closed-loop system will be stable, providing that all the elements in the left-hand column of the Routh array are positive. We see immediately that the first two elements and both b_1 and d_1 are positive, so that the only questionable term is

$$c_1 = \frac{1.815K^2 - 372.9K + 270.3}{90.1 + 0.605K} \qquad 7.4\text{--}17$$

Since the denominator of this expression is always positive, the sign of c_1 is determined by the quadratic expression in the numerator. We could simplify the numerator by dividing each coefficient by a constant without changing the result of the test. However, initially it is more important to gain a better appreciation of the kind of behavior that may be encountered. We immediately see that when $K = 0$, which corresponds to the open-loop configuration of three first-order systems in series, then $c_1 = 270$ and the systems is stable. When $K = 10$, we find that $c_1 = 181 - 3730 + 270 < 0$, so that the system becomes unstable. A similar result is obtained when $K = 100$. However, if we continue to increase K and let $K = 1000$, then $c_1 = 18.1 \times 10^5 - 3.73 \times 10^5 + 0.002 \times 10^5 > 0$, so that the closed-loop system becomes stable once more. Of course, the region where unstable operation is to be expected can be established by evaluating the roots of c_1. Calculating these roots, we find that unstable operation will occur approximately when $0.73 < K < 205$. Using Nyquist diagrams, Coughanowr and Koppel obtain the values $0.59 < K < 360$.

An interesting feature of the problem is that the range of loop gains corresponding to unstable operation is finite.

Bode Plots

As we mentioned in Chapter 4, a Bode plot provides a convenient way of representing the dynamic characteristics of linear systems. The Bode plot for any particular element in a control loop can be determined either experimentally or from a theoretical model of that element. Also, the plot for a series of elements can be obtained merely by adding the graphs for the individual components. Thus we are interested in seeing if Bode diagrams can be used to evaluate the stability of open and closed-loop systems.

Taking the open-loop case first, we choose a simple, unstable first-order system, Eq. 7.4–3, which has the transfer function

$$\frac{\tilde{y}}{\tilde{y}_f} = \frac{1}{\tau s - 1} \qquad 7.4\text{--}18$$

Unless we were careful, we might attempt to follow the procedure outlined in Example 4.4–3 for the pseudo-steady state output: substitute $j\omega$ for s, rationalize the complex number, and then find that the system gain and phase angle are given by the expressions

$$G = \frac{1}{\sqrt{1 + \tau^2\omega^2}} \qquad 7.4\text{--}19$$

$$\tan \theta = \tau\omega \qquad 7.4\text{--}20$$

The gain expression is identical to that for a stable first-order plant, and the phase angle turns out to be a lead rather than a lag. Of course, *this analysis is incorrect*, as can be seen by letting $\tilde{y}_f = A/(s^2+\omega^2)$ and inverting the transform of the output signal

$$\tilde{y} = \left(\frac{1}{\tau s - 1}\right)\left(\frac{A\omega}{s^2 + \omega^2}\right) \qquad 7.4\text{-}21$$

to obtain

$$y(t) = \left(\frac{A\tau\omega}{1 + \tau^2\omega^2}\right)e^{t/\tau} - \left(\frac{A}{1 + \tau^2\omega^2}\right)\sin \omega t - \left(\frac{A\tau\omega}{1 + \tau^2\omega^2}\right)\cos \omega t$$

$$7.4\text{-}22$$

Although the amplitude ratio and phase angle of the pseudo-steady state portion of this result agree with Eqs. 7.4–19 and 7.4–20, it is clear that the transient term dominates the solution and the system output tends to become unbounded. In other words, the pseudo-steady state solution, and therefore the Bode plot, has no meaning if the system is unstable. This agrees with our intuition that it would never be possible to measure experimentally the frequency response of any unstable plant.

However, Bode diagrams often provide a simple way of establishing when an initially stable third- or higher-order process becomes unstable because of an excessive gain in a feedback control loop. In order to illustrate this behavior, we consider the simplest case where there are three first-order systems in series

$$H(s) = \frac{0.00516}{(\tau_1 s + 1)(\tau_2 s + 1)(\tau_3 s + 1)} \qquad 7.4\text{-}23$$

For this case, the maximum slope of the gain curve on a Bode plot is -3 and the maximum phase lag is 270 deg. When we put a variable gain element K_c in series with these three first-order lags, the phase characteristics are not changed and the gain curve shifts upward as K_c increases; that is, since we multiply the transfer functions, we add a constant value K_c to every point on the logarithmic plot (see Figure 7.4–2). According to the analysis presented in Section 7.2, this procedure should make a closed-loop system become unstable.

The reason for the onset of instability becomes apparent if we consider the experimental problem outlined below. We set up a proportional feedback controller for our third-order plant, assume that there are no measurement, controller, or valve dynamics and no disturbances, and we install a switch in the feed-back loop (see Figure 7.4–3). Now from Figure 7.4–2 we find the frequency that corresponds to a phase lag of 180 deg ($\omega = 9.2$), and then we vary the controller set point sinusoidally at this frequency with the switch in the open position. After an initial transient period, we will obtain a sinusoidal signal at the position of the switch. The frequency of this signal will be identical with the driving frequency, but it will be exactly out of phase with the driving signal because there is a 180 deg phase shift as the original signal passes through

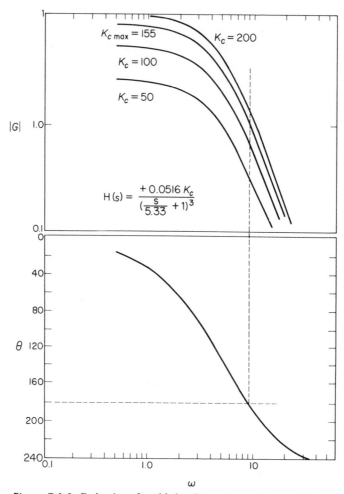

$$H(s) = \frac{+0.0516\,K_c}{\left(\dfrac{s}{5.33} + 1\right)^3}$$

Figure 7.4-2. Bode plot of a third-order system with a variable gain element, K_c.

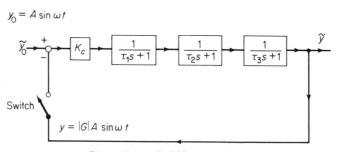

Figure 7.4-3. Stability experiment.

the system. The ratio of the amplitude of this signal to the driving signal depends on the controller gain, and at the $K_{c\,max}$ setting in Figure 7.4–2 ($K_{c\,max}$ = 155) the amplitudes will be identical. Therefore if we suddenly stop fluctuating the set point and simultaneously close the switch, the system will continue to oscillate because of the 180 deg phase change at the summing junction. In other words, when $K_c = K_{c\,max}$ the system itself generates oscillations, which means that one of the transfer functions in the loop must be quadratic with a zero damping coefficient or that there is a pair of pure imaginary roots. This result agrees with the assertion we made in Section 7.2 for the stability of third-order plants.

If we repeat the experiment for values of the proportional gain less than $K_{c\,max}$, the amplitude of the signal at the switch will be less than the driving signal, so that when the driving signal is terminated and the switch is closed, the error signal will gradually decay. Thus for a case where the loop gain is less than unity when the phase shift is -180 deg, the system will be stable. Similarly, if we performed the experiment for loop gains greater than unity, the amplitude of the signal at the switch would be larger than the driving signal. Then when the switch was closed, the amplitude of the oscillation would continue to grow in magnitude every time the signal passed around the loop, so that the output would tend to increase without a bound.

To summarize our results, we say that a closed-loop system will be unstable if the open-loop frequency response has an amplitude ratio greater than unity at the frequency corresponding to a phase lag of 180 deg. This frequency is usually called the *crossover* or *critical frequency*. It should be noted that this stability criterion is valid only for systems where the gain and phase curves continuously decrease with increasing frequency. For example, if the phase angle can be -180 deg for two values of frequency, as illustrated in Figure 7.4–4, the criteria are not applicable. For these cases, it is necessary to use a different approach, such as the Routh criterion or a Nyquist plot, to test for stability. One great advantage of the Bode analysis is that it can be applied to

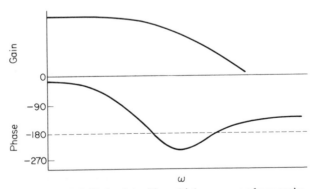

Figure 7.4-4. Bode plot with multiple crossover frequencies.

experimentally determined frequency response data of the controller and open-loop plant, and therefore is useful even when a theoretical model of the plant dynamics is not available.

Other Graphical Techniques

In addition to the stability methods just described, two other graphical methods are often used to develop stability criteria—the Nyquist diagram and the root-locus plot. The Nyquist diagram is essentially a polar plot of the same information presented on a Bode diagram; that is, the magnitude of the complex number obtained when $j\omega$ is substituted for the Laplace parameter s in the transfer function is plotted as a radius vector and the phase angle is plotted clockwise from the abscissa to establish the direction of the vector. Thus the magnitude and phase information appear as a single curve, with each point on the curve corresponding to a different frequency. This situation is in contrast to a Bode diagram, where the magnitude and phase were plotted separately against frequency. The main advantage of the Nyquist representation is that it is a fairly simple job to determine the closed-loop dynamics of a system from the graph for the open-loop plant. Also, by introducing some additional results from the theory of complex variables, it is possible to determine the process stability characteristics under very general conditions; for example, whether or not a closed-loop configuration with a controller will make an unstable open-loop plant become stable.

Without this background, however, the Nyquist method does not appear to offer many advantages over a Bode analysis. The asymptotic techniques that simplified the preparation of Bode diagrams are no longer applicable, and because log scales are not used, any modification of the system, such as the addition of an integral control mode, means that a new graph must be prepared. Thus the common approach used to construct a Nyquist diagram is first to make a Bode plot and then translate the information on to polar coordinates. A number of excellent references describing the use of Nyquist diagrams to determine system stability and to design control systems are available. The reader is encouraged to master this approach because it provides an alternate point of view from the one we shall pursue.

Another graphical technique that can be used to develop stability criteria is called a root-locus plot. This method is based on the fact that the transfer function of both open-loop and closed-loop, lumped parameter systems can be represented by the ratio of polynomials in the Laplace parameter s. By studying the behavior of functions that are the ratios of polynomials, it is possible to develop some general rules indicating how the zeros (the roots of the polynomial in the numerator) and the poles (the roots of polynomial in the denominator) of the transfer function change as some parameter (the controller gain) is varied. These rules can be used to prepare a graph of the locus of the characteristic roots (the poles) on a complex plane, with the controller gain as a parameter. On this graph the region of unstable operation corresponds to the right half of the complex plane, where the real part of any root is positive. The root-locus plot also provides a useful means for designing control systems, and therefore will be discussed in much greater detail in Section 7.6.

A third approach, similar to the Nyquist method, is to construct the Mikhailov curve.[3] After substituting $j\omega$ for s in the characteristic polynomial and writing the result as a single complex number, we can plot the complex number on a real and imaginary axis for various values of frequency between zero and infinity. It is easy to show that the angle between a vector connecting the origin to any point on the curve and the positive real axis must increase 90 deg for every root with a negative real part but decrease 90 deg for every root with a positive real part. Thus in order for the nth-order system to be stable, this angle must pass through n quadrants. The angle for an unstable system will only pass through $n - 2m$ quadrants, where m is the number of roots with a positive real part. The factor of two occurs, because the positive root decreases the angle, in addition to not adding an extra 90 deg. The approach applies to the characteristic polynomials describing both open-loop and closed-loop systems.

Example 7.4–3 System stability using a Mikhailow curve

The characteristic polynomial of a third-order plant with a proportional feedback control system is given by the equation

$$g(s) = s^3 + 16s^2 + 85.2s + 151.1 + 7.83K_c = 0 \qquad 7.4\text{–}24$$

Sketch the Mikhailov curves for $K_c = 100$ and $K_c = 200$ and determine the system stability for these two cases.

Solution

Substituting $j\omega$ for s and collecting terms, we obtain

$$g(j\omega) = (j\omega)^3 + 16(j\omega)^2 + 85.2(j\omega) + 151.1 + 7.83K_c$$
$$= (151.1 + 7.83K_c - 16\omega^2) + j\omega(85.2 - \omega^2) \qquad 7.4\text{–}25$$

Now we substitute one value of K_c, evaluate the real and imaginary quantities at a number of frequencies, and plot the real versus the imaginary parts (see Figure 7.4–5). For $K_c = 100$, we see that the angle between the vector and the positive real axis passes through three quadrants (it increases from 0 to $270°$), so that the plant with its feedback control loop is stable. However, for $K_c = 200$, the angle sweeps through -1 quadrants (from 0 to $-90°$). Hence we find that there should be two roots with positive real parts, $n - 2m = 3 - 2(2) = -1$, and that the system will be unstable.

---

[3] Additional information is available in D. D. Perlmutter, *Introduction to Chemical Process Control*, p. 111, Wiley, N. Y., 1965.

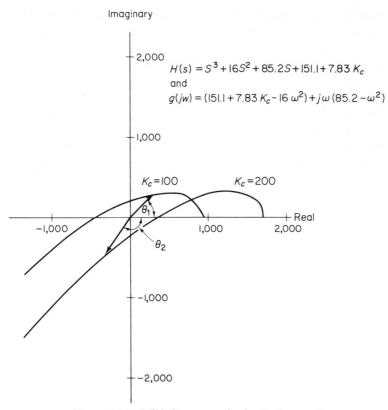

Figure 7.4-5. Mikhailov curves for feedback control systems.

Liapunov's Direct Method

All the stability techniques presented above are based on the linearization of the nonlinear dynamic equations in the neighborhood of some steady state operating point and a determination of the nature of the characteristic roots of the linearized equations. In other words, we have either quantitatively or qualitatively established the dynamic behavior of the linearized equations. The fact that the linearized equations provide a valid description of the nonlinear plant in a sufficiently small region around a steady state solution, providing that no pairs of pure imaginary roots are obtained, was proved by Liapunov. This approach is often called *Liapunov's indirect method*, for it requires an approximate solution of the system equations. There is a completely different approach, however, called *Liapunov's direct method*, where no attempt is made to solve the system equation.[4] Unfortunately, practical considerations seem to

[4] An excellent discussion of this approach was published by R. E. Kalman, and J. E. Bertram, *J. Basic Eng.*, **82**, 371, 394 (1960).

limit the usefulness of the method, but we include a brief description of it here because it provides some useful insight into the behavior of dynamic systems from a completely new point of view.

Perhaps the easiest way of understanding the direct method is first to consider the behavior of a ball in a bowl (see Figure 7.4–6). When we place the ball on the wall of the bowl, we are putting potential energy into the system. After we let the ball go, this energy is transformed into kinetic energy as it rolls down the side, and then back into potential energy as it moves up the opposite side of the bowl. Because of friction, the new value of potential energy is less than the original value, and the ball rolls back and forth until it finally comes to rest at the bottom of the bowl. Our intuitive notion of stability is

Ball

Bowl   **Figure 7.4-6.** A ball in a bowl.

satisfied, for the system tends to return to an equilibrium condition after a small perturbation. Also, at least for this system, stability is equivalent to a decrease in the total energy of the system (kinetic plus potential) with respect to time as the system approaches a state of minimum energy at the equilibrium condition.

It is possible to show that the same results are obtained for other simple mechanical systems, such as mass and spring combinations or the model of a manometer we presented in Example 4.5–1. Thus we form a tentative hypothesis that whenever the time rate of change of total energy of an isolated system is negative, the system will be stable. We hope that it might be possible to determine the sign of this energy change, and thereby determine the system stability, without solving the equations describing the system. Liapunov's direct method is a generalization of this basic idea, which is applicable to both linear and nonlinear systems and to cases where the system energy may either increase or decrease with time. Before stating his theorem, however, it is helpful to solve a simple problem in order to gain an appreciation for the quantitative aspects of the analysis.

The equations for simple mechanical systems can be written

$$m\ddot{x} + k_d\dot{x} + Kx = 0 \qquad\qquad 7.4\text{–}26$$

where the first term represents the inertial force, the second assumes that the frictional resistance is proportional to velocity, the third is a spring force, and we have assumed that the distance variable is a deviation from an equilibrium condition and that there is no forcing signal. The corresponding terms for a manometer are given in Example 4.5–1. The kinetic and potential energy can be written as

$$KE = \tfrac{1}{2} m\dot{x}^2 \qquad PE = \tfrac{1}{2} Kx^2 \qquad\qquad 7.4\text{-}27$$

so that the total energy of the system is

$$E = \tfrac{1}{2} m\dot{x}^2 + \tfrac{1}{2} Kx^2 \qquad\qquad 7.4\text{-}28$$

Before proceeding, we make a transformation of variables and let

$$y_1 = x \qquad y_2 = \dot{y}_1 = \dot{x} \qquad\qquad 7.4\text{-}29$$

In terms of these state variables, the system equation becomes

$$\dot{y}_1 = y_2 \qquad \dot{y}_2 = -\frac{K}{m} y_1 - \frac{k_d}{m} y_2 \qquad\qquad 7.4\text{-}30$$

and the total energy is

$$E = \tfrac{1}{2} Ky_1^2 + \tfrac{1}{2} my_2^2 \qquad\qquad 7.4\text{-}31$$

Now the time derivative of this energy expression is

$$\frac{dE}{dt} = Ky_1\dot{y}_1 + my_2\dot{y}_2 \qquad\qquad 7.4\text{-}32$$

or, after substituting the state equations

$$\frac{dE}{dt} = -K_d y_2^2 \qquad\qquad 7.4\text{-}33$$

It is clear that the time rate of change of the total energy will be negative, providing that $k_d > 0$. From our previous work we know that a positive damping coefficient, or $k_d > 0$, guarantees that the system will be stable, so that stability does correspond to $dE/_{dt} < 0$.

When we attempt to apply this approach to other kinds of second-order linear systems, we soon find out that it is necessary to develop a new interpretation for the analysis. For instance, if we consider the dynamic model of a thermometer placed in a stirred heater, which was described in Example 4.5–4, the system equations themselves are energy balances. Thus the total energy of the system may increase or decrease, and the sign of the change in the energy content can have no connection with stability. One way to make this point apparent is to consider the response when a thermometer is inserted into a stirred vessel operating at steady state conditions. The final temperature of the mixer is not changed (according to our model) by this perturbation, and the temperature of the thermometer will increase to the value of the fluid in the heater. This means that the total energy of the plant increases even though the system is stable. It is easy to show that the system must be stable by combining the equations for the deviations

$$\dot{x}_1 = -\frac{1}{\tau_1} x_1 \qquad\qquad 4.5\text{-}95$$

$$\dot{x}_2 = \frac{1}{\tau_2} x_1 - \frac{1}{\tau_2} x_2 \qquad\qquad 4.5\text{-}96$$

into a single second-order equation

$$\ddot{x}_2 + \left(\frac{1}{\tau_1} + \frac{1}{\tau_2}\right)\dot{x}_2 + \frac{1}{\tau_1\tau_2}x_2 = 0 \qquad 7.4\text{-}34$$

and recognizing that the damping coefficient is positive.

After some additional thought about the matter, we might attempt to generalize our original approach by introducing some arbitrary function, V, which has no relation to the energy of the system. By analogy, we let

$$V = \frac{c_1}{2}x_1^2 + \frac{c_2}{2}x_2^2 \qquad 7.4\text{-}35$$

where c_1 and c_2 are positive constants, which we shall specify later. We find that

$$\begin{aligned}\dot{V} &= c_1 x_1 \dot{x}_1 + c_2 x_2 \dot{x}_2 \\ &= -\frac{c_1}{\tau_1}x_1^2 + \frac{c_2}{\tau_2}x_1 x_2 - \frac{c_2}{\tau_2}x_2^2 \\ &= -\frac{c_2}{\tau_2}\left(\frac{c_1}{c_2}\frac{\tau_2}{\tau_1}x_1^2 - x_1 x_2 + x_2^2\right) \end{aligned} \qquad 7.4\text{-}36$$

If we want the function V to serve the same purpose as the total energy in the mechanical problem, we would expect that for a stable system $\dot{V}$ would be negative for all real values of x_1 and x_2 except the equilibrium values $x_1 = x_2 = 0$. It should not be suprising that the sign of $\dot{V}$ depends on the choice we make for the constants c_1 and c_2. For example, if we let $c_1 = c_2 = 1$, consider a case where the thermometer time constant is one-tenth of the vessels, $\tau_2 = 0.1\tau_1$, and choose $x_2 = 1$, we find that $\dot{V} < 0$ when $x_1 = 1$ and $\dot{V} > 0$ when $x_1 = 2$, despite the fact that the system is always stable. However, if we select the constants such that $4(c_1/c_2)(\tau_2/\tau_1) > 1$, we find that $\dot{V} < 0$ for all positive or negative values of x_1 and x_2. Thus we must be quite careful in our specification of the V function if we decide to use this approach to assess the stability of a system.

A clearer picture of the significance of our analysis can be obtained by using geometrical arguments. When lines of constant V in Eq. 7.4–35 are plotted on an x_1 versus x_2 plane for a case where $c_1 = c_2$, we generate a set of circles, with the largest circles corresponding to large values of V (see Figure 7.4–7). We can write $\dot{V}$ as

$$\dot{V} = \frac{\partial V}{\partial x_1}\dot{x}_1 + \frac{\partial V}{\partial x_2}\dot{x}_2 \qquad 7.4\text{-}37$$

or

$$\dot{V} = \dot{\mathbf{x}}_1^T \, \mathbf{grad} \, \mathbf{V} \qquad 7.4\text{-}38$$

where

$$\mathbf{grad} \, \mathbf{V} = \begin{pmatrix}\dfrac{\partial V}{\partial x_1} \\ \dfrac{\partial V}{\partial x_2}\end{pmatrix} \qquad 7.4\text{-}39$$

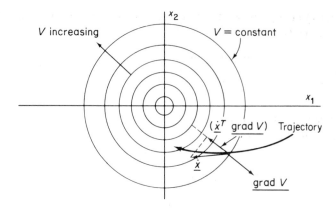

Figure 7.4-7. V functions.

Remembering that the direction of this gradient vector is always normal to a surface of constant V and that the positive direction is toward increasing values of V, we can sketch the gradient vectors on Figure 7.4-7. Also, recalling that the scalar or dot product of two vectors, $\mathbf{x}^T\,\mathbf{y}$ is simply the projection of one on the other, our criterion that $\dot{V} < 0$ means that the projection of the velocity vector, $\dot{\mathbf{x}}$, on the gradient must be negative. In other words, our criterion means that the projection of the tangent to the system trajectories must be directed toward the origin, which is an equilibrium point. With this background and motivation, we now present the theorem proved by Liapunov:

> If it is possible to find a function $V(\mathbf{x})$, which we call a Liapunov function, such that $V(\mathbf{x}) > 0$ for all $\mathbf{x} \neq 0$ and $V(\mathbf{x}) = 0$ for $\mathbf{x} = 0$, and, in addition, that $\dot{V} < 0$ for all $\mathbf{x} \neq 0$ in some closed region, then the system will be asymptotically stable in that region.

As might be anticipated from our stirred-heater example, it is always easy to find functions that satisfy the condition that $V(\mathbf{x}) > 0$ for $\mathbf{x} \neq 0$ and that are zero at the origin, but normally it is extremely difficult to also satisfy the condition that $\dot{V}(\mathbf{x}) < 0$ for $\mathbf{x} \neq 0$. This is the practical difficulty we referred to early in our discussion.

However, in certain cases it is possible to develop fairly general results. For example, for sets of linear equations

$$\dot{\mathbf{x}} = \mathbf{A}\mathbf{x} \qquad\qquad 7.4\text{-}40$$

we can choose V to be the quadratic form

$$V = \mathbf{x}^T\mathbf{Q}\mathbf{x} \qquad\qquad 7.4\text{-}41$$

We want V to be positive for all values of $\mathbf{x}$ except $\mathbf{x} = 0$; therefore we require that $\mathbf{Q}$ be a positive definite matrix. This is equivalent to requiring that the determinants

$$|q_{11}|, \quad \begin{vmatrix} q_{11} & q_{12} \\ q_{21} & q_{22} \end{vmatrix}, \quad \begin{vmatrix} q_{11} & q_{12} & q_{13} \\ q_{21} & q_{22} & q_{23} \\ q_{31} & q_{32} & q_{33} \end{vmatrix}, \cdots \qquad 7.4\text{-}42$$

are all positive. This restriction on $\mathbf{Q}$ is not sufficient, however, as can be seen by developing the expression for $\dot{V}$

$$\begin{aligned} \dot{V} &= \dot{\mathbf{x}}^T\mathbf{Q}\mathbf{x} + \mathbf{x}^T\mathbf{Q}\dot{\mathbf{x}} \\ &= \mathbf{x}^T\mathbf{A}^T\mathbf{Q}\mathbf{x} + \mathbf{x}^T\mathbf{Q}\mathbf{A}\mathbf{x} \\ &= -\mathbf{x}^T\mathbf{R}\mathbf{x} \qquad\qquad\qquad 7.4\text{-}43 \end{aligned}$$

where

$$-\mathbf{R} = \mathbf{A}^T\mathbf{Q} + \mathbf{Q}\mathbf{A} \qquad 7.4\text{-}44$$

In order for $\dot{V}$ to be negative for all values of $\mathbf{x}$ in Eq. 7.4–43, we must have that $\mathbf{R}$ is a positive definite matrix, where the relationship between $\mathbf{Q}$ and $\mathbf{R}$ is given by Eq. 7.4–44. In other words, we can choose any positive definite matrix $\mathbf{R}$ (simply by forming $\mathbf{S}^T\mathbf{S}$, where $\mathbf{S}$ is an arbitrary matrix), solve Eq. 7.4–44 for the elements of $\mathbf{Q}$, and if $\mathbf{Q}$ turns out to be positive definite, Eq. 7.4–44 will be a Liapunov function and we can guarantee that the origin is asymptotically stable.

Krasovskii's method[5] provides a general approach for many nonlinear systems of the form

$$\dot{\mathbf{x}} = \mathbf{f}(x) \qquad \mathbf{f}(0) = 0 \qquad 7.4\text{-}45$$

First we let

$$V = \mathbf{f}^T\mathbf{f} \qquad 7.4\text{-}46$$

and we note that

$$\dot{\mathbf{f}} = \mathbf{A}\dot{\mathbf{x}} = \mathbf{A}\mathbf{f} \qquad 7.4\text{-}47$$

where $\mathbf{A}$ is the matrix of first partial derivatives of $\mathbf{f}$ (the Jacobian matrix we use in the normal linearization procedure). Then

$$\dot{V} = \dot{\mathbf{f}}^T\mathbf{f} + \mathbf{f}^T\dot{\mathbf{f}} = \mathbf{f}^T\mathbf{A}^T\mathbf{f} + \mathbf{f}^T\mathbf{A}\mathbf{f} = -\mathbf{f}^T\mathbf{R}\mathbf{f} \qquad 7.4\text{-}48$$

where in this case

$$-\mathbf{R} = \mathbf{A}^T + \mathbf{A} \qquad 7.4\text{-}49$$

It is clear that V will always be a positive function, so that if $\mathbf{R}$ is a positive definite matrix (and therefore $\dot{V}$ is always negative), the system will be asymptotically stable.

[5] See footnote 3 on p. 58.

A number of other techniques for generating Liapunov functions are available.[6] Unfortunately, for most chemical process problems, the predicted region of asymptotic stability turns out to be much smaller than the actual stable domain; that is, Liapunov's method seems to be extremely conservative. Warden, Aris, and Amundson[7] describe a procedure for extending the method to cover larger regions, but they point out that the analysis is almost as difficult as solving the nonlinear equations numerically to obtain a complete description of the dynamic behavior of the system.

Example 7.4–4 Application of Krasovskii's method to a CSTR

Use Krasovskii's method to develop the stability criteria for a CSTR.

Solution

From Eq. 7.4–49 we know that the matrix

$$\mathbf{R} = -(\mathbf{A}^T + \mathbf{A})$$

must be positive definite. Using the definitions given in Eqs. 7.1–6, we have

$$\mathbf{A} = \begin{pmatrix} a_{11} & a_{12} \\ a_{21} & a_{22} \end{pmatrix} = \begin{pmatrix} -(1 + \alpha) & -\gamma \\ \alpha & -(1 + \beta - \gamma) \end{pmatrix}$$

so that

$$\mathbf{R} = \begin{pmatrix} 2(1 + \alpha) & \gamma - \alpha \\ \gamma - \alpha & 2(1 + \beta - \gamma) \end{pmatrix} \qquad 7.4\text{–}50$$

In order for **R** to be positive definite, we require that the determinants in Eq. 7.4–42 are positive, or

$$2(1 + \alpha) > 0 \qquad 7.4\text{–}51$$

$$4(1 + \alpha)(1 + \beta - \gamma) - (\gamma - \alpha)^2 > 0 \qquad 7.4\text{–}52$$

From the physics of the problem, Eq. 7.4–51 will always be satisfied. However, Eq. 7.4–52 is a new result.

The criteria we obtained using characteristic roots were

$$1 + \alpha + 1 + \beta - \gamma > 0 \qquad 7.1\text{–}9$$

$$(1 + \alpha)(1 + \beta) - \gamma > 0 \qquad 7.1\text{–}10$$

Since Eq. 7.4–52 can be put into the form

$$[(1 + \alpha)(1 + \beta) - \gamma] - \left[\frac{(\gamma - \alpha)^2}{4} + \alpha\gamma\right] > 0 \qquad 7.4\text{–}53$$

It is a more conservative restriction than Eq. 7.1–10. This is in contrast to Eq. 7.4–51, which is not as restrictive as Eq. 7.1–9.

[6] A useful list of references is given in D. M. Himmelblau and K. B. Bischoff, *Process Analysis and Simulation*, p. 153, Wiley, N. Y., 1968.

[7] R. B. Warden, R. Aris, and N. R. Amundson, *Chem. Eng. Sci.*, **19**, 149, 173 (1964).

A region of asymptotic stability can be plotted in the composition-temperature phase plane by plotting the locus of points where each of the stability criteria is violated and then determining the contour, $V = f^T f = $ constant, which is tangent to the closest one of these curves. An example of this approach has been published by Berger and Perlmutter.[8,9]

Other Definitions of Stability

In our discussions of stability we have emphasized local asymptotic stability: if a system at equilibrium is perturbed slightly, it will return to equilibrium conditions in an asymptotic manner. However, we could use a more general definition: if a system at equilibrium is perturbed slightly, it will always remain in some small neighborhood of the origin. This second definition would include the case of a harmonic oscillator, which does not meet the test of asymptotic stability. Whenever a system exhibits asymptotic stability for arbitrarily large perturbations, we say that it is *globally stable*. Although this case would seem to be the most desirable situation from an engineering standpoint, it is the most difficult to establish mathematically.

SECTION 7.5 DESIGN PROCEDURES USING BODE DIAGRAMS

We have developed a great deal of information concerning the dynamic behavior of chemical processes and the manner in which feedback control systems can be used to alter that behavior. It is obvious that what we want to do is select a feedback control system and the appropriate controller gains so that the controlled plant will have a more desirable dynamic response than the original uncontrolled plant; that is, we want to develop a procedure for synthesizing control systems. Before this step can be accomplished, however, we must be willing to make a quantitative statement of what we consider a "desirable response."

Performance Measures

The early work on the design of control systems was almost entirely in the province of electrical engineering. A majority of the plants of interest to them were single-input, single-output systems described by linear, ordinary differential equations. Therefore the techniques for synthesizing controllers were directed toward plants of this type. In addition, the primary consideration was the development of servomechanism control systems, although, as we indicated in Section 7.2, there are only minor differences between regulating control

[8] J. S. Berger, and D. D. Perlmutter, *AIChE Journal*, **10**, 233, 238 (1964).

[9] J. S. Berger, and D. D. Perlmutter, *Chem. Eng. Sci.*, **20**, 147 (1965).

problems and servomechanism systems. For this reason, the techniques that we will describe below are often referred to as classical, linear, servomechanism theory, even though our discussion will be directed toward the design of regulating controllers.

The basic hypothesis of all these synthesis techniques is that the most desirable response of a plant resembles that of a second-order underdamped system. This hypothesis was tested for a tremendous number of electrical systems and always gave a satisfactory performance. In other words, after a large number of experimental studies on various kinds of plants with various kind of controllers, a "rule of thumb" emerged that we wanted to obtain the fastest response of the controlled plant without having too many oscillations in the output signal. This same performance measure, plus the same design procedures, was adopted by the chemical industry.

Of course, high-order plants or nonlinear processes with feedback control loops will not behave exactly the same as a second-order system. Thus in order not to place too stringent a requirement on the design methods, it is common practice to define a region where the total response of the plant and controller is judged acceptable. Although this approach is eminently reasonable from an engineering point of view, it means that a number of different control systems might lead to an overall response that falls inside the acceptable region. Consequently, we cannot expect to obtain a unique design for a controller. This situation is in contrast to an optimal control system, which we will discuss in Chapter 9, but it turns out that an equivalent problem is encountered with that approach.

From our previous study of the dynamic characteristics of second-order systems, we know that we can obtain all the information of interest from the impulse, step, or frequency response. The standard practice is to choose the step and frequency response, because these provide a time domain and a frequency domain description. The step response is preferable to the impulse response because a step input is simple to introduce, provides a fairly drastic disturbance, activates the important dynamic modes, and tends to force the system to a new steady state operating condition. A sketch of the acceptable region for the dynamic response of a plant with its controller is shown in Figure 7.5-1 for the case of a step input and in Figure 7.5-2 for the case of a sinusoidal driving force.[1]

Now we see why we spent so much time in Section 4.5 introducing various special definitions that can be used to describe the response of second-order processes. Not all these definitions are independent, however. For example,

---

[1] These curves are taken from Olle I. Elgerd, *Control Systems Theory*, Ch. 7, McGraw-Hill, N. Y., 1967, although the results initially were published by J. E. Gibson et al., "Specification and Data Presentation in Linear Control Systems," *Report AFMDC-TR-61-5*, Purdue University, Lafayette, Ind., May, 1961, and AIEE Trans., **80**, Part 2 (Applications and Industry), No. 54, 65 (1961).

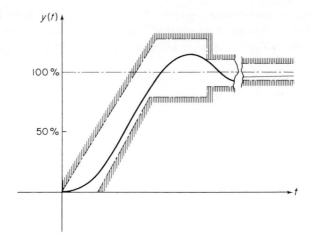

Figure 7.5-1. Desired step response of plant and controller. (Reperinted from Olle I. Elgerd, *Control Systems Theory,* Chap. 7, McGraw-Hill, N. Y., 1967, with permission of McGraw-Hill Book Co. See footnote 1, p. 67.)

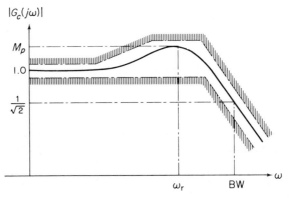

Figure 7.5-2. Desired frequency response of plant and controller. (Reprinted from Olle I. Elgerd, *Control Systems Theory,* Chap. 7, McGraw-Hill, N. Y., 1967, with permission of McGraw-Hill Book Co. See footnote 1, p. 67.)

if we specify that the decay ratio of the output (the ratio of the height of the second peak to the first, see Eq. 4.5–32) should be approximately $\frac{1}{4}$, we have implicitly required that the overshoot be $\frac{1}{2}$ (see Eq. 4.5–31). Typical design values for the decay ratio are in the range $\frac{1}{4}$ to $\frac{1}{3}$ and for the overshoot 0.3 to 0.5. In order to characterize the dynamic response completely, it is also necessary to specify a value for the rise time, response time, or the period of the oscillation. No "rules of thumb" are available to fix these quantities, for they are strongly dependent on the dynamics of the original plant. Thus we can hardly expect some process that has an apparent time constant of several hours to have a rise time in the order of milliseconds after we install a feedback controller. There are cases, however, where the settling time might be more important than the decay ratio. For example, a large distillation column that requires a 2– or 3–day start-up period might not be economical, whereas if a controller can be used to decrease this start-up time to a few hours, the plant might be profitable. Since cases of this kind are relatively rare, perhaps it is worth emphasizing that the synthesis problem for a controller, using a time domain representation, is usually based on a specification of the decay ratio or overshoot, as well as the offset.

With this background, we can now speculate about a procedure for designing control systems using a time domain representation. First, we would solve the dynamic equations for a step change in some input of importance. For most chemical processes, we expect this result to look like a second-order over-damped system. Then we would select a particular feedback control system and a particular set of gains for the control modes, modify the system equations to take into account the presence of the feedback loop, solve the equations again for the step response, and see if the decay ratio or overshoot is in the appropriate range and if the offset is acceptable. If not, we would change the controller gains and try again. If we are still unsuccessful, we would change the kind of controller we had selected and try various values for the gain settings until we eventually achieve a satisfactory result. When we envision applying this method to a plate, gas absorption unit, such as that described in Example 4.6–3, we quickly recognize that we need to develop a simpler approach. In Section 7.6 we will show that a root-locus method can be used for this purpose, although this technique has not been favored by chemical engineers. Instead, the most common procedure is to study the behavior of the system in the frequency domain.

We are hopeful that a frequency-domain synthesis technique will avoid the difficulties just discussed, for it will not require an inversion of complicated expressions in the Laplace parameter s. Also, it should be an easier job to account for modifications of the controller in the frequency domain because we are only considering their transfer functions. Before describing a few of the synthesis techniques, it is helpful to introduce some of the particular terms commonly used to describe the frequency response of a second-order underdamped system and to give the typical design values for these quantities. The definitions are analogous to those for the time domain representation presented in Chapter 4.

Consider the gain curve for a second-order underdamped system shown in Figure 7.5–2. The definitions of interest are listed below.

Bandwidth. For a first-order system, we characterize the process dynamics by a time constant. Similarly, we often use an apparent time constant to provide an order of magnitude description for higher-order systems, by determining the time it takes for the plant to reach 63 percent of its final value after a step change. In the frequency domain, the gain of the system at a frequency equal to the reciprocal of the time constant is $1\sqrt{2}$ (see Eq. 4.3–24). In our previous discussion we called this value, $\omega_c = 1/\tau$, the corner frequency, but it is also called the bandwidth of the first-order process. For higher-order plants, we define the bandwidth, BW, as the frequency where the gain curve has decreased to $1/\sqrt{2}$ times its low-frequency value. Thus high bandwidths correspond to low apparent time constants and a rapid response.

Resonance peak and resonance frequency. A second-order underdamped system may exhibit resonance. The gain at the resonance peak is closely related to the settling time, where high peak values correspond to long settling times. Dangerous situations may arise if the gain at resonance, M_p, is very large, for input signals having frequency components near the resonance frequency, ω_r, might produce such large output fluctuations that the system is damaged. Experience has shown that an acceptable performance is generally obtained if the gain is between 1.5 and 3.0.

Gain margin. In our previous discussion of the use of Bode plots to determine the stability of a feedback control system, we showed that the transition from stable to unstable operation occurred when the system gain was equal to unity at the frequency corresponding to a phase lag of 180 deg. Thus we can attempt to ensure that the feedback system will be stable by requiring that the loop gain is always less than unity at this frequency. This procedure is equivalent to defining the gain margin as

$$\text{Gain margin} = \frac{1}{|G|} \qquad\qquad 7.5\text{--}1$$

at the appropriate frequency and then specifying that the gain margin should be between 1.75 and 2.0. In other words, if for any reason the loop gain in-

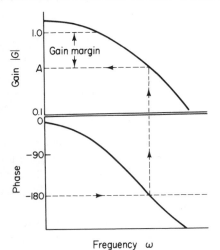

Frequency ω Figure 7.5-3. Gain margin.

creased by a factor less than 1.75 or 2.0, the system would still be stable. Figure 7.5–3 illustrates the gain margin on a Bode plot.

Phase margin. An alternate way of making certain that the closed-loop control system will be stable is to specify that the phase lag is less than 180 deg at the frequency where the system gain is equal to unity. Typical design values for the phase margin are in the range 45 to 60 deg. The graphical significance of the phase margin is presented in Figure 7.5–4. It should be noted that the

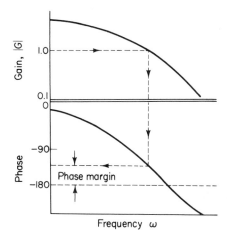

Figure 7.5-4. Phase margin.

phase margin can be used as a design criterion even for second-order systems, whereas the gain margin has no meaning, because the phase lag approaches 180 deg only asymptotically.

A number of other quantities are also used to provide performance measures for a control system. The quantities we choose to emphasize are the gain and phase margins. The design is accomplished by attempting to combine the transfer function of a controller with the transfer function of the plant, so that the resulting gain or phase margin is in the appropriate range. If we use a Bode diagram, we know that multiplication of the transfer functions only requires that we graphically add their characteristics on the Bode plot. Consequently, this process seems to provide a simple approach for synthesizing controllers, once the Bode diagrams for the various kinds of controllers have been established.

Bode Diagrams for Ideal Controllers

The Bode diagrams for ideal controllers are simple to determine. All we need do is substitute $j\omega$ for s in the transfer function, rationalize the complex number obtained, and plot both the amplitude and phase angle of this number against frequency. The equations and graphs follow.

Proportional controller. The transfer function of a proportional controller is simply

$$G_c(s) = K_c \qquad\qquad 7.5\text{-}2$$

Hence the system gain is K_c and the phase angle is equal to zero at all frequencies (see Figure 7.5–5). This means that the effect of a proportional controller is to add a constant value to every point on the open-loop gain curve; that is, it shifts it upward.

Proportional plus integral control. The transfer function for this case can be written

$$G_c(s) = K_c\left(1 + \frac{1}{T_i S}\right) \qquad\qquad 7.5\text{-}3$$

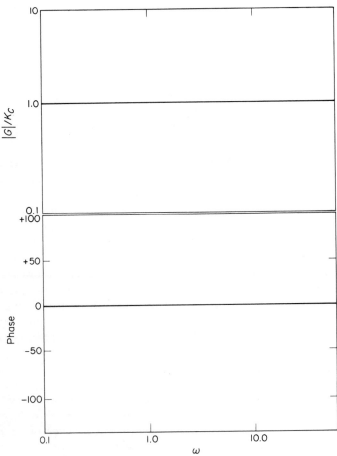

Figure 7.5-5. Bode plot for a proportional controller.

so that its frequency response is

$$G_c(j\omega) = K_c\left(1 + \frac{1}{T_i j\omega}\right) \qquad 7.5\text{-}4$$

Thus the gain and phase are

$$|G_c| = K_c\sqrt{1 + \left(\frac{1}{T_i\omega}\right)^2} \qquad 7.5\text{-}5$$

$$\phi = \tan^{-1}\left(\frac{-1}{T_i\omega}\right) \qquad 7.5\text{-}6$$

and these results are plotted on Figure 7.5-6. The corner frequency on the

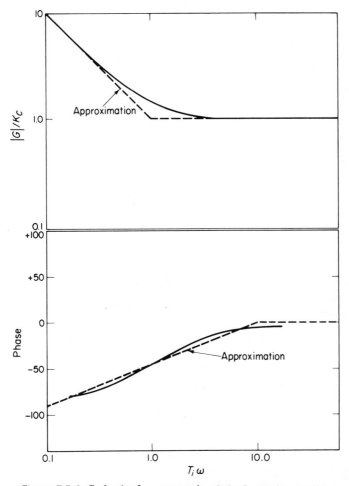

Figure 7.5-6. Bode plot for a proportional plus Integral controller.

graph is $\omega_c = 1/T_i$, where T_i is called the reset time. The straight-line, asymptotic approximation of the gain curve and an approximation of the phase curve (which we will discuss later) are also shown.

Proportional plus derivative control. For this case, the transfer function is

$$G_c(s) = K_c(1 + T_D s) \qquad\qquad 7.5\text{-}7$$

or in terms of frequency,

$$G_c(j\omega) = K_c(1 + jT_D\omega) \qquad\qquad 7.5\text{-}8$$

Proceeding as before, we obtain

$$|G_c| = K_c\sqrt{1 + (T_D\omega)^2} \qquad\qquad 7.5\text{-}9$$

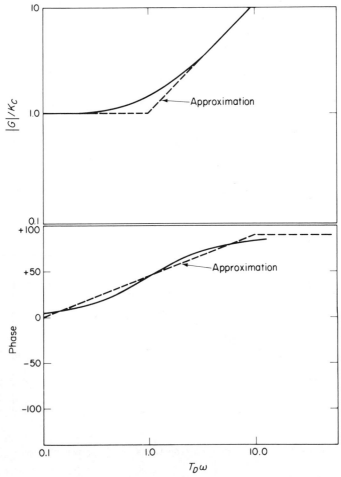

Figure 7.5-7. Bode plot for proportional plus derivative controller.

and

$$\phi = \tan^{-1}(T_D\omega) \qquad 7.5\text{--}10$$

These expressions, as well as an approximation of the curves, are plotted on Figure 7.5–7. It should be noted that the derivative mode introduces a phase lead into the system, and this behavior is often helpful in stabilizing a plant by increasing the phase margin.

Example 7.5–1 Statement of control problem

In order to illustrate the various techniques for designing simple feedback control systems, it is helpful to consider a particular plant. For this purpose, we choose to study an isothermal CSTR and the complex reaction scheme (see Figure 7.5–8),

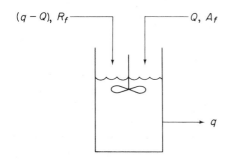

Figure 7.5-8. Isothermal CSTR.

$$A + R \longrightarrow B$$
$$B + R \longrightarrow C$$
$$C + R \rightleftharpoons D \qquad 7.5\text{--}11$$
$$D + R \longrightarrow E$$

We suppose that component R is in sufficiently large excess so that the reaction rates can be approximated by first-order expressions. The system equations are written as

$$V\frac{dA}{dt} = QA_f - qA - k_1VA \qquad 7.5\text{--}12$$

$$V\frac{dB}{dt} = -qB + k_1VA - k_2VB \qquad 7.5\text{--}13$$

$$V\frac{dC}{dt} = -qc + k_2VB - k_3VC + k_3'VD \qquad 7.5\text{--}14$$

$$V\frac{dD}{dt} = -qD + k_3VC - k_3'VD - k_4VD \qquad 7.5\text{--}15$$

where we have assumed that the feed compositions of components B, C, D, and E are zero and that the rate constants have the form

$$k_1 = k_AR, \quad k_2 = k_BR, \quad k_3 = k_cR, \quad k_4 = k_DR \qquad 7.5\text{--}16$$

Actually, for a complete description of the system behavior, we should also include a material balance equation for component R. However, if this component is in large excess and remains relatively constant, we can neglect its dynamic behavior in comparison with the other components.

At steady state conditions, the accumulation terms are all equal to zero; and the steady state compositions of the various components are

$$A_s = \frac{QA_f}{q + k_1V}, \qquad B_s = \frac{Qk_1VA_f}{(q + k_1V)(q + k_2V)}$$

$$C_s = \frac{Qk_1Vk_2V(q + k_3'V + k_4V)A_f}{(q + k_1V)(q + k_2V)[(q + k_3V)(q + k_4V) + qk_3'V]} \qquad 7.5\text{-}17$$

$$D_s = \frac{Qk_1Vk_2Vk_3VA_f}{(q + k_1V)(q + k_2V)[(q + k_3V)(q + k_4V) + qk_3'V]}$$

For a particular case, where $A_f = 0.001$ lb mole/ft^3, $Q = 100$ ft^3/hr, $q = 200$ ft^3/hr, $k_1 = 2.0$ hr^{-1}, $k_2 = 3.5$ hr^{-1}, $k_3 = 1.2$ hr^{-1}, $k_3' = 0.4$ hr^{-1}, and $k_4 = 1.0$ hr^{-1}, we can show that the reactor volume that gives the maximum amount of the desired component C is $V = 235$ ft^3. Even though this procedure represents a particular kind of an optimum steady state design problem, it is apparent that the true optimum design conditions should be established using economic considerations similar to those discussed in Chapter 2.

Nevertheless, here we consider the possibility of designing a control system that will maintain the composition of component C as close as possible to the steady state design value, despite the fact that disturbances enter the system. We attempt to accomplish this objective by measuring the actual composition of C, and using the difference between the desired and measured values to manipulate the inlet flow rate Q of component A in some way. Of course, when Q is varied, the feed rate of component R will also change according to Figure 7.5–8, so that neglecting the dynamic equation for component R might introduce an error in our analysis. In addition, for simplicity, we assume that the disturbances are due to changes in the composition of component R in the vessel, and we know that these changes could be predicted by including the extra material balance. Despite this apparent discrepancy, we will proceed with the analysis, recognizing that once we have selected a control system, we could develop the dynamic equation for component R, and evaluate the error introduced by our simplifying assumptions. Thus for small deviations from steady state operation, we can linearize the system equations around the steady state conditions to obtain

$$\dot{x} = a_{11}x_1 + b_{11}u + c_{11}v$$

$$\dot{x}_1 = a_{21}x_1 + a_{22}x_2 + c_{12}v$$

$$\dot{x}_3 = a_{32}x_2 + a_{33}x_3 + a_{34}x_4 + c_{13}v \qquad 7.5\text{-}18$$

$$\dot{x}_4 = a_{43}x_3 + a_{44}x_4 + c_{14}v$$

where

$$x_1 = \frac{A - A_s}{A_f}, \quad x_2 = \frac{B - B_s}{A_f}, \quad x_3 = \frac{C - C_s}{A_f}, \quad x_4 = \frac{D - D_s}{A_f}, \quad u = \frac{Q - Q_s}{Q_s}$$

$$v = \frac{R - R_s}{R_s}, \quad a_{11} = -\left(\frac{q}{V} + k_{1s}\right), \quad b_{11} = \frac{Q_s}{V}, \quad c_{11} = -\left(k_{1s}\frac{A_s}{A_f}\right)$$

$$a_{21} = k_{1s}, \quad a_{22} = -\left(\frac{q}{V} + k_{2s}\right), \quad c_{12} = \left(k_{1s}\frac{A_s}{A_f} - k_{2s}\frac{B_s}{A_f}\right), \quad a_{32} = k_{2s}$$

$$a_{33} = -\left(\frac{q}{V} + k_{3s}\right), \quad a_{34} = k_{3s}', \quad c_{13} = \left(k_{2s}\frac{B_s}{A_f} - k_{3s}\frac{C_s}{A_f}\right), \quad a_{43} = k_{3s}$$

$$a_{44} = -\left(\frac{q}{V} + k_{3s}' + k_{4s}\right), \qquad c_{14} = \left(k_{3s}\frac{C_s}{A_f} - k_{4s}\frac{D_s}{A_f}\right) \qquad \text{7.5-19}$$

As an alternate problem, we could have considered the reactor temperature to be the disturbance. This change would not affect the form of the linearized equations above, although the definitions of c_{11} through c_{14} would be different. For example, if we let $v = (T - T_s)/T_s$, then $c_{11} = k_{1s}A_sE_1/RT_s$. Therefore we will try to keep the analysis in general terms, so that we can apply the results to other problems.

Taking the Laplace transforms of the linearized equations and solving for the outputs in terms of the inputs gives the expressions

$$\tilde{x}_1 = \frac{b_{11}}{s - a_{11}}\tilde{u} + \frac{c_{11}}{s - a_{11}}\tilde{v}$$

$$\tilde{x}_2 = \frac{a_{21}b_{11}}{(s - a_{11})(s - a_{22})}\tilde{u} + \frac{a_{21}c_{11} + (s - a_{11})c_{11}}{(s - a_{11})(s - a_{22})}\tilde{v}$$

$$\tilde{x}_3 = \frac{a_{21}a_{32}b_{11}(s - a_{44})}{(s - a_{11})(s - a_{22})[(s - a_{33})(s - a_{44}) - a_{34}a_{43}]}\tilde{u}$$

$$+ \left[\frac{(s - a_{11})\{(s - a_{22})[a_{34}c_{14} + c_{13}(s - a_{44})] + a_{32}c_{12}(s - a_{44})\}}{(s - a_{11})(s - a_{22})[(s - a_{33})(s - a_{44}) - a_{34}a_{43}]} + \right.$$

$$\left. \frac{a_{21}a_{32}c_{11}(s - a_{44})}{(s - a_{11})(s - a_{22})[(s - a_{33})(s - a_{44}) - a_{34}a_{43}]}\right]\tilde{v} \qquad \text{7.5-20}$$

$$\tilde{x}_4 = \frac{a_{21}a_{32}a_{43}b_{11}}{(s - a_{11})(s - a_{22})[(s - a_{33})(s - a_{44}) - a_{34}a_{43}]}\tilde{u}$$

$$+ \frac{(s - a_{11})\{a_{32}a_{43}c_{12} + (s - a_{22})[a_{21}c_{11} + c_{12}(s - a_{11})]\} + a_{21}a_{32}a_{43}c_{11}}{(s - a_{11})(s - a_{22})[(s - a_{33})(s - a_{44}) - a_{34}a_{43}]}\tilde{v}$$

The next to last expression is the one of primary interest. In order to simplify the notation, we write it as

$$\tilde{x}_3 = G_u\tilde{u} + G_v\tilde{v} \qquad \text{7.5-21}$$

where G_u and G_v can be determined by inspection.

At this point we assume that the dynamics of the measuring instrument are negligible, and we consider a general expression for the transfer function of the controller

$$\tilde{u} = G_c(\tilde{x}_0 - \tilde{x}_3) \qquad\qquad 7.5\text{--}22$$

so that G_c could include any or all of proportional, derivative, or integral control modes. Substituting this general controller equation into Eq. 7.5–21 and rearranging, we find that

$$\tilde{x}_3 = \frac{G_c G_u}{1 + G_c G_u}\,\tilde{x}_0 + \frac{G_v}{1 + G_c G_u}\,\tilde{v} \qquad\qquad 7.5\text{--}23$$

Although we should direct our attention to the second term on the right-hand side, because we are interested in designing a regulating control system and want $\tilde{x}_0$ to be equal to zero, actually we will concentrate most of our effort on studying the effect of the controller transfer function on the first term. This means that, in reality, we are designing a servomechanism control system, rather than a regulator, which is a common practice in control system synthesis. Of course, we hope that a good servo system will also act as a good regulator, and for this simple problem we will be able to check this assumption. The major reason for considering what appears to be the wrong problem is that in a great number of practical situations the disturbances are unknown, or cannot be measured, so that G_v is unknown. Despite this fact, we still want the controller to compensate for the disturbances. We have previously shown that the stability characteristics of both the servo and regulating control systems depend only on the denominators of the respective transfer functions, which are the same for the two cases, and we hope that the other dynamic characteristics will be similar. Hence the control system synthesis becomes a problem of specifying a controller transfer function G_c that will make the outlet composition of component C, or $\tilde{x}_3$, relatively insensitive to various kinds of changes in the controller set point $\tilde{x}_0$ (which would correspond to variations in the desired steady state output of component C).

Controller Design Using a Bode Plot

From our previous discussions of performance measures, we know what kind of a system output or frequency response we would like to achieve. The actual frequency response characteristics of the process can be represented on a Bode plot, using either the theoretical model or experimental data obtained from an operating unit, and we want to add a control system that will make the controlled plant have a desirable response. Perhaps the simplest way of illustrating this synthesis procedure for the control system is to consider our specific control problem again.

Example 7.5–2 Reactor control systems

The relationship between the desired output and the control variable was given in Eq. 7.5–20 as

$$\tilde{x}_3 = \frac{a_{21}a_{32}b_{11}(s - a_{44})}{(s - a_{11})(s - a_{22})[(s - a_{33})(s - a_{44}) - a_{34}a_{43}]} \tilde{u} \qquad 7.5\text{–}24$$

or

$$\tilde{x}_3 = G_u \tilde{u} \qquad 7.5\text{–}25$$

To prepare a Bode plot of this transfer function, we must evaluate the parameters. Using the values given earlier, we find that the steady state compositions are

$$A_s = 1.493 \times 10^{-4}, \quad B_s = 6.86 \times 10^{-5}, \quad C_s = 1.306 \times 10^{-4}$$
$$D_s = 6.96 \times 10^{-5}, \ E_s = 8.18 \times 10^{-5}$$

Then the coefficients in the linearized equations become

$$a_{11} = -2.85, \quad a_{21} = 2.0, \quad a_{22} = -4.35, \quad a_{32} = 3.5, \quad a_{33} = -2.05$$
$$a_{34} = 0.40, \quad a_{43} = 1.20, \quad a_{44} = -2.25, \quad b_{11} = 0.426$$

The quadratic factor in the denominator can be written as

$$(s - a_{33})(s - a_{44}) - a_{34}a_{43} = (s - a)(s - b)$$

and we can see that

$$a = -1.451 \quad \text{and} \quad b = -2.85$$

It is interesting to note that this value of b is equal to a_{11}, so that there is a pair of repeated roots in the denominator. This fact is due to a fortuitous choice of the original system parameters. We can rearrange the terms in the transfer function and put it into the more standard form

$$\frac{\tilde{x}_3}{\tilde{u}} = \frac{K(\tau_0 s + 1)}{(\tau_1 s + 1)^2 (\tau_2 s + 1)(\tau_3 s + 1)} \qquad 7.5\text{–}26$$

where the gain and time constants are

$$K = \frac{b_{11}a_{21}a_{32}(-a_{44})}{(-a_{11})(-a_{22})(-a)(-b)} = 0.1315, \quad \tau_0 = \frac{1}{-a_{44}} = 0.445$$

$$\tau_1 = \frac{1}{-a_{11}} = 0.351, \quad \tau_2 = \frac{1}{-a_{22}} = 0.230, \quad \tau_3 = \frac{1}{-a} = 0.689$$

In order to develop a first estimate of the design of a control system, we can use the asymptotic approximations of the various terms in the transfer function to draw the Bode plot. Figure 7.5–9 shows the approximations used for both the gain and phase curves. The corner frequencies for the two first-order terms in the denominator are $\omega_c = 1.45$ and $\omega_c = 4.35$, and both of these lines have a slope of -1. The corner frequency for the repeated root is $\omega_c = 2.85$, and this line has a slope of -2 since it represents a second-order system. Obviously the asymptotic approximation for the second-order system has a larger error than that for a first-order plant, as can be seen by examining Figure 4.5–5 for the case of critical damping. In fact, from Eq. 4.6–16 we find that the maximum error occurs at the corner frequency and that the actual

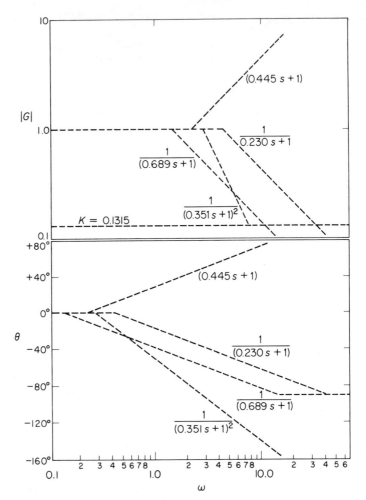

Figure 7.5-9. Asymptotic approximations for terms in the transfer function.

gains for the first- and second-order systems should be 0.707 and 0.5, respectively, rather than unity. The first-order term in the numerator has the same form as a proportional plus derivative controller, so that the line has a slope of $+1$ originating at the corner frequency $\omega_c = 2.25$.

Asymptotic approximations are also used for the phase curves. For the first-order terms in the denominator, these are drawn by fixing a phase lag of 45 deg at the corner frequency, a phase lag of 0 deg at $\omega_c/10$, and a phase lag of 90 deg at $10\omega_c$. The first-order term in the numerator is similar except that a phase lead is used. For the second-order term in the denominator, a 180 deg phase lag is distributed between $\omega_c/10$ and $10\omega_c$, with a 90 deg phase lag at the corner frequency.

The Bode plot is then obtained by adding the asymptotic curves, using the procedure described in Section 4.5. The results are shown in Figure 7.5–10. From this graph we find that the phase lag is 180 deg at a frequency of $\omega_{co} = 6.05$, and at this crossover frequency the system gain is $|G| = 0.0131$.

Proportional Control

Now if we consider the addition of a proportional feedback controller, we shift the gain curve upward without changing the phase. By setting the controller gain to $K_c = 1/|G| = 76.3$ the overall loop gain will be equal to unity when the phase lag is 180 deg, so that the closed-loop system will become marginally stable and start to generate continuous oscillations. We can ensure

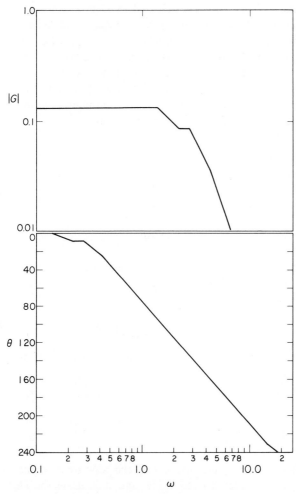

Figure 7.5-10. Bode plot for the reactor.

stable operation, and a satisfactory damped oscillatory response, by requiring that the gain margin is in the range from 1.75 to 2.0 or that the phase margin is between 45 and 60 deg. The gain margin criterion will be satisfied when the controller gain is

$$K_c = \frac{1}{1.75(0.0131)} = 43.6 \quad \text{to} \quad K_c = \frac{1}{2(0.0131)} = 38.1$$

while the phase margin criterion leads to the results

$$K_c = \frac{1}{[G|_{-135°}]} = 11.91 \quad \text{to} \quad K_c = \frac{1}{[G|_{-120°}]} = 11.43$$

For a conservative design, we would want to satisfy both criteria and thus might select a value of $K_c = 11.5$ as our first estimate.

However, we should notice that the system gains corresponding to phase lags of 135 and 120 deg—that is, $[G|_{-135°}] = 0.084$ and $[G|_{-120°}] = 0.0875$—occur in a region of the Bode plot where the slope of the asymptotic approximation is changing rapidly. We expect to obtain the greatest discrepancy between the approximate and the true Bode plots in a region of this type, for a slope change occurs at a corner frequency of one of the elements. Hence a Bode plot was prepared for the original transfer function. This is compared with the approximate results in Figure 7.5–11. Using this graph, we can calculate more accurate values for the controller gain

$$K_c = \frac{1}{1.75(0.0135)} = 42.3 \quad \text{to} \quad K_c = \frac{1}{2(0.0135)} = 37.0$$

and

$$K_c = \frac{1}{[G|_{-135°}]} = 25.3 \quad \text{to} \quad K_c = \frac{1}{[G|_{-120°}]} = 19.6$$

Thus we see that a controller gain setting of $K_c = 20$ is still a conservative value.

In order to verify that the closed-loop system performs properly for a step change in the input disturbance, we need to find the inverse transform of Eq. 7.5–23 with $\tilde{v} = A_v/s$.

$$\tilde{x}_3 = \left(\frac{G_v}{1 + G_c G_u} \right) \left(\frac{A_v}{s} \right) \qquad\qquad 7.5–27$$

After we substitute the appropriate expressions from Eq. 7.5–20, this becomes

$$\tilde{x}_3 = \frac{A_v}{s} \Bigg[\frac{(s - a_{11})\{(s - a_{22})[c_{14}a_{34} + c_{13}(s - a_{44})] + c_{12}a_{32}(s - a_{44})\}}{(s - a_{11})(s - a_{22})[(s - a_{33})(s - a_{44}) - a_{34}a_{43}] + K_c a_{21}a_{32}b_{11}(s - a_{44})}$$
$$+ \frac{c_{11}a_{21}a_{32}(s - a_{44})}{(s - a_{11})(s - a_{22})[(s - a_{33})(s - a_{44}) - a_{34}a_{43}] + K_c a_{21}a_{32}b_{11}(s - a_{44})} \Bigg]$$
$$7.5–28$$

The inverse is obtained by first evaluating the four roots of the denominator for $K_c = 20$, or any other desired value, and then using the Heaviside theo-

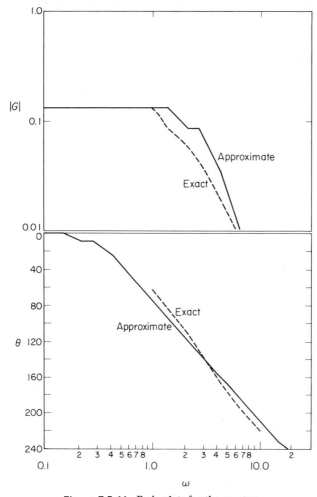

Figure 7.5-11. Bode plots for the reactor.

rems to find the inverse transform. With $K_c = 20$, it can be shown that the roots are

$$\lambda_1 = -2.240, \quad \lambda_2 = -7.137, \quad \lambda_{3,4} = -1.063 \pm 3.235j$$

The step response, $x_3(t)$, is shown in Figure 7.5–12 for a 15 percent change in v. In addition, the responses corresponding to controller gains of $K_c = 0$, 10, 30, and 40 are presented. It is apparent that the offset decreases as the gain increases, but the system also becomes more oscillatory. Remembering that another performance criterion we would like to satisfy is to achieve a decay ratio of about $\frac{1}{4}$, we see that a controller gain in the range $K_c = 20$ to 25 would be acceptable. It is interesting to note that the direction of the initial change in the output is opposite to that of the final curve. This situation is

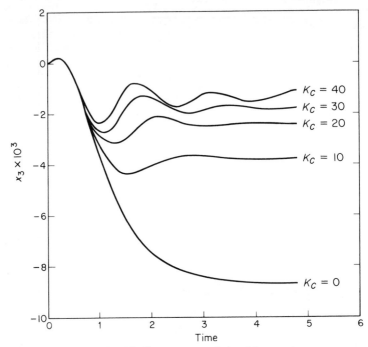

Figure 7.5-12. Step response of closed loop system.

due to the fact that the disturbance produces an immediate effect on x_3 because of the presence of the $c_{13} v$ term in the equation for $\dot{x}_3$ (see Eq. 7.5–18) and a slower effect because of the coupling of the equations.

Plants exhibiting this kind of behavior are often called non-minimum-phase systems. A *minimum-phase system* is defined as one where the transfer function is a ratio of polynomials and each root of both the polynomial in the numerator and denominator has a negative real part. For this case, there is a unique relationship between the gain and phase curves, and the phase angle asymptotically approaches the value $\phi = -90$ deg $(n - m)$, where n is the number of poles and m is the number of zeros of the transfer function. Since the roots of the numerator of Eq. 7.5–28 are

$$s_1 = +0.3612, \quad s_2 = -0.222 + 1.304j, \quad s_3 = -0.222 - 1.304j$$

it does not satisfy the minimum-phase criterion.

The amount of offset can be determined from the transformed expression for the closed-loop step response (Eq. 7.5–28) by applying the final value theorem. As t approaches infinity, we find that

$$x_3 = A_v \frac{c_{11}a_{21}a_{32}(-a_{44}) + c_{12}a_{32}(-a_{11})(-a_{44}) + c_{13}(-a_{11})(-a_{22})(-a_{44})}{(-a_{11})(-a_{22})[(-a_{33})(-a_{44}) - a_{34}a_{43}] + K_c a_{21}a_{22}b_{11}(-a_{44})}$$

$$+ \frac{c_{14}(-a_{11})(-a_{22})a_{34}}{(-a_{11})(-a_{22})[(-a_{33})(-a_{44}) - a_{34}a_{43}] + K_c a_{21}a_{22}b_{11}(-a_{44})} \qquad 7.5\text{–}29$$

A decision as to whether or not the amount of offset obtained with $K_c = 20$ (see Figure 7.5–12) is acceptable depends on the product quality specifications and the process economics. In order to present a more complete illustration of the control system synthesis technique, we will suppose that the product quality specifications require that we minimize the offset.

Proportional Plus Integral Control

Of course, we know that the offset can be eliminated by adding an integral mode to the controller, so that the controller transfer function becomes

$$G_c = K_c\left(1 + \frac{1}{T_i s}\right) \qquad \text{7.5–30}$$

Now we have two constants, K_c and T_i, which we hope to be able to choose so that the closed-loop system will have a more desirable dynamic response than the original reactor. The simplest approach for selecting these two constants is to consider the transfer function

$$\left(1 + \frac{1}{T_i s}\right)G_u$$

first and decide on an appropriate value for T_i, and then to select K_c so that we satisfy the gain and phase margin criteria. On a Bode plot this means that we add the transfer function for a proportional plus integral controller, without the proportional gain, $(1 + 1/T_i s)$, to the original open-loop transfer function for the plant, for various values of the reset time T_i. After we have determined which value of T_i will give us the best response characteristics, we can find the value of K_c that will make the closed-loop system have an overall gain of unity when the phase lag is 180 deg, so that it becomes marginally stable. Next, by specifying a gain or phase margin, we can fix the value of K_c such that we obtain a satisfactory, damped, oscillatory response.

Figure 7.5–13 shows the asymptotic approximations of the original reactor and the controller for the three reset times, $T_i = 1.0$, 0.5, and 0.2. When we add the curves, we obtain the results given in Figure 7.5–14. It is apparent from these graphs that the integral control mode introduces an additional phase lag. This means that a phase lag of 180 deg will be obtained at a lower frequency—that is, the crossover frequency is decreased—and that a lower value of K_c will make the overall system gain equal to unity. At the same time, the addition of the integral mode increases the system gain at low frequencies, which again has the effect of requiring a lower value of K_c in order to obtain an overall gain of unity when the phase lag is 180 deg. Both effects are detrimental because smaller values of K_c and the crossover frequency ω_{co} mean that the response will be slower.

Of course, we can minimize the effect of the integral mode on K_c and ω_{co} by choosing very large values of T_i; that is, we shift the dashed curves on

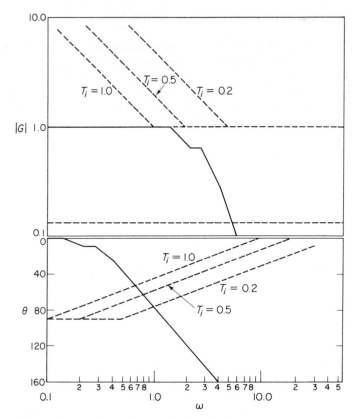

Figure 7.5-13. Asymtotic transfer function for process and $P + I$ controller.

Figure 7.5–13 far to the left. However, since the equation for the control action can be written as

$$u = -K_c \left(x_3 + \frac{1}{T_i} \int x_3 \, dt \right) \qquad 7.5\text{-}31$$

large values of T_i essentially mean that we are eliminating the integral action. Thus, it would take a very long time for the controller to remove the offset, and this factor would defeat our purpose in adding the integral mode. Now it is clear that we must be willing to make a compromise between rapid elimination of offset and high values of K_c and ω_{co}. Experience has shown that the best response is obtained when the integral mode just starts to have an appreciable effect on the critical frequency; that is, it adds about an extra 10 deg phase lag at the original crossover frequency. From Figure 7.5–6 we see that the phase lag of the controller is roughly 10 deg when $\omega T_i \approx 2\pi$, so that as a first estimate we set

$$T_i \approx \frac{2\pi}{\omega_{co}} \qquad 7.5\text{-}32$$

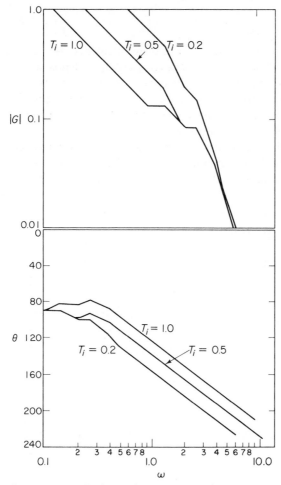

Figure 7.5-14. Bode plot for reactor and $P + I$ controller.

For our particular problem, this equation, or Figure 7.5–13, shows that we should make $T_i \approx 1.0$.

With this value of T_i, we learn from the Bode plot that the phase lag is 180 deg when the crossover frequency is $\omega_{co} = 4.65$ and the system gain is $|G| = 0.0275$. Then the value of K_c that will make the overall closed-loop gain unity is $K_c = 1/0.0275 = 36.4$. This gain will lead to undamped oscillations; therefore we must use a lower value. From the gain margin criterion, we find that

$$K_c = \frac{36.4}{1.75} = 20.8 \quad \text{or} \quad K_c = \frac{36.4}{2} = 18.2$$

while the phase margin criterion gives

$$K_c = \left[\frac{1}{|G|_{-135°}}\right] = 7.81 \quad \text{or} \quad K_c = \left[\frac{1}{|G|_{-120°}}\right] = 7.6$$

Again, some of the values used in these calculations occur in regions where the slope of the Bode diagram is changing rapidly—that is, a region where we expect to observe the maximum discrepancy between the exact transfer function and the asymptotic approximation. Thus, the original transfer functions of the plant and proportional plus integral controller were used to prepare a more accurate Bode plot in the region of interest. The curves are shown in Figure 7.5-15. Using the new values, we find that the crossover frequency is $\omega_{co} = 4.30$, and the gain at this frequency is $|G| = 0.020$; hence the controller gain corresponding to marginal stability is $K_c = 1/0.020 = 50$. From the gain mar-

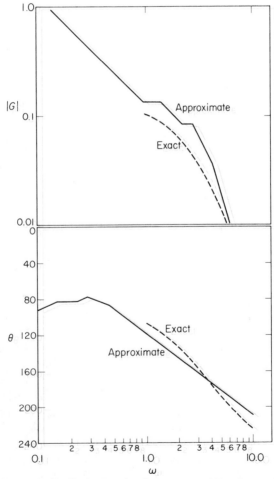

Figure 7.5-15. Bode plots for the reactor and $P + I$ controller.

gin criterion, we find that

$$K_c = \frac{50}{1.75} = 28.6 \quad \text{and} \quad K_c = \frac{50}{2} = 25$$

and from the phase margin criterion we obtain

$$K_c = \left[\frac{1}{G|_{-135°}}\right] = 15.9 \quad \text{and} \quad K_c = \left[\frac{1}{G|_{-120°}}\right] = 11.9$$

Thus we might select a value of $K_c = 12.0$ in order to satisfy all these criteria rather than a value of $K_c = 8.0$ as predicted by the asymptotic approximations. However, it should be noted that the integral mode adds more than a 10 deg phase lag at the original critical frequency, according to Figure 7.5-15. This means that we should increase the value of T_i somewhat; and after a few trial-and-error calculations, we would obtain slightly larger values for K_c. Nevertheless, the results of the calculations clearly show that the addition of the integral mode has forced us to decrease the proportional gain from the value obtained for just a proportional controller.

In order to determine the response of the closed-loop system to a step input, we must substitute the new controller transfer function into Eq. 7.5-27

$$\tilde{x}_3 = A_v \left[\frac{(s - a_{11})\{(s - a_{22})[c_{14}a_{34} + c_{13}(s - a_{44})] + c_{12}a_{32}(s - a_{44})\} + c_{11}a_{21}a_{32}(s - a_{44})}{s(s - a_{11})(s - a_{22})[(s - a_{33})(s - a_{44}) - a_{34}a_{43}] + K_c a_{21}a_{32}b_{11}(s - a_{44})(s + 1/T_i)}\right]$$

$$7.5\text{-}33$$

find the roots of the fifth-order polynomial in the denominator for the values of K_c and T_i of interest, and use the Heaviside theorems to determine the inverse transform. A fifth-order polynomial is obtained in this case because the

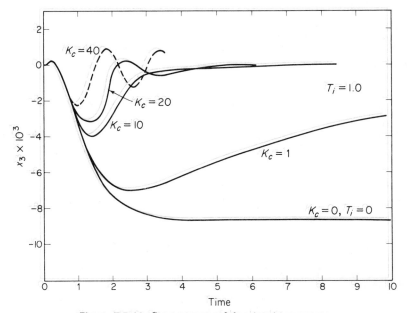

Figure 7.5-16. Step response of the closed loop system.

addition of the integral mode increases the order of the system. Some results for 15 percent changes in the disturbance, $A_v = 0.15$, are given in Figure 7.5–16 for $T_i = 0.1$ and values of $K_c = 1.0$, 10, 20, and 40.

It is apparent from the graph that the output becomes more oscillatory as the gain is increased. Also, the closed-loop response is slower than the corresponding system with just a proportional controller (see Figure 7.5–12) although the offset is completely eliminated. Judging from the step response data, it looks as if a gain setting in the range $K_c = 15$ to 25 will give a satisfactory response.

Proportional Plus Derivative Control

The preceding analysis demonstrates that the addition of an integral control mode will eliminate the possibility of offset, but it slows down the response of the system. If the product quality specifications are stringent, it would seem desirable to have a system that returns to steady state conditions very quickly. Our earlier discussions indicated that the addition of a derivative control mode often makes the system respond faster, for it introduces a large control action whenever the error signal is changing rapidly. Therefore we would like to explore the possibility of obtaining an improved response by including a derivative mode. Initially we will consider only a proportional plus derivative controller, in order to gain a good understanding of the effect of derivative action, and later we will study a three–mode controller.

The transfer function of the $P + D$ controller is

$$G_c = K_c(1 + T_D s) \qquad\qquad 7.5\text{–}34$$

Again there are two constants we can select, K_c and T_D. We make our decision by first considering the effect of the value of T_D on the combination of the open-loop transfer function of the plant and the controller

$$(1 + T_D s)G_u$$

Later we choose the controller gain, K_c, so that it satisfies the gain and phase margin criteria.

Figure 7.5–17 shows the Bode plot for the asymptotic approximations of the reactor and controller transfer functions for the four derivative times $T_D = 0.05$, 0.1, 0.2, and 0.5. The sum of the curves is given in Figure 7.5–18, and it should be noted that two scales are used for the ordinate of the gain plot. These graphs make it clear that a derivative mode introduces a phase lead into the system. This has the effect of shifting the crossover frequency to higher values, and thereby makes it possible to increase the value of K_c before stability considerations become important. However, as the amount of phase lead near the critical frequency is increased, the presence of the derivative mode also makes the overall gain of the system larger, so that the amount by which

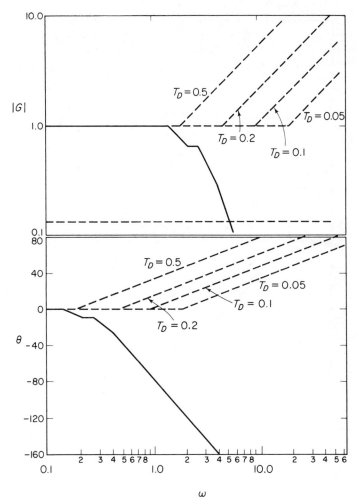

Figure 7.5-17. Asymptotic transfer function for process and $P + D$ controller.

K_c can be raised might start to fall off. Thus we normally attempt to select the derivative time T_D in such a way as to obtain the maximum value for K_c. From experience, it has been observed that this usually occurs when about a 45 deg phase lead is added at the original crossover frequency. At this point $\omega_{co} T_D = 1$, and as a first estimate we can let $T_D = 1/\omega_{co}$. For our particular problem, we find that we should take $T_D \approx 1/6.05 \approx 0.165$, but we will let $T_D = 0.1$. The reason for this choice will be discussed later.

When $T_D = 0.1$, we find from the graph that the crossover frequency is $\omega_{co} = 14.7$ and the system gain is $|G| = 0.001315$ when the phase lag is 180

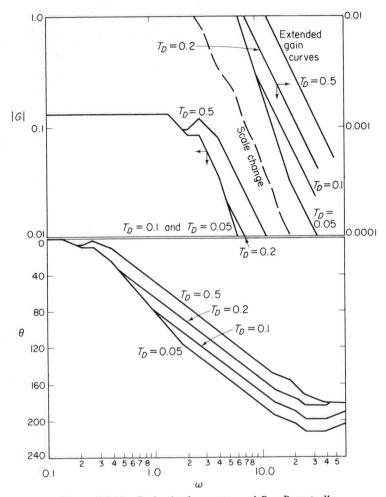

Figure 7.5-18. Bode plot for reactor and $P + D$ controller.

deg. Thus the controller gain corresponding to marginal stability is $K_c = 1/|G| = 760$, and using the gain margin criterion, we find that

$$K_c = \frac{760}{1.75} = 434 \quad \text{or} \quad K_c = \frac{760}{2} = 380$$

while the phase margin criterion gives

$$K_c = \frac{1}{[G|_{-135°}]} = 35.1 \quad \text{or} \quad K_c = \frac{1}{[G|_{-120°}]} = 14.1$$

For this case, the two criteria lead to very different results both because the slope of the gain curve is very large and because the slope of the phase curve is very small. Once again we see that we are in a region where the asymptotic approximations are changing slope, and therefore it seems wise to plot the exact transfer function. This result is shown in Figure 7.5-19. Now the crossover frequency is $\omega_{co} = 10.2$, the system gain at this frequency is $|G| = 0.00239$, the gain margin criterion gives

$$K_0 = \frac{419}{1.75} = 239 \qquad K_c = \frac{419}{2} = 210$$

and the phase margin criterion gives

$$K_c = \frac{1}{[G|_{-135°}]} = 16.4 \qquad K_c = \frac{1}{[G|_{-120°}]} = 11.0$$

If we plot the Bode diagrams of the exact transfer functions for $T_D = 0.15$ or $T_D = 0.2$, we find that the phase curve approaches -180 deg asymptoti-

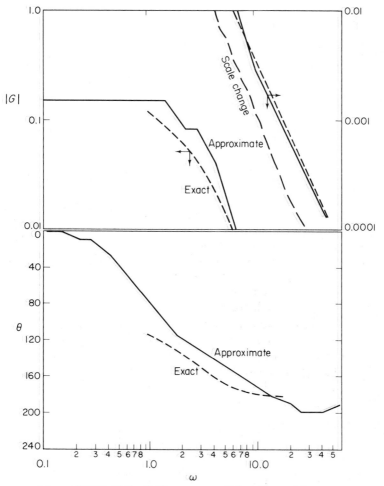

Figure 7.5-19. Bode plot for reactor and $P + D$ controller.

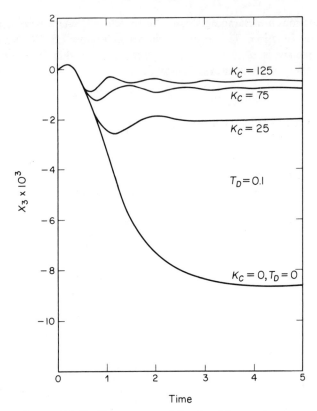

Figure 7.5-20. Step response of the closed loop system.

cally so that the crossover frequency occurs at $\omega_{co} = \infty$. This behavior arises because the transfer function consists of a second-order polynomial divided by a fourth-order polynomial, which may behave like a second-order system for certain values of the parameters. However, it means that the selection of the controller gain must be based only on the phase margin criterion because the gain margin has no meaning. In order to illustrate the synthesis procedure and the variation in results obtained by using the gain margin and phase margin criteria, it seemed to be better to select $T_D = 0.1$, rather than a somewhat higher value, which would probably give a better performance.

The step response of the closed-loop system can be determined by inverting the transfer function

$$\tilde{x}_3 = A_v \frac{(s - a_{11})\{(s - a_{22})[c_{14}a_{34} + c_{13}(s - a_{44})] + c_{13}a_{32}(s - a_{44})\} + c_{11}a_{21}a_{32}(s - a_{44})}{(s - a_{11})(s - a_{22})[(s - a_{33})(s - a_{44}) - a_{34}a_{43}] + K_c a_{21}a_{32}b_{11}(s - a_{44})(T_D s + 1)}$$

$$7.5\text{-}35$$

For $A_v = 0.15$, $T_D = 0.1$, and controller gains $K_c = 25$, 75, and 125. the response curves are shown in Figure 7.5–20. Comparison of these curves with our previous results reveals that the addition of the derivative control mode makes it possible to use higher controller gains so that the system has a faster response. Of course, if the controller gain is increased too much, the system response is too oscillatory. A gain setting in the range of $K_c = 25$ to 40 would seem to give a satisfactory performance.

After giving some thought to the foregoing analysis, we see that the potential advantages of a derivative mode can be quickly established simply by examining the slopes of the Bode plot in the neighborhood of the crossover frequency of the open-loop plant. Whenever the slope of the phase curve is small, derivative action will be beneficial, for the added phase lead will cause a large increase in the crossover frequency and a corresponding increase in the controller gain K_c. Similarly, derivative action will lead to an improved performance if the slope of the gain curve is large, because even slight changes in the crossover frequency make it possible to increase the controller gain significantly. However, for cases where the slope of the phase curve is large or the slope of the gain curve is small, a derivative mode will have little effect on the response. Series of first-order systems normally exhibit both small slopes of the phase curve and large slopes of the gain curve, whereas plants with a large dead time usually have the opposite characteristics.

Proportional Plus Integral plus Derivative Control, Three-Mode Control

When we combine all three control modes in a single controller, the transfer function is

$$G_c = K_c\left(1 + \frac{1}{T_i s} + T_D s\right) \qquad 7.5\text{–}36$$

Substituting $j\omega$ for s and rationalizing the complex numbers, we find that the gain and phase of this transfer function are

$$|G| = K_c\sqrt{\left(T_D\omega - \frac{1}{T_i\omega}\right)^2 + 1}$$

$$\phi = \tan^{-1}\left(T_D\omega - \frac{1}{T_i\omega}\right) \qquad 7.5\text{–}37$$

Providing that the derivative time T_D is small in comparison with the integral time T_i, we can draw the Bode plot for the controller by using the results of a proportional plus integral controller at low frequencies and the results for a proportional plus derivative controller at high frequencies. In other words, the gain curve will have a slope of -1 up to the corner frequency of $\omega = 1/T_i$, then a slope of zero up to the corner frequency $\omega = 1/T_D$, and, finally, a slope of $+1$. The phase curves can be added in a similar way. If the

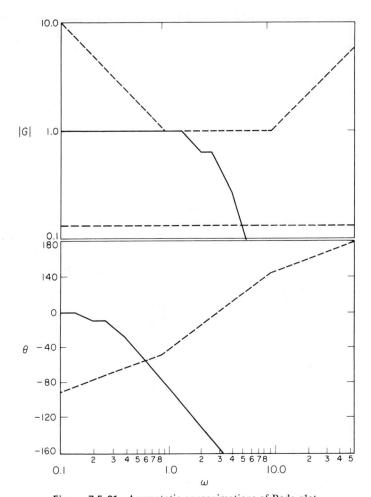

Figure 7.5-21. Asymptotic approximations of Bode plot.

derivative time is of the same order of magnitude as or larger than the integral time, it is necessary to plot the functions given by Eqs. 7.5-37.

A Bode plot of the asymptotic approximations of the reactor and the controller, excluding the controller gain K_c, are shown in Figure 7.5-21 for $T_i = 1.0$ and $T_D = 0.1$. The combined asymptotic transfer function, as well as the exact values, are plotted in Figure 7.5-22. Using the exact transfer function, we find that the crossover frequency is $\omega_{co} = 10.25$ and the gain at this frequency is $|G| = 0.00185$. Hence the gain margin criterion sets the controller gain as

$$\frac{1}{1.75(0.00185)} = 309 \quad \text{or} \quad K_c = \frac{1}{2(0.00185)} = 270$$

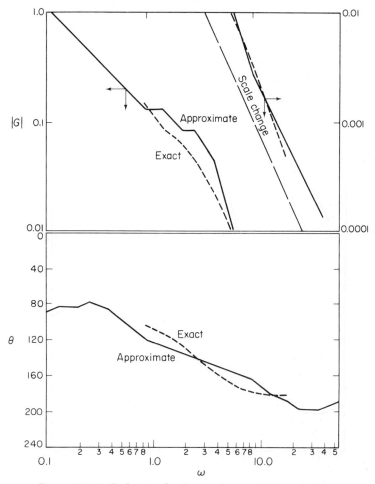

Figure 7.5-22. Bode plots for the reaction and *PID* controller.

whereas the phase margin criterion gives

$$K_c = \frac{1}{[G|_{-135°}]} = 20.8 \quad \text{or} \quad K_c = \frac{1}{[G|_{-120°}]} = 13.7$$

The step response of the closed-loop system is presented in Figure 7.5–23 for $K_c = 35$. The previous results for the uncontrolled plant, a proportional controller with $K_c = 25$, a proportional plus integral controller with $K_c = 20$ and $T_i = 1.0$, and a proportional plus derivative controller with $K_c = 40$ and $T_D = 0.1$, are also plotted on this graph, so that various kinds of control systems can be directly compared.

Perhaps it should be emphasized that no real attempt was made to evaluate the best values of the constants in the controller transfer function. Doing so

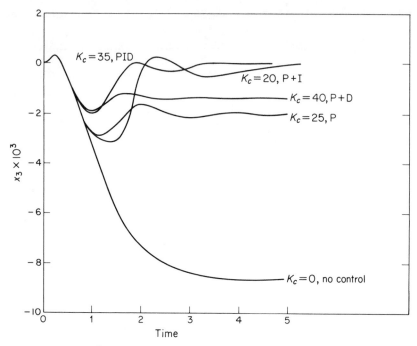

Figure 7.5-23. Step response of closed loop system.

would require a trial-and-error procedure. Our main purpose was to illustrate the synthesis approach rather than determine the best possible controller. In fact, the output deviations are so small, even for 15 percent changes in the disturbance (which we could have established, approximately, using the steady state model), that a controller would not be required unless the product quality specifications were very rigid or the product material was very valuable.

If we continue to presume that the control problem is meaningful, there are still a number of other factors that must be considered. For example, we based our selection of the controller constants on the gain and phase margin criteria and verified the settings by examining the step response. Thus one problem of interest would be to check the design further by developing the frequency response of the closed-loop system and seeing if the performance measures in the frequency domain are satisfied.

Closed-Loop Frequency Response from Open-Loop Data

In the control system synthesis problem just discussed, we found that the gain and phase margin criteria gave somewhat different values for the controller gain K_c. In addition, an examination of the closed-loop response of the system to step changes in the disturbance revealed that it might be advisable to make a further adjustment in the controller gain. The reason for this kind of behavior is that we are attempting to design a controller so that the closed-loop response of the plant will look a second-order underdamped system, and depending on which one of the performance measures we choose to consider, we get slightly different results. We hope to be able to satisfy all the different performance measures, as well as those for the closed-loop frequency response of the plant.

An evaluation of how well our previous designs satisfy the performance criteria for the closed-loop frequency response will require that we find some way of using our open-loop Bode diagrams to determine the amplitude ratios and phase angles of the closed-loop process. A simple technique for accomplishing this task is available. We note that the equation for the closed-loop system can be written

$$\tilde{x} = \frac{G_c G_u}{1 + G_c G_u} \tilde{x}_0 + \frac{G_v}{1 + G_c G_u} \tilde{v} \qquad 7.5\text{--}38$$

and recognize that our earlier studies were based on the open-loop transfer function of the plant and the controller, $G_c G_u$. If we represent this open-loop response by the expression

$$G_c G_u = A_0(\cos \theta_0 + j \sin \theta_0) \qquad 7.5\text{--}39$$

where A_0 is the amplitude, which now includes the controller gain K_c, and θ_0 is the phase, the closed-loop transfer function for set-point changes is

$$\begin{aligned} \frac{G_c G_u}{1 + G_c G_u} &= \frac{A_0(\cos \theta_0 + j \sin \theta_0)}{1 + A_0(\cos \theta_0 + j \sin \theta_0)} \\ &= \frac{A_0(A_0 + \cos \theta_0) + j A_0 \sin \theta_0}{(1 + A_0 \cos \theta_0)^2 + (A_0 \sin \theta_0)^2} \end{aligned} \qquad 7.5\text{--}40$$

Letting A_c be the magnitude of this complex number and θ_c its phase angle, we see that for the closed-loop system

$$A_c = \frac{A_0}{\sqrt{1 + 2A_0 \cos \theta_0 + A_0{}^2}} \qquad 7.5\text{--}41$$

and

$$\theta_c = \frac{\sin \theta_0}{A_0 + \cos \theta_0} \qquad 7.5\text{--}42$$

These are very general expressions that are applicable for any system. Thus by selecting arbitrary values for the open-loop amplitude ratio, A_0, and phase angle, θ_0, we can calculate the corresponding values for the closed-loop system, A_c and θ_c. These results can be plotted, and the final graph is called a Nichols chart (see Figure 7.5–24).

Now, for any particular system of interest, we can choose a frequency, find the amplitude ratio and phase angle of the open-loop system on a Bode plot, use these values to locate a particular point on the Nichols chart, and read the values of the closed-loop gain and phase angle from the Nichols chart. When we repeat this procedure for various frequencies, eventually we can obtain the information required to prepare a Bode plot for the closed-loop plant.

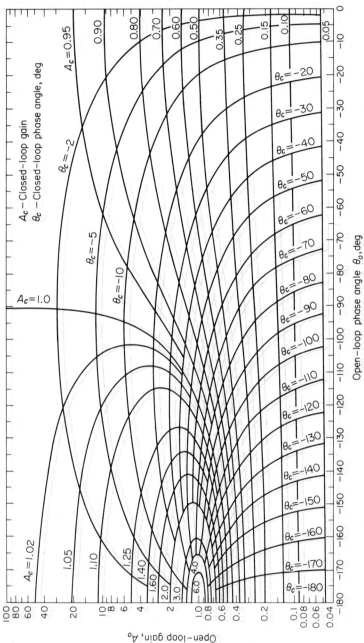

Figure 7.5-24. Nichols chart.

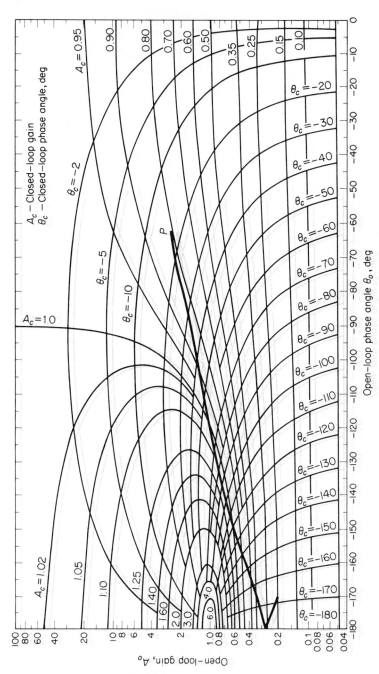

Figure 7.5-25. Nichols chart for reactor control system.

Only a slight modification of the procedure is necessary when we consider the transfer function for disturbances. In this case, we write

$$\frac{\tilde{x}}{\tilde{v}} = \frac{G_v}{1 + G_c G_u} = \frac{G_v}{G_c G_u}\frac{G_c G_u}{1 + G_c G_u}$$

7.5-43

Thus if we graphically add the Bode plot of the transfer function $(G_v/G_c G_u)$ to the plot of $G_c G_u/(1 + G_c G_u)$, which we can determine by using the procedure described above, we obtain the desired result.

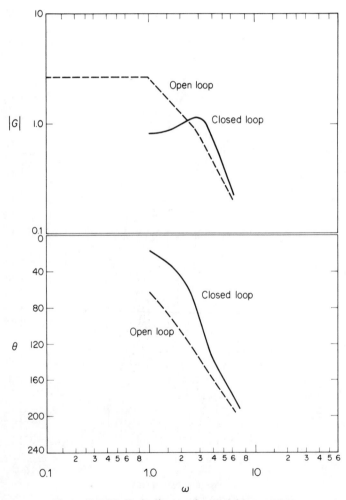

Figure 7.5-26. Bode diagram for closed loop system.

Example 7.5–3 Closed-loop frequency response

As an illustration of the use of the Nichols chart, let us consider the preceding reactor problem with just a proportional controller. The reactor transfer function, G_u, is given by Eq. 7.5–24, the controller transfer function is $G_c = K_c$, and for this example we let $K_c = 20$. We can select various values of frequency, read the amplitude ratio and phase from the Bode plot of the exact transfer function (given in Figure 7.5–11), multiply the amplitude ratio by the controller gain, $K_c = 20$, plot these points on the Nichols chart (see Figure 7.5–25), read the values of the closed-loop amplitude and phase from the Nichols chart, and, finally, plot these values on a new Bode diagram (see Figure 7.5–26). As a specific illustration, when $\omega = 1.0$, we find that $|G_u| = 0.130$ and $\theta_0 = -63$ deg from Figure 7.5–11. Thus $A_0 = K_c |G_u| = 2.60$. Point P on Figure 7.5–25 corresponds to these conditions, and we also find from this graph that $A_c = 0.81$ and $\theta_c = -17$ deg. Now we plot these values at $\omega = 1.0$ on Figure 7.5–26.

We see from the Bode diagram of the closed-loop plant that the resonance frequency is approximately $\omega_r = 3.0$, and the closed-loop gain at this frequency is $A_c = 1.15$. This value falls below the normal range of 1.2 to 3.0 mentioned in our discussion of performance criteria; therefore we would probably evaluate the effect of an additional slight increase in the controller gain on the performance specifications we used previously. In other words, we follow a trial-and-error approach to determine the controller gain that provides a good compromise between the various performance measures.

Perhaps we should mention that it is not really necessary to construct the Bode diagram for the closed-loop plant. Once we have selected a desirable value for the process gain at the resonance frequency, such as 2.0 or some other value, all we need do is keep adjusting the Bode diagram for the open-loop plant so that when the points are plotted on the Nichols chart, the curve just becomes tangent to the contour where the closed-loop gain is equal to 2.0.

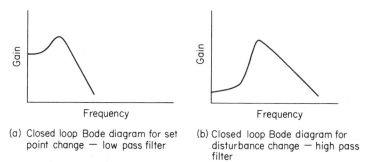

(a) Closed loop Bode diagram for set point change — low pass filter

(b) Closed loop Bode diagram for disturbance change — high pass filter

Figure 7.5-27. Closed loop Bode diagrams: (a) low pass filter; (b) high pass filter.

The closed-loop Bode diagrams for proportional plus integral, proportional plus derivative, or a three-mode controller can be obtained in a similar way. Clearly, the preparation of an open-loop Bode diagram for the plant and controller greatly simplifies the construction of a Nichols chart, and therefore it is also simpler to base the initial design of the controller on the gain and phase margin criteria. A Bode diagram for the closed-loop response to a disturbance can be obtained by adding the curves on Figure 7.5–26 to those for the transfer function G_v/G_cG_u. The gain curves for these two cases generally have characteristically different shapes (see Figure 7.5–27). We say that the closed-loop plant responds to set-point changes like a low-pass filter, because low-frequency signals are not attenuated significantly; whereas the closed-loop plant responds to disturbances like a high-pass filter, because both very low and very high frequency signals are damped.

Controller Design Using Nyquist Diagrams

When the magnitude and phase of a complex number are plotted on polar coordinates, with frequency as a parameter, we obtain a Nyquist diagram. The same information as presented in a Bode plot is then available, so that it would be possible to study the synthesis of control systems on Nyquist diagrams. Since logarithmic coordinates are not used, it is more difficult to see how the addition of a particular control mode modifies the dynamic behavior of the system. However, Nyquist plots have the advantage in that it is a simple matter to find the closed-loop characteristics from the open-loop dynamics. Moreover, by introducing some advanced theorems from the theory of complex variables, the Nyquist method can be applied to more complicated systems, including plants that are unstable without a controller. Unfortunately, space limitations preclude a discussion in depth of complex variables, and so we will not pursue a study of Nyquist diagrams. Several excellent treatments of the procedures are available,[2] however, and an interested reader is encouraged to master this technique.

SECTION 7.6 A TIME DOMAIN PROCEDURE: ROOT LOCUS

Early in the preceding section we noted that it would be a difficult task to develop a procedure for synthesizing control systems in the time domain, because this kind of an approach requires that we evaluate the characteristic roots corresponding to each particular set of gains we choose for the various control modes. Therefore we developed a frequency domain technique that we used to decide which control modes would be desirable and to obtain a first estimate of the gains of those modes. However, as an additional check of the perfor-

[2] For example, see Ch. 21 of the reference given in footnote 2 on p. 51.

mance of the controlled plant, we calculated the step response of the closed-loop system. This procedure again involved an evaluation of the characteristic roots and an application of the Heaviside theorems to invert the transfomed expression, but in this case we needed to consider only a few values of the controller gains.

Despite the fact that the ready access to digital computers makes it a relatively easy job to find the roots of the characteristic equation, it still would be helpful to have a simple procedure for determining the qualitative behavior of the roots as we change one of the system parameters (the controller gain). The root-locus method can be used for this purpose. Once the basic rules are mastered, it is possible to gain a good "feeling" for the behavior of the plant with little effort. Also, the technique can be used to synthesize control systems.

To develop the root-locus procedure, we consider the transfer function for a closed-loop process, which can be written

$$\tilde{x} = \frac{G_c G_u}{1 + G_c G_u} \tilde{x}_0 + \frac{G_v}{1 + G_c G_u} \tilde{v} \qquad 7.6\text{-}1$$

so that the stability of the system is determined by the condition that

$$1 + G_c G_u = 0 \qquad 7.6\text{-}2$$

which is just the characteristic equation. If we initially limit our attention to a proportional controller—that is, $G_c = K_c$, we would like to be able to determine the roots of the characteristic equation as a function of K_c. A root-locus plot is a graph showing how these roots change in the complex plane, with the value of K_c marked as a parameter on the curves.

Rules for Finding the Root Locus

The basic rules for constructing a root-locus diagram were developed by Evans.[1] The method is applicable whenever the transfer function $G_c G_u$ can be written in terms of a ratio of a polynomials, so that Eq. 7.6–2 with $G_c = K_c$ has the form

$$K_c \frac{(s - \alpha_1)(s - \alpha_2)\cdots(s - \alpha_m)}{(s - \lambda_1)(s - \lambda_2)\cdots(s - \lambda_n)} = -1 \qquad 7.6\text{-}3$$

Here $\alpha_1, \alpha_2, \cdots, \alpha_m$ are the zeros of the transfer function; $\lambda_1, \lambda_2, \cdots, \lambda_n$ are the poles of the transfer function (or the characteristic roots of the open-loop plant), and $m < n$. By inspection of Eq. 7.6–3, we see that as K_c approaches zero, G_u must approach infinity in order to maintain the relationship $K_c G_u = -1$. This means that the Laplace parameter s must approach the poles of the open-loop transfer function and that the root-locus diagram must originate at these open-loop poles. In addition, we see that there will be a branch of the

---

[1] W. R. Evans, *Trans. AIEE*, **67**, 547 (1948), and *Control-System Dynamics*, McGraw-Hill, N.Y., 1954.

curve corresponding to each pole. A formal statement of these results provides rules 1 and 2 below.

From the preceding expression we also see that as K_c approaches infinity, the equation $K_c G_u = -1$ can be satisfied only if G_u approaches zero or if s approaches the zeros of G_u. Thus m of the loci will terminate at the m open-loop zeros, and this fact provides the basis of rule number 3.

Another way of interpreting the characteristic equation, $K_c G_u = -1$, is that when the overall system gain is unity, the phase angle must be equal to -180 deg (or any odd multiple of π). In mathematical terms, we can write

$$K_c \frac{(s - \alpha_1)(s - \alpha_2) \cdots (s - \alpha_m)}{(s - \lambda_1)(s - \lambda_2) \cdots (s - \lambda_n)} = -1 \qquad 7.6\text{-}4$$

and

$$\measuredangle(s - \alpha_1) + \measuredangle(s - \alpha_2) + \cdots + \measuredangle(s - \alpha_m) - \measuredangle(s - \lambda_1)$$
$$- \measuredangle(s - \lambda_2) - \cdots - \measuredangle(s - \lambda_n) = (2k + 1)\pi \qquad 7.6\text{-}5$$

where k is any positive integer. By assuming various values for s and substituting them into the phase angle equation, we can use a trial-and-error procedure to establish the root-locus diagram eventually. Next, we can use the gain equation to find the value of K_c on any point on each branch of the curve. Although this procedure is often used to determine intermediate points, there are some additional asymptotic values that we can find and that are helpful in establishing the nature of the curves. For example, if we consider very large values of s (corresponding to high frequencies), our expression becomes

$$K_c \left(\frac{1}{s^{n-m}} \right) = -1 \qquad 7.6\text{-}6$$

Hence, for large values of K_c, the loci must approach asymptotes at angles of $(2k + 1)\pi/(n - m)$. Similarly, it can be shown that these $(n - m)$ asymptotes originate at the center of gravity of the pole-zero configuration, which is given in rule number 5.

The fact that the angle from the poles and zeros to any point on the locus must always be -180 deg means that the real axis will always be part of a locus whenever there are an odd number of poles and zeros to the right of any point on the axis (see rule number 4). Also, we can determine points where the loci enter or leave the real axis by considering the angle criterion for a point slightly above (a distance ϵ) the axis. For a case where there are three simple open-loop poles (see Figure 7.6–1), the angle criterion at point ϵ gives

$$\theta_1 + \theta_2 + \theta_3 = (2k + 1)\pi \qquad 7.6\text{-}7$$

Remembering that small angles are approximately equal to their tangents and recognizing that from geometrical considerations

$$\theta_1 = \pi - \frac{\epsilon}{\lambda_1 - \gamma} \qquad 7.6\text{-}8$$

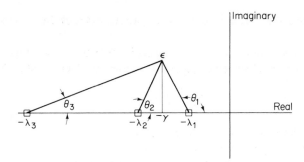

Figure 7.6-1. Root locus diagram.

the angle criterion becomes

$$\pi - \frac{\epsilon}{\lambda_1 - \gamma} + \frac{\epsilon}{\gamma - \lambda_2} + \frac{\epsilon}{\gamma - \lambda_3} = (2k + 1)\pi \qquad 7.6\text{-}9$$

Letting $k = 0$, subtracting π from both sides of the equation, and then letting ϵ approach zero, so that we approach the real axis, we find that

$$\frac{1}{\gamma - \lambda_1} + \frac{1}{\gamma - \lambda_2} + \frac{1}{\gamma - \lambda_3} = 0 \qquad 7.6\text{-}10$$

The more general result for breakaway or entering points is given in rules number 6 and 7.

The other rules can be derived by using similar procedures. A summary of the results is given below:

Rules for Root-Locus Diagram

1. *Number of curves.* The number of loci or branches of the curves is equal to the number of poles, n.

2. *Origin of curves.* The loci or branches of the curves originate at the n poles when $K_c = 0$, and an rth-order pole is the source of r loci.

3. *Termination of curves.* The n loci terminate, as K_c approaches infinity, at either the m zeros of the open-loop transfer function, or approach $(n - m)$ zeros at infinity along certain asymptotes which will be specified later. Also, r loci will terminate at an rth-order zero.

4. *Loci on the real axis.* The real axis will be part of a locus whenever the sum of the number of poles and zeros to the right of any point on the axis is odd. An rth-order pole and/or zero must be counted r times when applying this rule, and complex conjugate poles and/or zeros may be neglected.

5. *Asymptotes.* The $(n - m)$ asymptotes, which are approached by $(n - m)$ of the loci as K_c becomes very large, originate at the center of gravity of the poles and zeros of the open-loop transfer function, where the *center of gravity*, CG, is defined as

$$CG = \frac{\sum_{i+1}^{n} \lambda_i - \sum_{j=1}^{m} \alpha_j}{n - m} \qquad 7.6\text{-}11$$

and they leave the real axis at equally spaced angles of $2\pi/(n - m)$ to each other.

6. *Breakaway point.* Whenever the real axis between two adjacent poles is part of a locus, two loci leave the axis at a point γ which satisfies the equation

$$\sum_{i=1}^{n} \frac{1}{\gamma - \lambda_i} = \sum_{j=1}^{m} \frac{1}{\gamma - \alpha_j} \qquad 7.6\text{-}12$$

and the loci leave at right angles to the real axis.

7. *Entering point.* Whenever the real axis between two adjacent zeros is part of a locus, two loci enter the axis at a point where γ satisfies the equation above, and they enter perpendicular to the real axis.

8. *Angle of departure.* For simple poles on the real axis, the angle of departure of the locus from the pole is either 0 or π. For an rth-order pole, the angles of departure of the r loci satisfy the equation

$$\phi_k = \frac{1}{r}\left[(2k + 1)\pi + \sum_{\substack{i=1 \\ i \ne r}}^{n} \measuredangle(\lambda_r - \lambda_i) + \sum_{j=1}^{m} \measuredangle(\lambda_r - \alpha_j) \right] \qquad 7.6\text{-}13$$

where $k = 0, 1, \cdots, r - 1$ and λ_r is a pole of order r.

9. *Angle of approach.* For simple zeros on the real axis, the angle of approach of a locus to the zero will be either o or π. For an rth-order zero, the angles of approach of the r loci satisfy the equation

$$\phi = \frac{1}{r}\left[(2k + 1)\pi + \sum_{i=1}^{n} \measuredangle(\alpha_r - \lambda_i) + \sum_{\substack{j=1 \\ j \ne r}}^{m} \measuredangle(\alpha_r - \alpha_j) \right] \qquad 7.6\text{-}14$$

Application of these rules generally is adequate for sketching the locus of the roots as the controller gain changes from zero to infinity. Once the curves have been established, it is possible to locate the points corresponding to particular values of the controller gain. Before we describe this procedure, however, it is helpful to present a few additional criteria that are useful in the development of a root-locus plot and then to present an example illustrating the application of the foregoing rules.

10. We first note that since complex roots always appear as conjugate pairs, the root-locus diagram must be symmetrical with respect to the real axis. Actually, this criterion means that it is only necessary to plot the upper half of the root-locus diagram.

11. By examining the form of the characteristic equation, it is possible to show that the sum of the characteristic roots, which is equal to the coefficient of the s^{n-1} term, is a real constant that is independent of the controller gain whenever $n - m \ge 2$. This result implies that the root-locus diagram must exhibit a certain amount of balance with respect to a line parallel to the imaginary axis or that a locus moving to the left must be balanced by one or more loci moving to the right.

12. It is often convenient to visualize a locus as the path of a positively charged particle in an electrostatic field, noting that a locus will be repelled by a pole and attracted by a zero.

Example 7.6–1 Root-locus plot for a reactor with a proportional control system

The transfer function for the reactor described in Examples 7.5–1 through 7.5–3 was given as

$$\frac{\tilde{x}_3}{\tilde{u}} = G_u = \frac{a_{21}a_{32}b_{11}(s - a_{44})}{(s - a_{11})(s - a_{22})[(s - a_{33})(s - a_{44}) - a_{34}a_{43}]}$$

$$= \frac{2.98(s + 2.25)}{(s + 1.45)(s + 2.85)^2(s + 4.35)}$$

Consider a proportional controller in series with the plant and develop a root-locus diagram that shows how the characteristic roots of the closed-loop system change with controller gain.

Solution

The transfer function of interest is

$$K_c\frac{2.98(s + 2.25)}{(s + 1.45)(s + 2.85)^2(s + 4.35)} = -1$$

The root-locus diagram is developed by applying the rules listed above.

1 and 2. The poles of the open-loop transfer function are a simple pole at $\lambda_1 = -1.45$, a second-order pole at $\lambda_2 = \lambda_3 = -2.85$, and a simple pole at $\lambda_4 = -4.35$. Thus there will be four loci originating at these poles. We indicate the simple open-loop poles on the root-locus diagram (see Figure 7.6–2) by the symbol □ and the double pole by the symbol ⊠.

3. The zero of the open-loop transfer function is $\alpha_1 = -2.25$. Hence one of the four loci will terminate at this point, and $(4 - 1) = 3$ loci will approach zeros at infinity along asymptotic lines which will be defined later. We indicate the zeros of the transfer function on the root-locus diagram by the symbol ○.

4. The real axis will be a portion of a locus between the pole at $\lambda_1 = -1.45$ and the zero at $\alpha_1 = -2.25$, and also everywhere to the left of the pole at $\lambda_4 = -4.35$, because in these regions the sums of the number of poles and zeros to the right of any point on the real axis are one and five, respectively. Thus one locus goes from $\lambda_1 = -1.45$ to $\alpha_1 = -2.25$ along the real axis as K_c increases from zero to infinity. Similarly, one loci goes from $\lambda_4 = -4.35$ to $-\infty$ (see Figure 7.6–2).

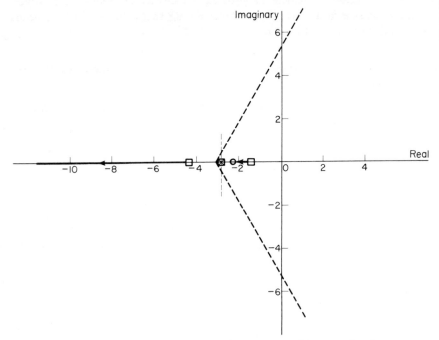

Figure 7.6-2. Root locus diagram for a reactor with proportional control.

5. The center of gravity of the system is

$$CG = \frac{\sum\limits_{i=1}^{n} \lambda_i - \sum\limits_{j=1}^{m} \alpha_j}{n - m}$$

$$= \frac{-1.45 - 2.85 - 2.85 - 4.35 + 2.25}{4 - 1} = -3.08$$

The angles the asymptotes make with each other are

$$\frac{2\pi}{n - m} = \frac{2\pi}{4 - 1} = \frac{2\pi}{3} = 120°$$

It is apparent from rule number four that one asymptote lies along the real axis to the left of $\lambda_4 = -4.35$; therefore the others must be located at $60°$ and $300°$

6. The real axis is never part of a locus between two adjacent poles; thus there are no breakaway points.

7. The real axis is never part of a locus between two adjacent zeros; thus there are no entering points.

8. The angles of departure from the simple poles are π, and the angles of

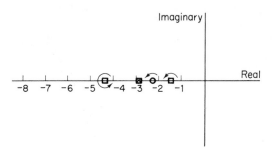

Figure 7.6-3. Root locus.

departure from the multiple pole at $\lambda_2 = \lambda_3 = -2.85$ are given by the equation

$$\phi_k = \frac{1}{2}\left[(2k+1)\pi + \sum_{\substack{i=1 \\ i \neq r}}^{n} \measuredangle(\lambda_r - \lambda_i) + \sum_{j=1}^{m} \measuredangle(\lambda_r - \alpha_j)\right]$$

From the sketch given in Figure 7.6-3, we see that

$$\measuredangle(-2.85 + 1.45) = \pi$$
$$\measuredangle(-2.85 + 4.35) = 2\pi = 0$$
$$\measuredangle(-2.85 + 2.25) = \pi$$

Thus

$$\phi_0 = \frac{1}{2}[\pi + (\pi + 0) - \pi] = \frac{\pi}{2}$$

$$\phi_1 = \frac{1}{2}[3\pi + (\pi + 0) - \pi] = \frac{3\pi}{2}$$

so that the two loci leave at right angles ($\pm 90°$) to the real axis.

9. By inspection, we know that the angle of approach to the single zero is 0 deg.

Using the preceding results, we can fairly well sketch in the locus of the roots as K_c varies from zero to infinity (see Figure 7.6-2). In order to locate exactly the curves leaving the double pole and approaching the asymptotes crossing the real axis, it would be necessary to use a trial-and-error method to determine pairs of complex conjugate roots that satisfy the angle criterion given by Eq. 7.6-5. Once these loci have been established, it is possible to find the controller gain corresponding to any point on any of the four loci by substituting the value of the root at that point into Eq. 7.6-4. Figure 7.6-4 shows the complete root-locus diagram with some values of the controller gain marked as a parameter.

Control System Synthesis Using Root-Locus Diagrams

We know that we want the response of the closed-loop plant to resemble that of a second-order underdamped system. Thus by comparing the root-locus

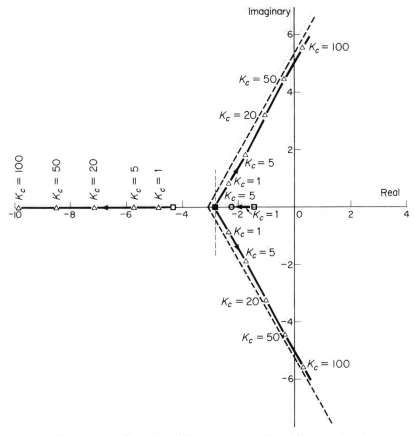

Figure 7.6-4. Root locus diagram for a reactor with proportional control.

diagram we obtain for a design problem with the desired response of a second-order process, we should be able to use this information to select the correct controller gain. The standard form for the transfer function of a second-order plant was given by Eq. 4.5-10 as

$$H(s) = \frac{1}{\tau^2 s^2 + 2\gamma \tau s + 1} \qquad 4.5\text{-}10$$

or, in terms of the characteristic roots, as

$$H(s) = \frac{1}{\tau^2 [s + (\gamma - \sqrt{\gamma^2 - 1})/\tau][s + (\gamma + \sqrt{\gamma^2 - 1})/\tau]} \qquad 4.5\text{-}12$$

When the damping coefficient γ is less than unity, we know that the roots will be complex conjugate, which is the case of interest. It is a simple job to sketch a root-locus diagram of this transfer function for various values of γ. Letting $\gamma = 0$, we see that both roots will be imaginary and will have the value $\pm j/\tau$. Similarly, letting $\gamma = 1$, we see that we obtain two identical real roots $-1/\tau$. At intermediate values of γ, the real part of the root is $-\gamma/\tau$ and the imaginary part is $\pm j\sqrt{1 - \gamma^2}/\tau$. Hence the roots always lie on a circle of radius

$$\left[\left(-\frac{\gamma}{\tau}\right)^2 + \left(\frac{\sqrt{1 - \gamma^2}}{\tau}\right)^2\right]^{1/2} = \left[\frac{\gamma^2}{\tau^2} + \frac{1 - \gamma^2}{\tau^2}\right]^{1/2} = \frac{1}{\tau} \qquad 7.6\text{-}15$$

and the angle between the negative real axis and a vector from the origin to the root is

$$\cos\theta = \left[\frac{\gamma/\tau}{1/\tau}\right] = \gamma \qquad 7.6\text{-}16$$

These results are shown in Figure 7.6-5.

According to Eq. 4.5-32, the decay ratio is uniquely related to the damping coefficient

$$\text{Decay ratio} = \exp\left(\frac{-2\pi\gamma}{\sqrt{1 - \gamma^2}}\right) \qquad 4.5\text{-}32$$

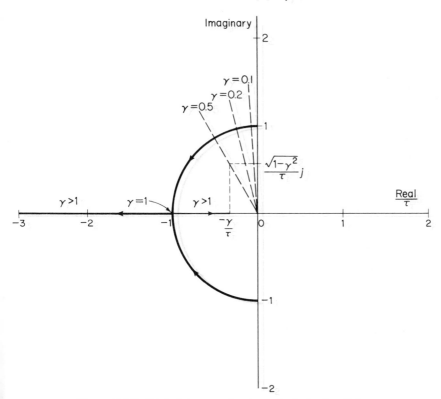

Figure 7.6-5. Roots for a second-order underdamped system.

Thus our "rule of thumb" for designing control systems such that the decay ratio should be in the range of $\frac{1}{4}$ to $\frac{1}{3}$ means that the damping coefficient must be between about 0.10 and 0.20. In addition, from Eq. 7.6–16, we know that the appropriate angular region on a root-locus plot is between approximately 75 and 85 deg. Thus we can draw lines at these angles until they intersect one of the loci of the roots of the closed-loop plant; and by interpolating the values of the controller gain marked as a parameter along the loci, we can find the range of gains that should give a satisfactory closed-loop performance. This kind of a control-system synthesis procedure is essentially equivalent to that based on Bode plots. However, it is expected to give slightly different results, for we are attempting to match the dynamic response of a higher-order plant to a second-order underdamped system in a different way. One great advantage of the root-locus approach is that the characteristic roots of the plant are always available; therefore the Heaviside theorems can be used to calculate the response in the time domain.

Example 7.6–2 Design of a proportional controller for a reactor problem

Use a root-locus plot to design a proportional controller for the reactor problem described in Examples 7.5–1 through 7.5–3.

Solution

A root-locus diagram for the reactor and proportional controller was given in Figure 7.6–4. Drawing lines through the origin at angles of 75 and 85 deg to the negative real axis, we can find the intersection of these lines with a locus at points A and B on Figure 7.6–6. By interpolating the values of the controller gain marked on this locus, we see that appropriate values of the gain should be between $K_c = 24$ and $K_c = 50$. The design method using Bode diagrams gave us values in the range from 11.0 to 44.0.

Example 7.6–3 Proportional plus integral control system design for a reactor problem

Determine the effect of adding an integral mode to the controller in the previous problem. In particular, consider the values of integral time, $T_i = 0.2$, 0.5, and 1.0.

Solution

The transfer function of the controller is

$$G_c = K_c\left(1 + \frac{1}{T_i s}\right) = \frac{K_c}{s}\left(s + \frac{1}{T_i}\right) \qquad \text{7.5–30}$$

so that the characteristic equation becomes

$$G_c G_u = -1$$

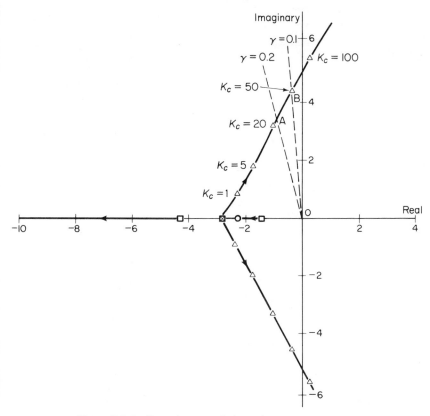

Figure 7.6-6. Control system design using a root locus plot.

or

$$K_c \frac{2.98(s + 2.25)(s + 1/T_i)}{s(s + 1.45)(s + 2.85)^2(s + 4.35)} = -1$$

Thus the integral control mode adds an additional pole at the origin and an additional zero at $s = -1/T_i$.

Case 1: $T_i = 0.2$. The added zero occurs at $s = -5$. Now there are five branches that must be determined, starting at the poles shown on Figure 7.6–7. Two of the curves must terminate at the two zeros and the other three must approach infinity. The real axis is part of the loci between 0 and -1.45, -2.25 and -2.85, -2.85 and -4.35, and to the left of -5. From this information we see that one locus starts at the double pole at -2.85 and proceeds to the zero at -2.25. The center of gravity of the asymptotes is $(-11.5 + 7.25)/3 = -1.42$, and the asymptotes leave the center of gravity at angles of 60, 180, and 300 deg. There are breakaway points between the adjacent poles at 0 and -1.45 and those at -2.85 and -4.35. Similarly, there is an entering point between the adjacent zeros at -5.0 and $-\infty$.

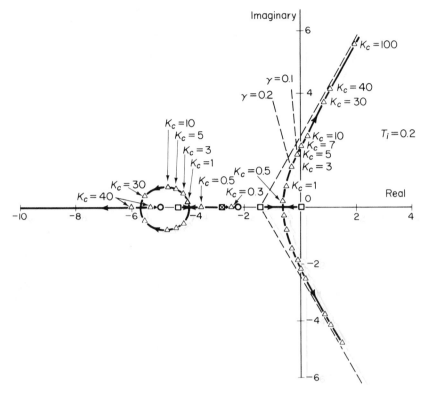

Figure 7.6-7. Root locus plot of reactor and $P + I$ controller.

This information is adequate to get a clear picture of the way in which the roots change as the controller gain is varied between zero and infinity. However, for the design of control systems, it is necessary to find some intermediate points on the loci, particularly in the region between the breakaway point nearest the origin and the asymptote leaving the center of gravity at an angle of 60 deg. Equation 7.6-5 is solved by trial and error to locate these points and then the controller gain at the points is calculated by using Eq. 7.6-4. The complete diagram is shown in Figure 7.6-7. Now, we could draw in the lines corresponding to the appropriate decay ratios and find the range of acceptable controller gains. However, we are more interested in determining the effect of the value of T_i on the root-locus diagram at this point, so we will proceed to the other cases.

Case 2: $T_i = 0.5$. For this case, the additional zero is located at -2.0. The real axis is part of a locus between 0 and -1.4, -2.0, and -2.25, and to the left of -4.35. The center of gravity of the pole-zero system moves to $(-11.5 + 4.25)/3 = -2.42$, and the asymptotes leave at the same angles as before. There are breakaway points between the adjacent poles at 0 and -1.45 and at the double pole at -2.85. An entering point exists between the adjacent zeros at -2.0 and -2.25.

Again we can fairly well establish the behavior of the system from this information. Since the asymptote leaving at 60 deg crosses the imaginary axis at a larger value than before, we expect the system to be somewhat more oscillatory, and we might expect that higher-controller gains would be obtained before the system became unstable. The root-locus plot is shown in Figure 7.6–8.

Case 3: $T_i = 1.0$. With $T_i = 1.0$, the additional zero is located at $s = -1.0$. Loci are located on the real axis between 0 and -1, -1.45 and -2.25 and to the left of -4.35. The center of gravity for the three asymptotes leaving at angles of 60, 180, and 300 deg is $(-11.5 + 3.25)/3 = -2.75$. There are no breakaway or entering points for this case except for the double pole at -2.85.

With this information, we can determine three of the loci immediately. Then we use the trial-and-error technique to locate points on the remaining two loci, which approach the asymptotes leaving at 60 and 300 deg. Again we expect to observe more oscillations in the response and to be able to use higher-controller gains before stability considerations become important. The complete plot is shown in Figure 7.6–9. The angular lines based on our performance

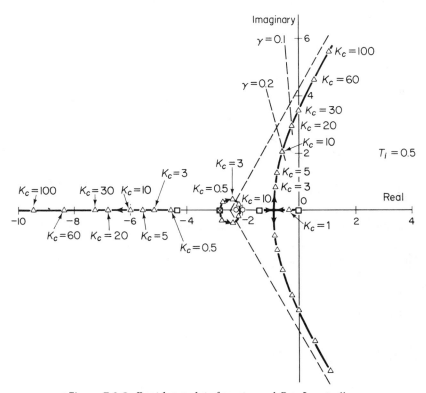

Figure 7.6-8. Root locus plot of reactor and $P + I$ controller.

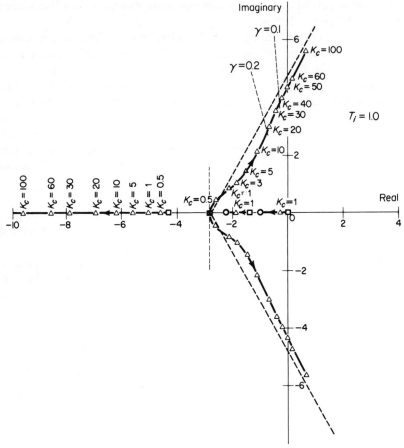

Figure 7.6-9. Root locus plot of reactor and $P + I$ controller.

index are also shown, and we find that values of K_c in the range from 18 to 37 should be acceptable.

Example 7.6–4 Proportional plus derivative control for a reactor problem

Design a proportional plus derivative control system for the reactor problem discussed above. Consider only the values $T_D = 0.2$ and $T_D = 0.1$.

Solution

The controller transfer function is

$$G_c = K_c(1 + T_D s)$$

so that the characteristic equation becomes

$$G_c G_u = -1$$

or

$$K_c \frac{0.298(s + 2.25)(1 + T_D s)}{(s + 1.45)(s + 2.85)^2(s + 4.35)} = -1$$

Thus the only difference between a proportional controller and a proportional plus derivative controller is the addition of a new zero at $s = -1/T_D$.

Case 1: $T_D = 0.2$. The new zero is located at $s = -5$. The real axis is part of a locus between -1.45 and -2.25 and between -4.35 and -5; in fact, the path between these poles and zeros must each be a loci. Two asymptotes leave the real axis from a center of gravity located at $(-11.5 + 7.25)/2 = -2.125$. There are no breakaway or entering points except at the double pole at -2.85.

Using this information to sketch in the root locus is a simple job (see Figure 7.6-10). The results show that the system will never become unstable, no matter how large a value is used for the controller gain (of course, we have neglected the dynamics of the measuring instrument, control valve, etc., so that this conclu-

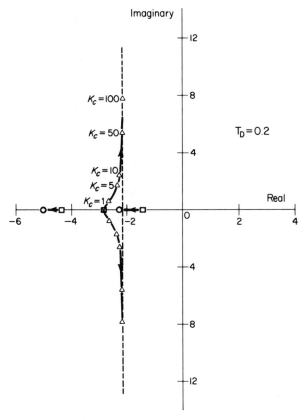

Figure 7.6-10. Root locus plot of reactor and $P + D$ controller.

sion will break down at some high value of K_c). This is the same kind of behavior we observe when the phase curve on a Bode plot does not cross -180 deg.

Case 2: $T_D = 0.1$. When T_D is decreased to 0.1, the new zero is shifted to $s = -10$. The real axis is part of a locus between -1.45 and -2.25 and between -4.35 and -10, and this information establishes two of the loci. Again there are only two asymptotes that leave at right angles to the real axis, but now the center of gravity of the system becomes $(-11.5 + 12.25)/2 = +0.375$. The only breakaway point is at the double pole at -2.85, and there are no entering points. The need for determining intermediate points on the loci that approach the asymptotes is apparent from Figure 7.6–11.

After comparing the results for the two values we used for the derivative time T_D, we find that decreasing the derivative time shifts the new zero far enough to the left so that the center of gravity of the asymptotes shifts to the

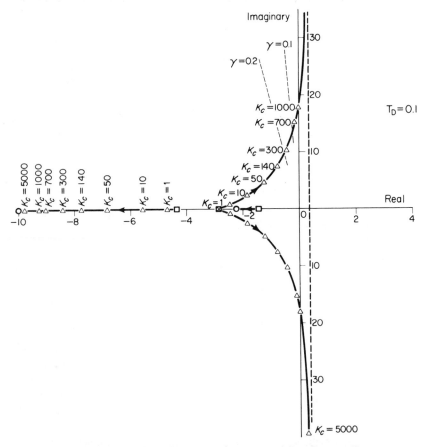

Figure 7.6-11. Root locus plot of reactor and $P + D$ controller.

right and becomes positive. This means that the system will have roots with positive real parts and become unstable at some finite value of the controller gain K_c. The angular lines corresponding to our performance measure are shown on Figure 7.6–11, and we find that we should set the controller gain to be in the range from 250 to 520. These values are similar to those we calculated using Bode plots.

SECTION 7.7 EMPIRICAL DESIGN TECHNIQUES

In the foregoing discussion we presented in some detail two alternate methods that can be used to design simple control systems for single-input, single-output plants. Both methods are based on a theoretical model of the process. We know that it is necessary to perform a few critical dynamic experiments on an existing plant (if available) in order to verify the theoretical model, although we never expect to obtain a perfect match between the data and the model predictions. Similarly, it would be unrealistic to expect that our control system designs will ever be exact, because of the assumptions used in the analysis. This problem of achieving a balance between the amount of effort required to make a calculation and the accuracy of the results is commonly encountered in engineering. Part of the information required to resolve this apparent difficulty is concerned with the simplicity of adjusting the constants appearing in the controller transfer function after the controller has been built and installed on a plant—that is, the flexibility of the design and the difficulty of gaining new information about the actual dynamic behavior of the real plant. Thus we need to explore the possibility of tuning the controller after it has been installed. In addition, we would like to find an even simpler approach, one enabling us to make a first estimate of the design of the controller, even if this ignores some of the information that we possess about the system dynamics.

Methods of this type are available. They are based primarily on a tremendous amount of empirical evidence. In fact, they have been found to be so powerful and useful that most industrial control systems were selected in this way. The results were developed by Ziegler and Nichols[1] and by Cohen and Coon.[2]

Reaction Curve Method

Cohen and Coon recognized that the step response of most process units had a sigmoidal shape (see Figure 7.7–1) that could be adequately represented by the transfer function of a pure time delay and a first-order lag.

$$G(s) = \frac{Ke^{-t_d s}}{\tau s + 1} \qquad 7.7\text{–}1$$

[1] J. G. Ziegler and N. B. Nichols, *Trans. ASME*, **64**, 759 (1942).
[2] G. H. Cohen and G. A. Coon, *Trans. ASME*, **75**, 827 (1953).

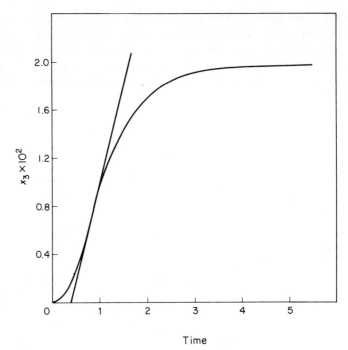

Time

Figure 7.7-1. Reaction curve.

The procedure for determining the constants in this expression (i.e., the gain K, the dead time t_d, and the time constant τ) from a plot of the step response, either calculated from a theoretical model or measured experimentally, was described in Section 5.3. Using this simple model as a description of the plant and considering a wide range of parameters in the model, they studied the effect of adding various feedback control systems. The choice of the control settings was based on a $\frac{1}{4}$ decay ratio, minimum offset, the minimum area under the error curve, and other performance measures. Fortunately they were able to generalize the results of their investigation and to find simple relationships between the parameters in the controller transfer function and the parameters in the plant transfer function, Eq. 7.7–1. The expressions of interest are

Proportional

$$K_c = \frac{1}{K} \frac{\tau}{t_d}\left(1 + \frac{t_d}{3\tau}\right) \qquad 7.7\text{–}2$$

Proportional plus integral

$$K_c = \frac{1}{K} \frac{\tau}{t_d}\left(0.9 + \frac{t_d}{12\tau}\right)$$

$$T_i = t_d\left(\frac{30 + 3t_d/\tau}{9 + 20t_d/\tau}\right) \qquad 7.7\text{–}3$$

Proportional plus derivative

$$K_c = \frac{1}{K}\frac{\tau}{t_d}\left(1.25 + \frac{t_d}{6\tau}\right)$$

$$T_D = t_d\left(\frac{6 - 2t_d/\tau}{22 + 3t_d/\tau}\right)$$

7.7-4

Proportional plus integral plus derivative

$$K_c = \frac{1}{K}\frac{\tau}{t_d}\left(1.333 + \frac{t_d}{4\tau}\right)$$

$$T_i = t_d\left(\frac{32 + 6t_d/\tau}{13 + 8t_d/\tau}\right)$$

7.7-5

$$T_D = t_d\left(\frac{4}{11 + 2t_d/\tau}\right)$$

The application of these results saves us the effort of using a trial-and-error approach to find the controller settings that best satisfy the performance criteria. Also, it should be noted that the approach is really valid only for the transfer function given by Eq. 7.7-1; and from Section 5.3 we know that this expression does not provide a very good fit of a sigmoidal-shaped curve. Nevertheless, experience has shown that the controller settings obtained in this way give a fairly good closed-loop response. When the reaction curve is determined experimentally, all the elements in the feedback loop should be included in the test.

Example 7.7-1 Design of a reactor control system
using the reaction curve method

Use the reaction curve method to design a control system for the reactor problem described in Examples 7.5-1 and 7.5-2.

Solution

The transfer function relating the desired output $\tilde{x}_3$ to the control variable $\tilde{u}$ is given by Eq. 7.5-24.

$$\tilde{x}_3 = \frac{a_{21}a_{32}b_{11}\,(s - a_{44})}{(s - a_{11})(s - a_{22})[(s - a_{33})(s - a_{44}) - a_{34}a_{43}]}\,\tilde{u}$$

7.7-6

In terms of the system parameters, this becomes

$$\tilde{x}_3 = \frac{0.1315(0.445s + 1)}{(0.230s + 1)(0.351s + 1)^2(0.689s + 1)}\,\tilde{u}$$

7.7-7

For a step input, $\tilde{u} = 0.15/s$, we can use the Heaviside theorems to find the inverse transform. The results are shown in Figure 7.7-1. From this reaction curve, we find that the apparent dead time is $t_d = 0.39$; the system gain (the final steady state value divided by the magnitude of the input) is $K = 1.96 \times$

$10^{-2}/0.15 = 0.1308$; and the slope of the tangent through the inflection point is, slope $= 1.96 \times 10^{-2}/(1.59 - 0.39) = 1.631 \times 10^{-2}$, so that the system time constant is $\tau = (1.96 \times 10^{2})/(1.631 \times 10^{-2}) = 1.20$. Hence the approximate transfer function can be written as

$$G(s) = \frac{0.1308\, e^{-0.39s}}{(1.2s + 1)} \qquad\qquad 7.7\text{-}8$$

Substituting these parameters into Eqs. 7.7-2 through 7.7-5, we find that the controller settings should be

Proportional: $K_c = 26.8$

Proportional plus integral: $K_c = 23.4$, $T_i = 0.732$

Proportional plus derivative: $K_c = 31.6$, $T_D = 0.0909$

Proportional plus integral plus derivative: $K_c = 34.2$, $T_i = 0.85$, $T_D = 0.134$

These values agree quite well with those obtained in Example 7.5-2.

Loop Tuning

After the control system has been installed, normally it is necessary to adjust the constants in the controller transfer function in order to make certain that we obtain the best possible response. One possible approach is first to set the integral time of a three-mode controller to infinity and the derivative time to zero, or the lowest possible value, and then to introduce a number of step changes into the closed-loop system with the controller gain set to various values. In this way we can find the gain that corresponds to the generation of continuous oscillations with a constant amplitude in the feedback loop—that is, the condition of marginal stability. We call this gain the *ultimate gain*, K_u, and we call the period of the oscillations the *ultimate period*, P_u. Ziegler and Nichols suggested using these values to calculate the controller settings according to the relationships listed below.

Proportional

$$K_c = 0.5 K_u \qquad\qquad 7.7\text{-}9$$

Proportional plus integral

$$K_c = 0.45 K_u$$

$$T_i = \frac{P_u}{1.2} \qquad\qquad 7.7\text{-}10$$

Proportional plus integral plus derivative

$$K_c = 0.6 K_u$$

$$T_i = \frac{P_u}{2} \qquad\qquad 7.7\text{-}11$$

$$T_D = \frac{P_u}{8}$$

Of course, we could also use these Ziegler-Nichols criteria to design control systems. The Bode diagram of the open-loop system can be used to find the frequency ω_{co} when the phase lag is 180 deg, as well as the system gain at this frequency. Marginal stability occurs when the closed-loop gain is equal to unity at this crossover frequency, so that the ultimate gain, K_u, is merely the reciprocal of the open-loop gain. Similarly, the ultimate period is just $P_u = 2\pi/\omega_{co}$. Now we see that the Ziegler-Nichols expression for the proportional controller gain is based on a gain margin of 2.0. The addition of an integral mode, which adds additional phase lag to the system, means that we must decrease the gain somewhat. Also, we are using a slightly smaller value of the integral time than that given by Eq. 7.5–32. When a derivative mode is added, we introduce some phase lead, and thus can increase the controller gain. In our previous discussion in Example 7.5–2, we considered values of the derivative time in the neighborhood of $T_D = 1/\omega_{co}$, whereas the Ziegler-Nichols setting is $T_D = \pi/4\omega_{co}$.

Example 7.7–2 Design of a reactor control system using the Ziegler-Nichols criteria

Use the Ziegler-Nichols results to design a control system for the reactor problem discussed in Examples 7.5–1 and 7.5–2.

Solution

From the open-loop Bode diagram of the reactor (Figure 7.5–11), we find that the crossover frequency is $\omega_{co} = 5.2$ and the system gain at this frequency is $|G| = 0.0135$. Hence the ultimate gain is $K_u = 1/0.0135 = 74.0$ and the ultimate period is $P_u = 2\pi/5.2 = 1.208$. Using the Ziegler-Nichols criteria, we obtain the following controller settings:

Proportional: $K_c = 37.0$

Proportional plus integral: $K_c = 33.3$, $T_i = 1.006$

Proportional plus integral plus derivative: $K_c = 44.4$, $T_i = 0.604$,

$$T_D = 0.151$$

Again we see that these results are quite similar to those obtained previously.

Damped Oscillation Method of Loop Tuning

In some industrial situations it is potentially dangerous to attempt to carry out a dynamic test that leads to continuous oscillations. Thus an alternate procedure[3] for tuning the control loop is to use only proportional control initially and to keep on increasing the controller gain until the step response of the sys-

[3] Taylor Instrument Companies, "Instructions for Transcope Controller," *Bull. 1B 404*, 1961.

tem exhibits a decay ratio of $\frac{1}{4}$. When this value has been obtained, we measure the period of the oscillations, P, which we know from Section 4.5 will always be greater than the ultimate period, P_u. Then we let

$$T_i = \frac{P}{1.5} \qquad T_D = \frac{P}{6} \qquad\qquad 7.7\text{-}12$$

Once we have set the integral and derivative times to these values, we readjust the proportional gain until we again achieve a decay ratio of $\frac{1}{4}$.

SECTION 7.8   OTHER CONSIDERATIONS IN CONTROL SYSTEM DESIGN

The three procedures we have described for synthesizing simple feedback control systems are based on a number of idealizations. For example, we have assumed that the dynamics of each of the elements in the feedback loop are negligible, that the control variable and the output we want to maintain constant have been specified, that we can actually implement our ideal control laws, that the process nonlinearities are negligible, and so on. Before actually implementing our best design, we certainly want to check and make sure that these assumptions are valid. Some procedures that can be used for this purpose are discussed below.

Effect of the Dynamics of the Elements in the Feedback Loop

Up to this point we have tried to ignore any effects of the dynamics of the measuring instrument, the controller itself, the control valve and actuator, the transmission lines, and so forth. However, we did note in several places that the dynamics of these other elements in the feedback loop always become significant for a first- or second-order plant, for they impose an upper bound on the controller gain before the system becomes unstable. The procedure we have followed is always the most desirable, but it would be helpful to have a simple criterion that we could use to predict whether or not the dynamics of these elements are actually negligible.

The simplest way of developing a criterion of this type is to reexamine our control-system synthesis technique using Bode plots. First we plotted the asymptotic approximation of the open-loop transfer function of the plant. Then we used the graph to find the frequency corresponding to a phase lag of 180 deg, and also the system gain at this crossover frequency. For a proportional control system, where we base the design on a gain margin criterion, we multiplied the system gain by a factor between 1.75 and 2.0 and took the reciprocal of this product to determine the controller gain setting, K_c. When we added an integral mode, in order to eliminate offset, we added additional phase lag to the system, so that the crossover frequency decreased; whereas the addition of a derivative mode introduced some phase lead, so that the crossover frequency increased.

Now, we want to consider the possibility of having one or more additional dynamic elements in the feedback loop, and we want to determine under what conditions they will have a negligible effect on the synthesis procedure described above. The answer to this problem is quite simple, for an inspection of the Bode plot indicates that neither the gain nor the phase curves in the neighborhood of the crossover frequency will be affected by any additional first-order elements, providing that the corner frequency of these elements is greater than 10 times the crossover frequency

$$\omega_c \gg 10\omega_{co} \qquad \text{7.8-1}$$

Inasmuch as the corner frequency of an element is the reciprocal of its time constant, the preceding result can also be written as

$$\tau_m \ll \frac{1}{10\omega_{co}} \qquad \text{7.8-2}$$

If this criterion is not satisfied, then the dynamic performance of the plant could be improved by installing a faster measuring instrument, control valve, or whatever provides the largest time constant in the feedback loop.

Example 7.8-1 Effect of measurement dynamics

Determine the maximum value of the time constant of a composition-measuring device such that the measuring instrument dynamics would not have any significant effect on the reactor control systems described in Examples 7.5-1 and 7.5-2.

Solution

For a proportional control system, we see from the asymptotic approximations of the Bode plot (Figure 7.5-10) that the crossover frequency is $\omega_{co} = 6.05$. If we add a new first-order element to the system in such a way that the phase angle is zero at the crossover frequency, increases to a phase lag of -45 deg at a frequency of 60.5, and, finally, adds a 90 deg lag at $\omega = 605$, we will not change our previous results (see Figure 7.8-1). The corner frequency of this element is $\omega_c = 60.5$; thus its time constant is $\tau_m = 1/60.5 = 0.0166$. Alternately, using Eq. 7.8-2, we obtain the same result

$$\tau_m \ll \frac{1}{6.05(10)} = 0.0166$$

This value is much less than the other time constants of the system (the values are listed after Eq. 7.5-26). A better estimate can be obtained by using the graph of the exact transfer function, Figure 7.5-11, to locate the crossover frequency. However, since we are primarily interested in order of magnitude calculations, normally it is not necessary to refine the estimates.

For a proportional plus integral controller, we see from Figure 7.5-14 that

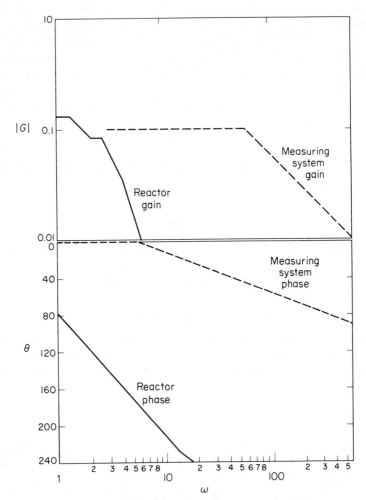

Figure 7.8-1. Bode diagram including measurement lag.

$\omega_{co} = 4.65$, so that

$$\tau_m \ll \frac{1}{4.65(10)} = 0.0215$$

Figure 7.5–18 shows that $\omega_{co} = 14.7$ for a proportional plus derivative controller, so that

$$\tau_m \ll \frac{1}{14.7(10)} = 0.0068$$

Also, according to Figure 7.5–22, $\omega_{co} = 10.25$ for a three-mode controller, so that

$$\tau_m \ll \frac{1}{10.25(10)} = 0.00975$$

Location and Selection of the Measured and Control Variables

Even though we have mentioned several times that generally a case-study approach must be used to decide which input variable to choose as a control variable and which output variable should be measured and used to generate the error signal for the controller, we hope that it might be possible to develop some guidelines to help in making this decision. Intuitively we might expect that we want to correct for the effects of disturbances as quickly as possible. After some reflection, it seems as if the best approach would be to choose to measure any output variable that appears in the same equation with the most significant disturbances; that is, the one leading to the largest steady state gain. In an analogous way, our first choice for a control variable probably would be one that appears in the equation for an output variable of interest. However, it should be recognized that there are so many interactions in some plants that these simple rules are not applicable. In addition, we know that some outputs can be measured more simply and economically than others (e.g., temperature as opposed to composition) and that some control variables and outputs have a greater effect on the plant economics than others. Thus it is difficult to make comprehensive generalizations concerning this type of question; perhaps the best thing to do is to work out some specific problems.

Example 7.8–2 Control of a battery of stirred-tank reactors

Suppose that we consider a battery of five, isothermal, stirred-tank reactors in series (see Figure 7.8–2), and we want to maintain the composition leaving the last reactor as constant as possible despite fluctuations in the feed com-

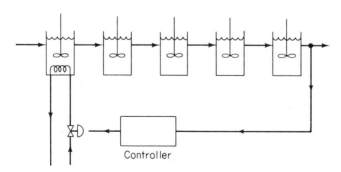

Figure 7.8-2. Control of a CSTR battery.

position. We attempt to compensate for the disturbances by adjusting the temperature (or the rate constant) in the first vessel, but we assume that this temperature change has no effect on the conditions in the remaining vessels.

The system equations are

$$V\frac{dA_1}{dt} = q(A_f - A_1) - k_1 VA_1$$

$$V\frac{dA_2}{dt} = q(A_1 - A_2) - k_2 VA_2$$

$$V\frac{dA_3}{dt} = q(A_2 - A_3) - k_3 VA_3 \qquad\qquad 7.8\text{-}3$$

$$V\frac{dA_4}{dt} = q(A_3 - A_4) - k_4 VA_4$$

$$V\frac{dA_5}{dt} = q(A_4 - A_5) - k_5 VA_5$$

Letting

$$x_1 = \frac{A_1 - A_{1s}}{A_{fs}}, \quad x_2 = \frac{A_2 - A_{2s}}{A_{fs}}, \quad x_3 = \frac{A_3 - A_{3s}}{A_{fs}}$$

$$x_4 = \frac{A_4 - A_{4s}}{A_{fs}}, \quad x_5 = \frac{A_5 - A_{5s}}{A_{fs}}, \quad v = \frac{A_f - A_{fs}}{A_{fs}} \qquad 7.8\text{-}4$$

$$u = \frac{k_1 - k_{1s}}{k_{1s}}$$

Subtracting the steady state equations, and dividing by VA_{fs}, we obtain

$$\dot{x}_1 = -\left(\frac{q}{V} + k_{1s}\right)x_1 + \left(\frac{q}{V}\right)v - \left(k_{1s}\frac{A_{1s}}{A_{fs}}\right)u - (k_{1s})ux_1$$

$$\dot{x}_2 = \left(\frac{q}{V}\right)x_1 - \left(\frac{q}{V} + k_2\right)x_2$$

$$\dot{x}_3 = \left(\frac{q}{V}\right)x_2 - \left(\frac{q}{V} + k_3\right)x_3 \qquad\qquad 7.8\text{-}5$$

$$\dot{x}_4 = \left(\frac{q}{V}\right)x_3 - \left(\frac{q}{V} + k_4\right)x_4$$

$$\dot{x}_5 = \left(\frac{q}{V}\right)x_4 - \left(\frac{q}{V} + k_5\right)x_5$$

Now we neglect the terms containing the product of small deviations, ux_1, to linearize the equations, let

$$k_{1s} = k_2 = k_3 = k_4 = k_5 = k \qquad \alpha_1 = \left(\frac{q}{V} + k\right)$$

$$\alpha_2 = \frac{q}{V} \qquad \beta = -k_{1s}\frac{A_{1s}}{A_{fs}} \qquad\qquad 7.8\text{-}6$$

take the Laplace transform of the set of linearized equations, and solve the equations for the state variables in terms of the disturbance and control variables, to obtain

$$\tilde{x}_1 = \left(\frac{\beta}{s + \alpha_1}\right)\tilde{u} + \left(\frac{\alpha_2}{s + \alpha_1}\right)\tilde{v}$$

$$\tilde{x}_2 = \frac{\alpha_2}{s + \alpha_1}\tilde{x}_1 = \frac{\alpha_2\beta}{(s + \alpha_1)^2}\tilde{u} + \frac{\alpha_2^2}{(s + \alpha_1)^2}\tilde{v}$$

$$\tilde{x}_3 = \frac{\alpha_2}{s + \alpha_1}\tilde{x}_2 = \frac{\alpha_2^2\beta}{(s + \alpha_1)^3}\tilde{u} + \frac{\alpha_2^3}{(s + \alpha_1)^3}\tilde{v} \qquad 7.8\text{-}7$$

$$\tilde{x}_4 = \frac{\alpha_2}{s + \alpha_1}\tilde{x}_3 = \frac{\alpha_2^3\beta}{(s + \alpha_1)^4}\tilde{u} + \frac{\alpha_2^4}{(s + \alpha_1)^4}\tilde{v}$$

$$\tilde{x}_5 = \frac{\alpha_2}{s + \alpha_1}\tilde{x}_4 = \frac{\alpha_2^4\beta}{(s + \alpha_1)^5}\tilde{u} + \frac{\alpha_2^5}{(s + \alpha_1)^5}\tilde{v}$$

In a conventional feedback control system, we measure the composition in the last vessel and use the difference between this value and its steady state value to activate the controller in some way. For a proportional controller, we write

$$\tilde{u} = -K_c(\tilde{x}_{50} - \tilde{x}_5) \qquad 7.8\text{-}8$$

so that the closed-loop equation of interest becomes

$$\tilde{x}_5 = \frac{\alpha_2^4\beta}{(s + \alpha_1)^5 + (-\beta)\alpha_2^4 K_c}\tilde{x}_{50} + \frac{\alpha_2^5}{(s + \alpha_1)^5 + (-\beta)\alpha_2^4 K_c}\tilde{v} \qquad 7.8\text{-}9$$

where we have used a negative sign in the control law and written $(-\beta)$ because, according to Eq. 7.8-6, β is negative.

One procedure for selecting the controller gain is to use the asymptotic approximations to develop the Bode diagram of the transfer function of the open-loop system

$$\frac{\tilde{x}_5}{\tilde{u}} = \frac{\alpha_2^4\beta}{(s + \alpha_1)^5} = \frac{\beta\alpha_2^4}{\alpha_1^5}\frac{1}{(\tau s + 1)^5} \qquad 7.8\text{-}10$$

where

$$\tau = \frac{1}{\alpha_1}$$

However, a relatively large error will be obtained in the neighborhood of the corner frequency if this approach is followed, for the system has five equal roots. Hence a simpler approach is to use Eqs. 4.6-16 and 4.6-17 to relate the fifth-order system to a first-order process

$$|G| = \frac{\alpha_2^4\beta}{\alpha_1^5}\left(\frac{1}{\sqrt{1 + \tau^2\omega^2}}\right)^5 \qquad 4.6\text{-}16$$

$$\theta = n\tan^{-1}(-\tau\omega) \qquad 4.6\text{-}17$$

and then use the Bode plot given in Figure 4.3-4 to determine the quantities of interest.

For a particular problem where $A_{fs} = 0.1$ g mole/liter, $k = 0.333$ hr^{-1}, $V = 20$ liters., $q = 100$ liters/hr, we find that

$$A_{5s} = \left(\frac{q}{q + kV}\right)^5 A_{fs} = 0.0726 \qquad \alpha_1 = \left(\frac{q}{V} + k\right) = 5.33$$

$$\alpha_2 = \frac{q}{V} = 5.0 \qquad \beta = -k_{1s}\frac{A_{1s}}{A_{fs}} = -k_{1s}\left(\frac{q}{q + kV}\right) = -0.313$$

and

$$\frac{\beta \alpha_2^4}{\alpha_1^5} = -0.0455$$

The crossover frequency occurs when the phase lag is 180 deg or, from Eq. 4.6–17, when each of the first-order systems contributes a lag of $-180°/5 = -36°$. From Figure 4.3–4 we see that a 36 deg phase lag corresponds to a value of $\tau\omega = 0.71$; and since $\tau = 1/\alpha_1 = 1/5.33$, the crossover frequency must be

$$\omega_{co} = 0.71(5.33) = 3.78$$

Thus, using Eq. 4.6–16, we find that the system gain must be

$$|G| = 0.0455\left(\frac{1}{\sqrt{1 + 0.71^2}}\right)^5 = 0.01639$$

so that for a gain margin of 2.0 the controller gain is

$$K_c = \frac{1}{2|G|} = \frac{1}{0.03278} = 30.5$$

Our analysis of the previous reactor control problem indicated that the addition of a derivative control mode would make it possible to increase the crossover frequency significantly, as well as the controller gain for this series of first-order systems. Rather than pursue the development of the "best" control system, however, we are interested in studying the effect of changing the measuring point to some other location in the system. For example, we know the desired composition in each tank for steady state conditions; and therefore we suppose that we measure the composition in the third vessel, rather than the fifth, and use this measurement to generate the error signal for the controller (see Figure 7.8–3). Since the measured variable is now closer to where the disturbance enters the plant, we might expect to obtain a more rapid response of the output variable (although larger deviations may be observed at the measuring point). Substituting the new control law

$$\tilde{u} = -K_c(\tilde{x}_{30} - \tilde{x}_3) \qquad\qquad 7.8\text{–}11$$

into the third expression in Eqs. 7.8–7

$$\tilde{x}_3 = \frac{\alpha_2^2 \beta}{(s + \alpha_1)^3}\tilde{u} + \frac{\alpha_2^3}{(s + \alpha_1)^3}\tilde{v} \qquad\qquad 7.8\text{–}12$$

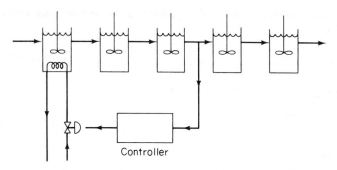

Figure 7.8-3. Control of a CSTR battery.

and then solving the remaining equations for $\tilde{x}_5$, we obtain

$$\tilde{x}_5 = \frac{\beta\alpha_2^4}{(s + \alpha_1)^2[(s + \alpha_1)^3 + (-\beta)\alpha_2^2 K_c]} \tilde{x}_{30}$$

$$+ \frac{\alpha_2^5}{(s + \alpha_1)^2[(s + \alpha_1)^3 + (-\beta)\alpha_2^2 K_c]} \tilde{v} \qquad 7.8\text{-}13$$

The design of the proportional controller is based on the transfer function given in Eq. 7.8–12

$$\frac{\tilde{x}_3}{\tilde{u}} = \frac{\beta\alpha_2^2}{(s + \alpha_1)^3} = \frac{\beta\alpha_2^2}{\alpha_1^3} \frac{1}{(\tau s + 1)^3} \qquad 7.8\text{-}14$$

where again

$$\tau = \frac{1}{\alpha_1}$$

Evaluating the parameters, we find that

$$\frac{\beta\alpha_2^2}{\alpha_1^3} = -0.0516 \qquad \tau = \frac{1}{5.33}$$

When the total phase lag is 180 deg, the three first-order systems in series will each contribute 60 deg phase lag according to Eq. 4.6–17, and from Figure 4.3–4 we see that $\tau\omega = 1.73$. Thus,

$$\omega_{co} = 1.73(5.33) = 9.22$$

This is a threefold increase over our previous value. The gain at this frequency is calculated by using Eq. 4.6–16 and is found to be

$$|G| = 0.0516\left(\frac{1}{\sqrt{1 + 1.73^2}}\right)^3 = 0.00646$$

With a gain margin of 2.0, the controller gain becomes

$$K_c = \frac{1}{2(0.00646)} = 77.4$$

which is twice the previous value.

Comparing the offset for $\tilde{v} = 0.15/s$ for the two systems, we obtain from Eq. 7.8-9 that

$$x_5 = \frac{\alpha_2^5(0.15)}{\alpha_1^5 + (-\beta)\alpha_2^4 K_c} = \frac{(5)^5(0.15)}{(5.33)^5 + (0.313)(5)^4(30.5)} = 0.0456$$

for the first case and from Eq. 7.8-13 that

$$x_5 = \frac{\alpha_2^5(0.15)}{\alpha_1^5 + (-\beta)\alpha_1^2\alpha_2^2 K_c} = \frac{(5)^5(0.15)}{(5.33)^5 + (0.313)(5.33)^2(5)^2(77.4)} = 0.0230$$

for the second case.

Obviously, shifting the measuring point closer to the source of the disturbance has led to an improvement. However, it should be recognized that this advantage is obtained only by sacrificing part of the feedback nature of the control system. In other words, if some other kind of disturbance, which is not considered in our model, causes a change in the first three vessels, the controller in Figure 7.8-3 will still provide a correction for the disturbance. Alternately, if some other disturbance causes a change in either of the last two vessels, no compensation will be provided. Hence the feedback controller shown in Figure 7.8-2 has some advantages, even though it does not have as good a performance as the second system for the particular problem under consideration. Also, it should be noted that an attempt to remove offset by adding an integral mode to the controller shown in Figure 7.8-3 might not lead to a zero offset at the output.

Perhaps at this point many readers will guess that we can improve on both these control systems by combining them in some way. If we measure the composition in the last vessel and use this deviation to change the set point $\tilde{x}_{30}$ of the controller shown in Figure 7.8-3, we call the result a *cascade control system*. Other multivariable systems can be designed, and this approach is treated in the next chapter. Before leaving this problem, however, we should note that an inspection of the system equations, Eqs. 7.8-3, indicates that we might obtain a better response if we choose either the flow rate q or the temperature in the last vessel k_5 as the control variable, because they will have a more immediate effect on the variable we hope to maintain constant, A_5. In fact, the linearized equations with k_5 chosen as the control variable will appear to be first order, so that the magnitude of the controller gain will be limited only by the dynamics of the elements in the feedback loop.

Example 7.8-3   *Control of a complex reacting mixture*

At first glance the results obtained for the preceding problem might appear to be a very special case, because we had a simple sequence of reactors. A more detailed consideration, however, reveals that this kind of system is similar to the one described in Example 7.5-1, where we had a sequence of reactions in a single vessel. Thus it might be better to measure the composition of a component that appears earlier in the sequence, such as B, rather than measure the desired component C. In order to explore this possibility, we let

$$u = K_c(x_{20} - x_2) \qquad 7.8\text{-}15$$

in Eqs. 7.5-18, take the Laplace transform of the equations, and solve them for the variables of interest. With this procedure, we obtain

$$\tilde{x}_2 = \frac{a_{21}b_{11}}{(s - a_{11})(s - a_{22})}\tilde{u} + \frac{(a_{21}c_{11}) + [c_{12}(s - a_{11})]}{(s - a_{11})(s - a_{22})}\tilde{v} \qquad 7.8\text{-}16$$

$$\tilde{x}_3 = \frac{a_{32}(s - a_{44})}{(s - a_{33})(s - a_{44}) - a_{34}a_{43}}\tilde{x}_2 + \frac{c_{13}(s - a_{44}) + a_{34}c_{14}}{(s - a_{33})(s - a_{44}) - a_{34}a_{43}}\tilde{v} \qquad 7.8\text{-}17$$

The control system design is based on the first of these equations; and after they are combined, we find that the expression for the output becomes

$$\tilde{x}_3 = \frac{a_{32}a_{21}b_{11}(s - a_{44})}{[(s - a_{11})(s - a_{22}) + K_c a_{21}b_{11}][(s - a_{33})(s - a_{44}) - a_{34}a_{43}]}\tilde{x}_{20}$$

$$+ \left\{\left[\frac{c_{11}a_{21} + c_{12}(s - a_{11})}{(s - a_{11})(s - a_{22}) + K_c a_{21}b_{11}}\right]\left[\frac{a_{32}(s - a_{44})}{(s - a_{33})(s - a_{44}) - a_{34}a_{43}}\right]\right.$$

$$+ \left.\frac{c_{13}(s - a_{44}) + c_{14}a_{34}}{(s - a_{33})(s - a_{44}) - a_{34}a_{43}}\right\}\tilde{v} \qquad 7.8\text{-}18$$

The asymptotic approximations of the transfer function given in Eq. 7.8-16

$$\frac{\tilde{x}_2}{\tilde{u}} = \frac{a_{21}b_{11}}{(s - a_{11})(s - a_{22})} = \frac{0.0687}{(0.351s + 1)(0.230s + 1)} \qquad 7.8\text{-}19$$

are shown in Figure 7.8-4 for the same system parameters listed in Example 7.5-1. This is a second-order transfer function, so that the crossover frequency approaches infinity and the gain margin cannot be used as a design criterion. However, if we use the phase margin criterion, we obtain

$$K_c = \frac{1}{[G|_{-135°}]} = \frac{1}{(0.0687)(0.1)} = 145.5$$

or

$$K_c = \frac{1}{[G|_{-120°}]} = \frac{1}{(0.0687)(0.224)} = 65.1$$

Our previous value for a proportional controller, using the phase margin as a performance measure, was about 11.5. Thus we expect to obtain a faster response and less offset if we measure the composition of component B rather than C, even though both measurements are made at the exit of the reactor. However, this advantage is compensated for by the fact that we no longer have a real feedback control system in the sense that any disturbances entering Eqs. 7.5-14 and 7.5-17 only will not activate the controller—for example, a change in the rate constants k_3, k_3', or k_4. This limitation can be removed by the use of a cascade control system.

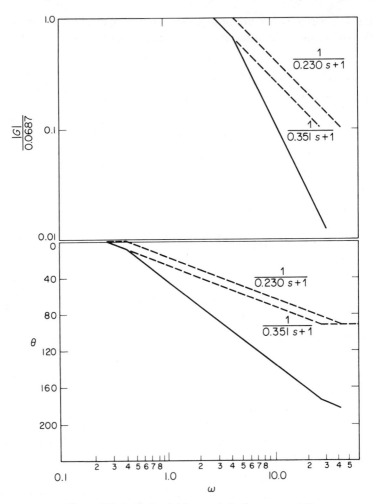

Figure 7.8-4. Bode plot for control of component B.

The same procedure can be used to evaluate the effect of adding other control modes. From an inspection of the system equations, Eqs. 7.5–12 through 7.5–15, we might also expect to obtain an improved dynamic response by using the flow rate q as a control variable, because this appears in the state equation describing the desired component, Eq. 7.5–14.

Nonideal Controllers

In our discussions of control system synthesis techniques, we have assumed that it was possible actually to implement the equation describing a three-mode controller

$$u = -K_c \left[(x_0 - x) + T_D \frac{d(x_0 - x)}{dt} + \frac{1}{T_i} \int (x_0 - x)dt \right] \quad 7.8\text{-}20$$

This assumption is valid, for all practical purposes, if we use a small digital computer as a controller, but neither an analog computer nor a pneumatic controller will act exactly in this way. Instead, a more realistic expression, which is normally used to describe a three-mode pneumatic system, is

$$G_c = K \left(\frac{\tau_1 s + 1}{\tau_2 s + 1} \right) \quad 7.8\text{-}21$$

As τ_2 approaches zero, the expression reduces to

$$G_c = K(\tau_1 s + 1) \quad 7.8\text{-}22$$

which is just the result for an ideal proportional plus derivative controller; whereas if τ_2 becomes very large, we find that

$$G_c = K \left(\frac{\tau_1 s + 1}{\tau_2 s} \right) = K \frac{\tau_1}{\tau_2} \left(1 + \frac{1}{\tau_1 s} \right) \quad 7.8\text{-}23$$

which is simply the result for a proportional plus integral controller. In the intermediate range of values of τ_2, the main difference between Eq. 7.8–21 and the ideal three-mode controller transfer function is that the actual controller gain always remains finite, both at very large values of s (high frequencies) and very small values of s (low frequencies). However, the basic design procedure used to select the constants in the controller transfer function is essentially the same as that described previously.

Nonlinear Effects

Our procedures for synthesizing control systems have been based on the linearized system equations. It seems as if it would be a good idea to evaluate the importance of the nonlinearities before an attempt is made to implement the control system. This is particularly true for nonisothermal reactor problems where there is an exponential nonlinearity present. In order to establish the effect of the nonlinear terms on the dynamic response of a closed-loop system, we can again use perturbation theory. All that is required is substitution of the control action (determined by the linear design procedure) into the Taylor series expansion of the original nonlinear system equations; manipulation of the equations, if necessary, to put them in the proper form; application of the perturbation method described in Chapter 6; and then an evaluation to see if the first-order correction functions are small in comparison with the solutions of the linearized equations.

Example 7.8-4 Perturbation analysis of a reactor control problem

Show how it would be possible to determine the response of the system described in Example 7.5-1 to a step change in the composition of component

R when the proportional plus integral controller described in Example 7.5-2 is installed on the reactor.

Solution

Using the definitions of the deviation variables and system parameters presented in Eqs. 7.5-19 and expanding the original system equations, Eqs. 7.5-12 through 7.5-15, in a Taylor series around the steady state operating condition, we obtain

$$
\begin{aligned}
\dot{x}_1 &= a_{11}x_1 + b_{11}u + c_{11}v - (a_{21}vx_1) \\
\dot{x}_2 &= a_{21}x_1 + a_{22}x_2 + c_{12}v + (a_{21}vx_1 - a_{32}vx_2) \\
\dot{x}_3 &= a_{32}x_2 + a_{33}x_3 + a_{34}x_4 + c_{13}v + (a_{32}vx_2 - a_{43}vx_3) \\
\dot{x}_4 &= a_{43}x_3 + a_{44}x_4 + c_{14}v + (a_{43}vx_3 - a_{45}vx_4)
\end{aligned}
\qquad 7.8\text{-}24
$$

where the terms in parentheses represent all of the nonlinear terms. The control law is

$$
u = -K_c\left(x_3 + \frac{1}{T_i}\int x_3\,dt\right) \qquad 7.8\text{-}25
$$

where we have taken the controller set point as $x_{30} = 0$. Substituting this expression into the equation for $\dot{x}_1$ and differentiating, we find that

$$
\ddot{x}_1 = a_{11}\dot{x}_1 - b_{11}K_c\dot{x}_3 - b_{11}K_c\left(\frac{1}{T_i}\right)x_3 + c_{11}\dot{v} - a_{21}\dot{v}x_1 - a_{21}v\dot{x}_1 \qquad 7.8\text{-}26
$$

For a step change in the disturbance, $\dot{v} = 0$. Thus if we drop these terms and simultaneously eliminate $\dot{x}_3$ by substituting the third equation in the set, our expression becomes

$$
\begin{aligned}
\ddot{x}_1 = a_{11}\dot{x}_1 - b_{11}K_c[a_{32}x_2 &+ \left(a_{33} + \frac{1}{T_i}\right)x_3 + a_{34}x_4 + c_{13}v \\
&+ a_{32}vx_2 - a_{43}vx_3] - a_{21}v\dot{x}_1
\end{aligned}
$$

Now if we introduce a new variable

$$
x_0 = \dot{x}_1 \qquad 7.8\text{-}27
$$

we can put the equations into the desired form

$$
\begin{aligned}
\dot{x}_0 = \ddot{x}_1 &= a_{11}x_0 - a_{32}b_{11}K_cx_2 - \left(a_{33} + \frac{1}{T_i}\right)b_{11}K_cx_3 - a_{34}b_{11}K_cx_4 - c_{13}b_{11}K_cv \\
&\quad + (-a_{21}vx_0 - a_{32}b_{11}K_cvx_2 + a_{43}b_{11}K_cvx_3) \\
\dot{x}_1 &= x_0 \\
\dot{x}_2 &= a_{21}x_1 + a_{22}x_2 + c_{12}v + (a_{21}vx_1 - a_{32}vx_2) \\
\dot{x}_3 &= a_{32}x_2 + a_{33}x_3 + a_{34}x_4 + c_{13}v + (a_{32}vx_2 - a_{43}vx_3) \\
\dot{x}_4 &= a_{43}x_3 + a_{44}x_4 + c_{14}v + (a_{43}vx_3 - a_{45}vx_4)
\end{aligned}
\qquad 7.8\text{-}28
$$

The appearance of a fifth-order system results from the introduction of the integral control mode. The initial conditions are $x_1 = x_2 = x_3 = x_4 = 0$, which means that the system starts from a steady state operating condition, and $x_0 = c_{11}v$, which is obtained by combining Eqs. 7.8–27 and the first expression in Eq. 7.8–24. The perturbation equations are developed by multiplying the product terms in parenthesis on the right-hand sides of the equations above by a small parameter μ, assuming a solution having the form

$$x_i = y_{i1} + \mu y_{i2} + \mu^2 y_{i3} + \dots \qquad 7.8\text{–}29$$

and proceeding as described in Chapter 6. Actually, since the disturbance for this problem is a step input, v is a constant and Eqs. 7.8–28 are linear. Therefore they can be solved directly, and a perturbation analysis is not necessary, although it will give the correct results.

On-Off Control and Describing Functions

When the gain of a proportional controller is made very large, even a small deviation of an output variable from its desired value will call for a large amount of control action. Of course, in any real system the amount of available control action will be limited to the region between where the control valve is fully opened and where it is completely closed. At the extremes of the range, we say that the control action is saturated. In the limiting case, where even the slightest positive error calls for the maximum amount of control available and the slightest negative error calls for the minimum control, we call the result an *on-off*, or a *relay* controller. This type of controller is commonly encountered in industry and elsewhere. One common example is the household thermostat, which turns the furnace on (maximum control) when the room temperature drops below a certain value and shuts off the furnace (minimum contol) when the room temperature returns to the desired setting. Another example is the simple float system in a toilet tank.

It is apparent from Figure 7.8–5 that the output signal from the relay (the control settings $u_{\max}$ and $u_{\min}$) are not a linear function of the input signal (the measured error) and, therefore, that relay control systems are not amenable to the linear analysis we described above. Moreover, it should be recognized that a plant containing a relay controller can never operate at steady state conditions, for the control variable will always be switching between its maximum and minimum values. An additional consequence of this behavior is that we cannot expand the input-output relationship in a Taylor series and cannot use perturbation theory to estimate the effect of the nonlinearity on the dynamic response of the system. Another situation where this type of difficulty is encountered is hysteresis in valves (the flow rate through the valve is a different function of the valve stem position depending on whether the valve is being opened or closed).

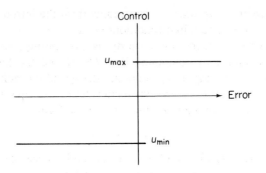

Figure 7.8-5. Relay controller.

The fact that perturbation techniques are not applicable to these kinds of problems means that we need an alternate approach. One method that has been demonstrated to be successful is to characterize the nonlinear element by its *describing function*. This is developed by considering a sinusoidal input to the element. For example, for the case of a relay, we let the input signal be

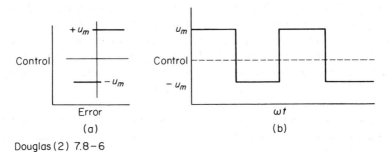

Douglas (2) 7.8−6

Figure 7.8-6. Discontinuous operation: (a) relay characteristic; (b) output signal.

$A \sin \omega t$. Then, as shown in Figure 7.8–6, the output will be a square wave. We can approximate this square wave signal by a Fourier series.

$$u(t) = A_0 + \sum_{n=1}^{\infty} A_n \sin n\theta + \sum_{n=1}^{\infty} B_n \cos n\theta \qquad 7.8\text{–}30$$

where

$$A_n = \frac{1}{\pi} \int_0^{2\pi} f(\theta) \sin n\theta \, d\theta \qquad B_n = \frac{1}{\pi} \int_0^{2\pi} f(\theta) \cos n\theta \, d\theta \qquad 7.8\text{–}31$$

Since the square wave is an odd function, $B_n = 0$, and since it has a zero mean value, $A_0 = 0$. Also,

$$A_n = \frac{1}{\pi} \int_0^{2\pi} f(\theta) \sin n\theta \, d\theta = \frac{2}{\pi} \int_0^{\pi} u_m \sin n\theta \, d\theta = \frac{2u_m}{n\pi} \qquad 7.8\text{-}32$$

Thus

$$u(t) = \frac{4u_m}{\pi}(\sin \omega t + \frac{1}{3}\sin 3\omega t + \frac{1}{5}\sin 5\omega t + \cdots) \qquad 7.8\text{-}33$$

In other words, the response of a relay to a sinusoidal input is a periodic signal that contains a fundamental component as well as higher harmonics.

If we consider the relay as part of a closed-loop configuration (see Figure 7.8-7), we want to find some way of characterizing the frequency response of the relay. Although we could use the complete expression given by Eq. 7.8-33, it would be advantageous to look for a simpler approximation that does not

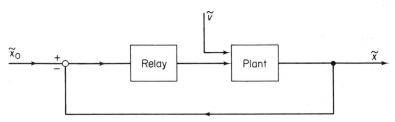

Figure 7.8-7. Relay control system.

contain an infinite number of terms. This step can be accomplished by first noting that the coefficients of the higher harmonic terms are smaller than the fundamental and that they continue to decrease as the order of the harmonic increases.

In addition, we know that if this output signal is used as the input to other dynamic systems, the higher harmonic terms will normally be damped more than the fundamental; that is, the amplitude ratio on a Bode plot usually decreases as the frequency of the input signal increases. Because of these two factors, it is common practice to assume that the frequency response of the nonlinear element can be adequately represented by its fundamental component and that the higher harmonics are completely damped by the plant dynamics. Hence the describing function for a relay (the relationship between the output and, input sine waves) is simply

$$G = \frac{4u_m}{\pi A} \qquad 7.8\text{-}34$$

For most nonlinear elements of this type,[1] the describing function depends on both the amplitude and the frequency of the input signal.

---

[1] An extensive amount of material is available in a number of electrical engineering texts on control theory.

SECTION 7.9 FEEDFORWARD CONTROL

One of the major considerations in our previous discussion of the design of feed-back control systems was that we could provide corrections for disturbances entering the system, even when the disturbances were unknown. This is an extremely important result, for the disturbances cannot be measured in a great number of cases. However, it is necessary to pay a price for ignoring these inputs in that we must wait until they produce a deviation in the measured output variable before we attempt to make any correction by manipulating the control variable. Thus it seems as if it should be possible to design control systems having a better performance than a feedback controller, whenever we can measure the disturbances. Indeed, doing so is possible, and in some cases we can achieve a perfect regulating controller. The approach of synthesizing controllers based on a knowledge of the disturbances and the system dynamics is called feedforward control.

Design Procedure

In this chapter we are restricting our attention to plants having a single desired output and a single control variable. Hence the transformed system equations can always be written

$$\tilde{y} = G_u \tilde{u} + G_v \tilde{v} \qquad\qquad 7.9\text{-}1$$

where G_u and G_v represent the transfer functions for the control and disturbance variables, respectively. If we can measure the disturbances, so that $\tilde{v}$ is known, we can manipulate the control variable in such a way that it is related to these input fluctuations, rather than the error in the output signal. Thus we let

$$\tilde{u} = G_c \tilde{v} \qquad\qquad 7.9\text{-}2$$

where G_c is the unknown transfer function of the controller. Substituting this expression into the system equation, we obtain

$$\tilde{y} = (G_v + G_u G_c)\tilde{v} \qquad\qquad 7.9\text{-}3$$

Now it is clear that if we select the controller transfer function so that

$$G_v + G_u G_c = 0 \qquad\qquad 7.9\text{-}4$$

or

$$G_c = -\frac{G_v}{G_u}$$

the output signal will be completely independent of the disturbances. In other words, whenever we can measure the disturbances and can select a controller transfer function that satisfies Eq. 7.9-4, we can manipulate the control variables to compentate for the disturbances exactly, and the system output will always remain constant at its optimum steady state value.

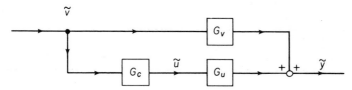

Figure 7.9-1. Feedforward control.

A block diagram of this approach is shown in Figure 7.9-1. The feedforward nature of the controller is apparent from the sketch. That is, we are measuring one of the inputs to the plant and using this signal to change another input. This situation is in contrast to a feedback controller, where a measured value of the output is used to change the control variable.

Limitations

At first glance the technique of feedforward control seems to provide a complete answer to all regulating control problems; therefore one starts to wonder why feedback control is ever utilized. However, the feedforward analysis contains a number of implicit assumptions, which must be satisfied if perfect control is to be achieved. These are

1. All disturbances must be measurable.
2. All the system transfer functions and the system parameters must be known exactly.
3. The required transfer function of the feedforward controller must be realizable.

The first of these requirements needs little elaboration. If there is some disturbance that is not recognized or that cannot be measured, the output will vary; and if the effects of the disturbance are cumulative, the results may be disastrous. For example, the effect of wind gusts, potholes, and so forth on the blind man's car during his trip between two cities will lead to his failure in reaching his objective. Similarly, pump failures, the breakdown of a measuring instrument, or the sticking of a valve might lead to a catastrophe.

The limitations imposed by the second requirement are similar to those discussed above. From our discussion in Chapters 3 through 5, we recognize that we seldom have an exact description of the process dynamics. This means that a feedforward controller cannot be expected to provide perfect regulation. In addition, if the errors in the system dynamics are cumulative, we again get into serious trouble. Our example of the blind man driving the car makes this point apparent.

The objections raised by the first two requirements can be overcome by augmenting a feedforward controller with a feedback system. In this way the feedforward system will compensate for most of the effects of the disturbances, and the feedback controller acts as a safeguard for our ignorance. Each control

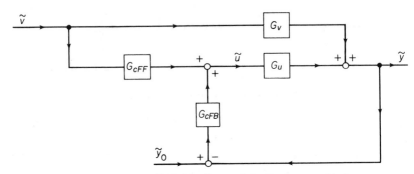

Figure 7.9-2. A combined feedforward–feedback control system.

system is designed as an independent entity, and they are then connected as shown in Figure 7.9–2.

The problem of the realizability of the feedforward controller sometimes provides a much more serious limitation. For a very simple example where

$$G_v = K_1 \quad \text{and} \quad G_u = \frac{K_2}{\tau s + 1} \qquad\qquad 7.9\text{-}5$$

the transfer function of the feedforward system becomes (see Eq. 7.9–4)

$$G_c = -\frac{G_v}{G_u} = -\frac{K_1}{K_2}(\tau s + 1) \qquad\qquad 7.9\text{-}6$$

which is just a proportional plus derivative controller.[1] However, for a case where

$$G_u = \frac{K_2}{(\tau_1 s + 1)(\tau_2 s + 1)} \qquad\qquad 7.9\text{-}7$$

the controller transfer function must be

$$G_c = -\frac{G_v}{G_u} = -\frac{K_1}{K_2}(\tau_1 s + 1)(\tau_2 s + 1) \qquad\qquad 7.9\text{-}8$$

which means that we need to generate the second derivative of the error signal. Normally this is not possible in practice because the differentiation of signals containing even slight amounts of noise (i.e., random disturbances) leads to wild fluctuations and meaningless results. In other words, while integration is a smoothing or error-reducing operation, differentiation is an operation that increases the error. This phenomenon can easily be demonstrated by considering a function such as that illustrated in Figure 7.9–3 and attempting to calculate the integral and the first and second derivatives.

[1] Actually, it is not possible to implement derivative action exactly, because doing so would require an infinite output for a step input. In other words, as we mentioned in Volume 1, Chapter 4, the denominator of the transfer function of any physical system must be of higher (or the same) order than the numerator. Nevertheless, we can approximate a first derivative action fairly well for smooth input signals, but we can seldom compute a second derivative accurately.

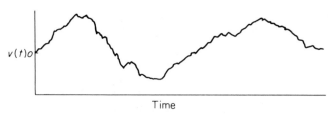

Figure 7.9-3. Disturbance signal.

An even more profound illustration of the difficulty encountered in the construction of a feedforward controller is obtained when we consider a case where G_u is a dead-time element. Thus we let

$$G_u = K_2 e^{-t_d s} \qquad\qquad 7.9\text{-}9$$

and find that

$$G_c = -\frac{G_v}{G_u} = \frac{K_1}{K_2} e^{+t_d s} \qquad\qquad 7.9\text{-}10$$

The positive sign on the exponential means that the controller must be able to predict future values of the disturbances, which is not possible under ordinary circumstances. (In some situations it might be possible to measure the disturbances far enough upstream of the unit to make the feedforward controller practical).

Example 7.9–1 Feedforward control of a reactor

Design a feedforward controller for the reactor described in Example 7.5–1.

Solution

The transfer functions relating the output of the desired component C, or x_3, to the control and disturbance variables are given in Eq. 7.5–19 as

$$\tilde{x}_3 = \frac{a_{21} a_{32} b_{11}(s - a_{44})}{(s - a_{11})(s - a_{22})[(s - a_{33})(s - a_{44}) - a_{34} a_{43}]} \tilde{u}$$

$$+ \left[\frac{c_{13}(s - a_{11})(s - a_{22})(s - a_{44}) + a_{34} c_{14}(s - a_{11})(s - a_{22})}{(s - a_{11})(s - a_{22})[(s - a_{33})(s - a_{44}) - a_{34} a_{43}]} \right.$$

$$\left. + \frac{a_{32} c_{12}(s - a_{11})(s - a_{44}) + a_{21} a_{32} c_{11}(s - a_{44})}{(s - a_{11})(s - a_{22})[(s - a_{33})(s - a_{44}) - a_{34} a_{43}]} \right] \tilde{v} \qquad 7.9\text{-}11$$

or

$$\tilde{x}_3 = G_u \tilde{u} + G_v \tilde{v} \qquad\qquad 7.9\text{-}12$$

For feedforward control

$$G_c = -\frac{G_v}{G_u} \qquad\qquad 7.9\text{-}13$$

or

$$G_c = \frac{(s - a_{11})[c_{13}(s - a_{22}) + a_{32}c_{12}] + a_{21}a_{32}c_{11}}{a_{21}a_{32}b_{11}}$$
$$+ \frac{a_{34}c_{14}(s - a_{11})(s - a_{22})}{a_{21}a_{32}b_{11}(s - a_{44})}$$

7.9-14

The first term in this transfer function is equivalent to the expression in Eq. 7.9–8, so that the feedforward controller is not realizable.

Perhaps it should be noted that even if the controller was realizable, it could not be expected to provide perfect regulation, for the development of the transfer functions was based on the assumptions that the feed composition of components R and A always remained constant, that changes in the flow rate Q had no effect on the outlet flow rate q, and that the linearized equations could be used to describe the dynamics. Thus it would be necessary to use the feedback controller discussed in the previous section, together with a feedforward controller.

Nonlinear Systems

The idea of feedforward control can also be applied to nonlinear plants. For these cases, the concept of a transfer function has no meaning, and we must work directly with the original set of nonlinear equations describing the system. The procedure is to look for the relationship between the control variable and the disturbance, or disturbances, which will maintain the desired output constant at its desired optimum steady state value. Of course, the problem of realizing the control system becomes much more difficult, as can be seen in the following simple example.

Example 7.9-2 Nonlinear feedforward control of a nonisothermal CSTR

Design a feedforward control system for a reactor described by the set of equations

$$V\frac{dA}{dt} = q(A_f - A) - kVA \qquad\qquad 7.9\text{-}15$$

$$VC_p\rho\frac{dT}{dt} = qC_p\rho(T_f - T) + (-\Delta H)kVA + \frac{UA_H Kq_H}{1 + Kq_H}(T_H - T) \qquad 7.9\text{-}16$$

which will manipulate the flow rate of heating fluid q_H in such a way as to maintain the effluent composition of A constant despite disturbances in A_f, T_f, and T_H.

Solution

If we are to maintain A constant at its optimum steady state value, we must also have

$$\frac{dA}{dt} = 0 \qquad\qquad 7.9\text{-}17$$

Then in order to satisfy the material balance expression, we require that

$$T = \frac{-E/R}{\ln{(q/k_0 V)[(A_f(t) - A_s)/A_s]}}$$ 7.9-18

Now we can use this expression and its time derivative, which will introduce a dA_f/dt term, to eliminate temperature from the energy balance. In this way, we can obtain a relationship between the control variable q_H and the disturbances A_f, T_f, and T_H, as well as dA_f/dt. This feedforward control law is highly nonlinear, and it would require special equipment, such as a small analog computer, in order to make it realizable. In addition, the feedforward controller should be augmented by a feedback system, because the model does not include the heat exchanger dynamics and is based on the assumption that the arithmetic-mean temperature driving force across the heating coil is equivalent to the log mean.

SECTION 7.10   DISTRIBUTED PARAMETER SYSTEMS

In order to complete our discussion of the synthesis of simple control systems for plants having single inputs and single outputs, we need to consider the control of distributed parameter processes. We might expect that much of our analysis of lumped parameter plants can be extended to the more complicated systems described by partial differential equations in a straightforward way. However, a quick survey of the literature indicates that this assumption is completely unwarranted. In fact, despite the great number of papers describing the dynamics of distributed parameter plants, there are almost no examples illustrating the addition of a feedback controller to these systems.

One possible reason for this situation is that the characteristic equations of the closed-loop plant are normally transcendental functions, which often have an infinite number of roots. As a result, discovering how these roots change as the controller gain is varied is quite a tedious task. Similarly, the job of inverting the transformed equations to find the time response, for cases where there are an infinite number of roots, appears to be one that will require an infinite amount of patience. Another major factor, which probably has impeded this type of control system design, is that until the recent advent of hybrid computers, it was quite difficult to generate numerical solutions of partial differential equations. Thus it was a major job to simulate the behavior of a controller on any new plant.

Perhaps the best way of illustrating the magnitude of some of these problems is to describe the approach used by Cohen and Coon[1] to develop their rules for determining the controller settings from information about the process reaction curve—that is, a transfer function containing a dead time and a first-order lag.

We can then attempt to make some general statements about the possibility of applying the control-system synthesis techniques described earlier to distribut-

[1] See footnote 2, p. 121.

ed parameter systems. The final portion of our discussion will be devoted to the stability analysis of a few plants. Hopefully, a study of these stability problems, where our purpose is only to determine if the real part of any root is positive, will provide a better insight into the kinds of characteristic equations encountered for distributed parameter plants and the mathematical complexities to be expected when control systems are installed in an attempt to change the characteristic roots.

Reaction Curve Method

In actual practice, the design of most control systems for distributed parameter plants is accomplished by using either the reaction curve method of Cohen and Coon[1] or the loop-tuning procedure of Ziegler and Nichols.[2] In the former case, it is assumed that any plant can be represented by the transfer function

$$G(s) = \frac{Ke^{-t_d s}}{\tau s + 1} \qquad\qquad 7.10\text{--}1$$

The procedure for determining the constants in this model from step response information, either experimental or numerical, was presented in Section 5.3, and the suggested controller settings for plants described by this transfer function are listed in Eqs. 7.7–2 through 7.7–5. Our purpose here is to study the technique used by Cohen and Coon to develop these rules. For simplicity, we limit our attention to the case of a proportional controller.

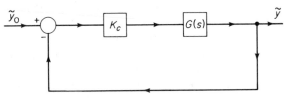

Figure 7.10-1. Feedback loop.

A sketch of the feedback control system is given in Figure 7.10–1. Using the results given in Section 7.2, we find that the closed-loop transfer function for the plant is simply

$$\frac{\tilde{y}}{\tilde{y}_0} = \frac{K_c G}{1 + K_c G} = \frac{K_c K e^{-t_d s}}{1 + \tau s + K_c K e^{-t_d s}} \qquad\qquad 7.10\text{--}2$$

The stability of the closed-loop system and the nature of the dynamic response depend on the roots of the characteristic equation, which is determined by setting the denominator of this closed-loop transfer function equal to zero

$$1 + \tau s + K_c K e^{-t_d s} = 0 \qquad\qquad 7.10\text{--}3$$

[2] See footnote 2, p. 121.

Unfortunately, this is a transcendental equation in the Laplace parameter and it has an infinite number of roots.

Before attempting to determine these roots, and to study the way in which the roots change as we vary the controller gain K_c, we try to simplify the problem by introducing the new variables

$$p = t_d s, \quad \tau_0 = \frac{t_d}{\tau}, \quad K_0 = \frac{K_c K t_d}{\tau} \qquad 7.10\text{-}4$$

With this transformation of variables, the characteristic equation becomes

$$\tau_0 + p + K_0 e^{-p} = 0 \qquad 7.10\text{-}5$$

Now if we suppose that the real roots of this equation occur when $p = -\lambda$, the real roots must satisfy the equation

$$K_0 e^{+\lambda} = \lambda - \tau_0 \qquad 7.10\text{-}6$$

For a particular case where $\tau_0 = 0.5$ and $K_0 = 0.2$ (see Figure 7.10-2), we find that there are two real roots, $\lambda_1 = 1.11$ and $\lambda_2 = 2.05$. A more general plot of the real roots was developed by Cohen and Coon, and this is presented as Figure 7.10-3.

In order to determine the complex conjugate roots, we let

$$p = \alpha \pm j\omega \qquad 7.10\text{-}7$$

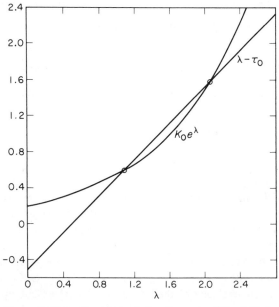

Figure 7.10-2. Graphical solution for roots of Eq. 7.10-6.

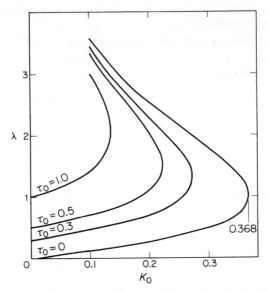

Figure 7.10-3. Real roots of characteristic equation. [Reproduced from G. H. Cohen and G. A. Coon, *Trans. ASME*, **75**, 827 (1953), by permission.]

substitute this expression into Eq. 7.10–5, and collect the real and imaginary expressions

$$(\tau_0 - \alpha + K_0 e^\alpha \cos \omega) + j(\omega - K_0 e^\alpha \sin \omega) = 0 \qquad 7.10\text{–}8$$

This equation can be satisfied only when both the real and the imaginary parts are equal to zero

$$\tau_0 - \alpha + K_0 e^\alpha \cos \omega = 0 \qquad \omega - K_0 e^\alpha \sin \omega = 0 \qquad 7.10\text{–}9$$

After some manipulation, these equations can be put into the form

$$\frac{\sin \omega}{\omega} \exp\left(\frac{\omega}{\tan \omega}\right) = \frac{1}{K_0 \exp \tau_0} \qquad 7.10\text{–}10$$

$$\alpha = \ln\left(\frac{\omega}{K_0 \sin \omega}\right) \qquad 7.10\text{–}11$$

Once values of K_0 and τ_0 have been specified, Eq. 7.10–10 can be solved, either numerically or graphically, for an infinite number of values of ω. Then Eq. 7.10–11 can be used to calculate the corresponding values of α. A sketch of the left-hand side of Eq. 7.10–10 is given in Figure 7.10–4 for values of ω between 0 and 5π. We are interested only in the positive values on this graph, for the right-hand side of Eq. 7.10–10 is always positive. Cohen and Coon prepared a more general plot of the first pair of complex conjugate roots (see

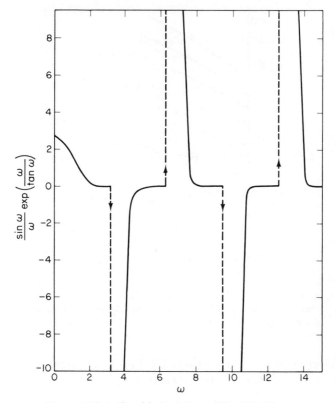

Figure 7.10-4. Graphical solution of Eq. 7.10-10.

Figure 7.10–5). It should be noted, however, that they use the parameters

$$\frac{P}{L} = \frac{2\pi}{\omega} \quad \text{and} \quad a = \exp\left(\frac{-2\pi\alpha}{\omega}\right) \qquad 7.10\text{-}12$$

rather than α and ω.

Once the characteristic roots have been determined, the step response of the system can be found by using the Heaviside expansion theorems. The general form of the solution will be

$$y(t) = \left(\frac{1}{\tau_0 + K_0}\right) + \sum_{n=0}^{\infty} A_n e^{-\alpha_n t / t_d} \cos\left(\frac{\omega t}{t_d} - \phi_n\right) + \sum_{n=1}^{\alpha} B_n e^{-\lambda_n t / t_d} \qquad 7.10\text{-}13$$

We want to adjust the controller gain so that the response resembles an under-damped, second-order system that has a decay ratio of approximately $\frac{1}{4}$. By evaluating the first three pairs of complex conjugate roots, Cohen and Coon were able to show that the amplitudes of the higher-harmonic terms in Eq. 7.10–13 were at least an order of magnitude less than the fundamental, providing that τ_0 was less than unity. Thus they decided to neglect these terms and to fix the decay ratio of the fundamental component as $\frac{1}{4}$. However, the decay

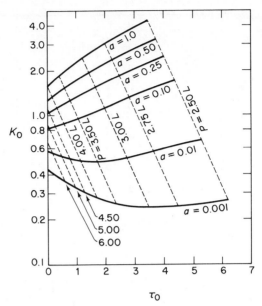

Figure 7.10-5. Complex conjugate roots. [Reproduced from G. H. Cohen and G. A. Coon, *Trans. ASME*, **75**, 827 (1953), by permission.]

ratio is identical to the parameter a in Eq. 7.10–12, and from Figure 7.10–5 we see that we can maintain $a = 0.25$ whenever

$$K_0 = 1.03 + 0.35\tau_0 \qquad\qquad 7.10\text{--}14$$

Modifying the constants slightly and substituting Eq. 7.10-4 to obtain our original system parameters, we find that

$$K_c = \frac{1}{K}\frac{\tau}{t_d}\left(1 + \frac{t_d}{3\tau}\right) \qquad\qquad 7.10\text{--}15$$

which is identical with Eq. 7.7–2.

Control-System Synthesis Techniques for Distributed Parameter Plants

In our discussion of the synthesis of control systems for the lumped parameter process, we emphasized design procedures based on Bode plots, root-locus diagrams, and empirical methods, such as the reaction curve. As mentioned previously, we know from experience that the empirical procedures usually lead to a satisfactory control system. However, since this approach generally requires either experimental measurement of the step response or the ultimate gain and period of a closed-loop system, it is not applicable to the development of control systems for new process units when they are still in the design stage. Of course, we can also apply the empirical techniques to numerical solutions of the partial differential equations describing the plant, but we would hope that it would be possible to find a simpler approach. Thus it seems to be worthwhile to consider briefly design methods based on Bode plots or root-locus diagrams.

Bode Plots

If we consider systems described by the approximate transfer function described above—that is, a dead time combined with a first-order lag, we can develop the Bode diagram for the plant by graphically combining the curves for each separate transfer function; see Figures 4.3–3 and 5.1–1. The procedure is straightforward and quite simple. Once this Bode diagram has been obtained, we can use the trial-and-error procedure described in Section 7.5 to synthesize the controller. The results obtained in this way will be essentially the same as those calculated by using the empirical method.

When we attempt to extend this approach to transfer functions obtained from the partial differential equations describing a plant, we sometimes run into trouble. For example, an inspection of the Bode diagram for a steam-heated exchanger with a wall capacity term shown in Figure 5.1–4 reveals that the phase angle curve may exhibit oscillations in the neighborhood of the crossover frequency. Similarly, the gain curve may fluctuate. Thus if we try to select a control system based on a gain margin or a phase margin criterion, even very small errors in the Bode diagram can give significantly different results. This kind of oscillatory behavior is often observed in the Bode diagrams for distributed parameter processes, and in these situations it is preferable to base the controller design on a Nyquist plot. With the Nyquist diagrams, it is easier to evaluate the stability of the closed-loop plant, and the difficulty of multiple crossings of the −180 deg line on a Bode diagram can be avoided.

Despite the fact that distributed parameter processes are normally said to be of infinite order and that there are often an infinite number of roots of the characteristic equation, not all of these systems have a crossover frequency. The Bode diagram for a countercurrent, double-pipe heat exchanger was presented in Figure 5.3–6, and we see that the phase curve approaches −90 deg asymptotically. This implies that we could install a proportional control system having an infinite gain, although it really means that the maximum controller gain is limited only by the dynamics of the elements in the feedback loop.

Root-Locus Diagrams

In our review of Cohen and Coon's analysis of the control of a plant approximately described by a dead time and a first-order lag, we found that even though the system had an infinite number of characteristic roots, only the first pair of complex conjugate roots had a major effect on the response. In a situation such as this, we often say that there is a *dominant* pair of *roots*. We could plot the locus of this pair of roots, with controller gain as a parameter, on a diagram similar to those used for lumped parameter plants and ignore all the other roots. Then we could use the same techniques described previously to select a satisfactory control system.

Some effort has been devoted to extending the root-locus method to more general partial differential equations,[3] but, in general, the simplicity of the approach is lost because the characteristic equations are transcendental. In order to gain a better understanding of the difficulty this situation introduces, it is necessary to look at some additional examples. For simplicity, we will restrict our attention to stability problems, where we only attempt to see if any of the roots can ever have a positive real part, rather than study control problems, where we are interested in the effect of the controller gain on the values of the roots.

Stability of Distributed Parameter Plants

Our previous study of stirred-tank reactor ploblems indicated that in certain situations it was possible to obtain steady state designs which were unstable. Since this kind of behavior is somewhat contrary to our intuition, perhaps it is worth demonstrating that distributed parameter processes can behave in the same way. The analysis of these of these problems also serves to illustrate the stability techniques that are applicable to both open-loop and closed-loop systems.

Example 7.10–1 Stability of a tubular reactor with recycle

In Example 5.3–5 we discussed the dynamic response of a nonisothermal tubular reactor, and now we want to consider the behavior of the system if we add a recycle loop (see Figure 7.10–6).[4] We let the fresh feed rate be q, the recycle flow rate be Q, and the recycle ratio be R:

$$R = \frac{Q}{q + Q} \qquad 7.10\text{–}16$$

The recycle flow has no effect on the partial differential equations describing the

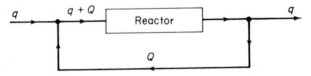

Figure 7.10-6. Tubular reactor with recycle.

---

[3] An illustration of the application of root locus to a dead-time element in a transfer function is presented on p. 200 of the reference given in footnote 2 on p. 51.

[4] The analysis is taken from O. Bilous and N. R. Amundson, *AIChE Journal*, **2**, 117 (1956).

conditions within the reactor, providing that the flow velocity v is now interpreted as the total flow rate, $q + Q$, divided by the cross-sectional area of the reactor. Assuming that no reaction takes place in the recycle line and that there is no heat loss from this line, the inlet conditions for the reactor are determined by making a material and energy balance at the mixing point

$$x_i(0) = (1 - R)x_f(0) + Rx(L)$$
$$T_i(0) = (1 - R)T_f(0) + RT(L)$$

7.10-17

In terms of the transformed deviation variables, these equations become

$$\tilde{y}_{c_i} = (1 - R)\tilde{y}_{c_f} + R\tilde{y}_c$$
$$\tilde{y}_{T_i} = (1 - R)\tilde{y}_{T_f} + R\tilde{y}_T$$

7.10-18

Thus the system response is approximately described by the differential equations given in Eqs. 5.3-81 and 5.3-82, together with the boundary conditions above. Using the same matrizant approach we developed in Example 5.3-5, we find that the solution of the equation is

$$\tilde{\mathbf{y}} = R[(1 + R)\mathbf{I} - \boldsymbol{\Omega}]^{-1}\boldsymbol{\Omega}\tilde{\mathbf{y}}_f$$

7.10-19

where

$$\boldsymbol{\Omega} = e^{-sL/v}\begin{pmatrix} b_{11} & b_{12} \\ b_{21} & b_{22} \end{pmatrix}$$

7.10-20

and the values of b_{ij} are identical to those given previously; see Eqs. 5.3-89, 5.3-92, and 5.3-93.

The stability of the recycle system only depends on the poles of the transfer function; and if there are any poles with a positive real part of a root, we say that the plant is unstable. From Eq. 7.10-19 we see that the only terms in the denominator of the transfer function are obtained when we calculate the inverse of the matrix $[(1 + R)\mathbf{I} - \boldsymbol{\Omega}]$. In other words, the poles will be the zeros of the determinant of this expression,

$$(1 + R)^2 e^{2sL/v} - (b_{11} + b_{22})(1 + R)e^{sL/v} + b_{11}b_{22} - b_{12}b_{21} = 0 \qquad \text{7.10-21}$$

where we have multiplied the determinant by the factor $e^{2sL/v}$. If we let

$$\lambda = (1 + R)e^{sL/v}$$

7.10-22

the characteristic equation becomes

$$\lambda^2 - (b_{11} + b_{22})\lambda + b_{11}b_{22} - b_{12}b_{21} = 0$$

7.10-23

However, in order to ensure that all the roots s satisfying Eq. 7.10-21 have negative real parts, we must examine the behavior of the function given by Eq. 7.10-22. By recognizing that s can be a complex number, we can write

$$\lambda = (1 + R)e^{\alpha + j\beta} = (1 + R)e^{\alpha}(\cos \beta + j \sin \beta)$$

7.10-24

The function

$$e^{j\beta} = \cos \beta + j \sin \beta \qquad 7.10\text{-}25$$

describes a circle of unit radius in the complex plane. Thus the largest value that we can ever obtain for λ is

$$\lambda_{\max} = (1 + R)e^{\alpha}\sqrt{\cos^2 \beta + \sin^2 \beta} = (1 + R)e^{\alpha} \qquad 7.10\text{-}26$$

For stable operation, we want all the roots to have negative real parts—that is, $\alpha < 0$. For this case, $e^{\alpha} < 1$, and the preceding equation can have a solution only if

$$\frac{\lambda_{\max}}{1 + R} < 1 \qquad 7.10\text{-}27$$

Then, in terms of Eq. 7.10-23, we see that the system will be stable if the magnitudes of the two roots of Eq. 7.10-23 are less than $\pm(1 + R)$ and if the product of the roots lies inside a circle of radius $(1 + R)^2$ centered at the origin. This is the stability criterion we hoped to derive. For a control problem it would be necessary actually to solve for the roots as a function of controller gain, to invert the Laplace transform, and to select the controller gain that corresponds to a decay ratio of $\frac{1}{4}$.

It should be noted that there is an alternate approach to the stability problem. Reilly and Schmitz[5] recognized that the stability information of interest can be obtained simply by following the progress of a plug of material as it flows through the system. That is, we can consider a particular plug of material at the inlet, calculate its composition and temperature change as it flows through

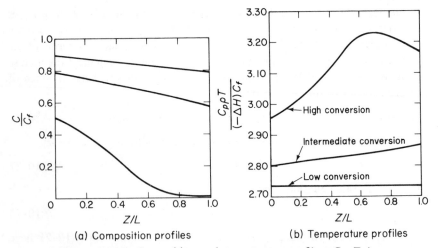

(a) Composition profiles (b) Temperature profiles

Figure 7.10-7. Composition and temperature profiles; $C_p \rho T_f/(-\Delta H)C_f = 2.737$, $4UL/C_p \rho v D = 0.45$. [Reproduced from M. J. Reilly and R. A. Schmitz, *AIChE Journal*, **12**, 153 (1966), by permission.]

[5] M. J. Reilly and R. A. Schmitz, *AIChE Journal*, **12**, 153 (1966).

the reactor, add a fraction, $Q/(q + Q)$, of this material to a fraction, $q/(q + Q)$, of fresh feed to find the inlet conditions for a second pass through the reactor, and repeat the procedure. If the system is stable, the inlet conditions for one plug will eventually become identical to those for the preceding plug. Also, after rewriting the system equations and boundary conditions so that they describe the conditions for the $(n + 1)$st pass through the reactor in terms of the nth pass, we obtain a set of finite-difference equations, and can use finite-difference techniques to determine the stability of the plant.

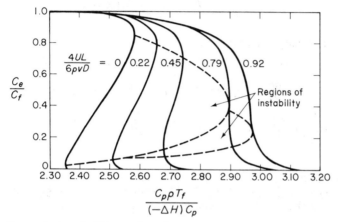

Figure 7.10-8. Stability regions for tubular reactor with recycle. [Reproduced from M. J. Reilly and R. A. Schmitz, *AIChE Journal*, **12,** 153 (1966), by permission.]

Reilly and Schmitz carried out a fairly extensive numerical study showing that sometimes there could be three steady state solutions (composition and temperature profiles) of the system equations (see Figure 7.10–7). They proved that the intermediate solution was always unstable. In addition, they found that it was possible to obtain a single steady state solution which was unstable. Their stability regions are given in Figure 7.10–8. In a later paper, Luss and Amundson[6] extended the results to complex kinetic mechanisms.

Example 7.10–2 Stability of reaction on a spherical particle

As a second illustration of the stability analysis of a distributed parameter system, we consider a problem where there is a surface reaction on a spherical particle. This might correspond to a case where the particle is burning in a fluid stream, or where it acts as a catalyst for reacting components that are present in a fluid flowing past the particle. In either case it is important to know the composition and temperature near the surface in order to calculate

[6] D. Luss and N.R. Amundson, *AIChE Journal*, **13,** 279 (1967).

the reaction rate. The most common approach for estimating the surface conditions is to introduce film heat-transfer and film mass-transfer coefficients, but, for some simple problems, it should be possible to undertake a more rigorous analysis. For simplicity, we will consider a single particle having an infinite thermal conductivity in an infinite supply of stationary fluid. The reaction is considered irreversible and first order.

The material and energy balances for the fluid phase are

$$\frac{\partial c}{\partial t} = D \frac{1}{r^2} \frac{\partial}{\partial r} \left(r^2 \frac{\partial c}{\partial r} \right)$$

$$\rho_f C_{pf} \frac{\partial T}{\partial t} = k_f \frac{1}{r^2} \frac{\partial}{\partial r} \left(r^2 \frac{\partial T}{\partial r} \right)$$

$$7.10\text{--}28$$

and the boundary conditions are

$$c(\infty, t) = c_f, \quad T(\infty, t) = T_f, \quad T(r_0, t) = T_p(r_0, t)$$

$$D \frac{\partial c}{\partial r} = c k_0 \exp \left(-\frac{E}{RT} \right) \quad \text{at } r = r_0 \text{ and any } t \qquad 7.10\text{--}29$$

$$4\pi r_0^2 \left[k_f \frac{\partial T}{\partial r} + (-\Delta H) c k_0 \exp \left(-\frac{E}{RT} \right) \right] = \frac{4}{3} \pi r_0^3 \rho_p C_{pp} \frac{\partial T_p}{\partial r}$$

$$\text{at } r = r_0 \text{ and any } t$$

The initial conditions will be introduced later.

At steady state conditions, the accumulation terms are equal to zero. The solutions of the material and energy balances have the form

$$c = \frac{c_1}{r} + c_2 \qquad T = \frac{c_3}{r} + c_4 \qquad 7.10\text{--}30$$

From the boundary conditions we find that

$$c_2 = c_f \qquad c_4 = T_f$$

$$\frac{-c_1 D}{r_0^2} = \left(\frac{c_1}{r_0} + c_f \right) k_0 \exp \left(\frac{-E/R}{T_f + c_3/r_0} \right) \qquad 7.10\text{--}31$$

$$\frac{c_3 k_f}{r_0^2 (-\Delta H)} = \left(\frac{c_1}{r_0} + c_f \right) k_0 \exp \left(\frac{-E/R}{T_f + c_3/r_0} \right)$$

By combining the last two expressions, we see that

$$c_1 = \frac{-k_f}{D(-\Delta H)} c_3 \qquad 7.10\text{--}32$$

so that we can write

$$\left[\frac{k_f}{(-\Delta H)} \right] c_3 = \left[c_f - \frac{k_f c_3}{D(-\Delta H) r_0} \right] k_0 \exp \left(\frac{-E/R}{T_f + c_3/r_0} \right) \qquad 7.10\text{--}33$$

This transcendental equation can be solved, either graphically or numerically, for c_3. Under most circumstances, there are either one or three roots. This means that either one or three sets of steady state profiles are described by Eqs. 7.10–30.

The stability of each of these profiles can be determined by studying the behavior of the linearized system equations. Actually, since the material and energy balances are linear, the only linearization required occurs in the nonlinear boundary conditions containing the reaction rate terms. These expressions are expanded in a Taylor series around one of the steady state solutions, and the series is truncated after the first-order terms. Following this procedure and letting

$$x_c = \frac{c - c_f}{c_f} - \frac{c_s - c_f}{c_f}, \quad x_T = \frac{T - T_f}{T_f} - \frac{T_s - T_f}{T_f}, \quad \tau = \frac{Dt}{r_0^2}, \quad R = \frac{r}{r_0}$$

$$\text{Le}_f = \frac{k_f}{\rho C_p D}, \quad \text{Le}_p = \frac{k}{\rho_p C_{pp} D}, \quad a_{11} = \frac{k_s r_0}{D}, \quad a_{12} = \frac{k_s r_0}{D} \frac{c_s}{c_f} \frac{T_f}{T_s} \frac{E}{RT_s}$$

$$a_{21} = \frac{k_s r_0 (-\Delta H) c_f D}{D} \frac{1}{k_f T_f}, \quad a_{22} = \frac{k_s r_0 (-\Delta H) c_f D}{D} \frac{1}{k_f T_f} \frac{c_s}{c_f} \frac{T_f}{T_s} \frac{E}{RT_s} \qquad 7.10\text{–}34$$

where the constants a_{ij} are evaluated at the particle surface, we obtain the linearized perturbation equations and boundary conditions

$$\frac{\partial x_c}{\partial \tau} = \frac{1}{R^2} \frac{\partial}{\partial R} \left(R^2 \frac{\partial x_c}{\partial R} \right)$$

$$\frac{1}{\text{Le}_f} \frac{\partial x_T}{\partial \tau} = \frac{1}{R^2} \frac{\partial}{\partial R} \left(R^2 \frac{\partial x_T}{\partial R} \right)$$

$$x_c(\infty, \tau) = 0 \qquad x_T(\infty, \tau) = 0 \qquad 7.10\text{–}35$$

$$\frac{\partial x_c(1, \tau)}{\partial R} = a_{11} x_c(1, \tau) + a_{12} x_T(1, \tau)$$

$$\left(\frac{1}{3\text{Le}_p} \right) \frac{\partial x_T(1, \tau)}{\partial \tau} = \frac{\partial x_T(1, \tau)}{\partial R} + a_{21} x_c(1, \tau) + a_{22} x_T(1, \tau)$$

Similarly, we can write general expressions for the initial conditions as

$$x_c(R, 0) = x_{ci} \qquad x_T(R, 0) = x_{T_i} \qquad 7.10\text{–}36$$

One method for solving these equations is first to take the Laplace transforms with respect to the time variable.

$$\frac{1}{R^2} \frac{d}{dR} \left(R^2 \frac{d\tilde{x}_c}{dR} \right) = s\tilde{x}_c - x_{ci}$$

$$\frac{1}{R^2} \frac{d}{dR} \left(R^2 \frac{d\tilde{x}_T}{dR} \right) = \frac{1}{\text{Le}_f} s\tilde{x}_T - \frac{1}{\text{Le}_f} x_{T_i}$$

$$\tilde{x}_c(\infty, s) = 0 \qquad \tilde{x}_T(\infty, s) = 0 \qquad 7.10\text{–}37$$

$$\frac{d\tilde{x}_c(1, s)}{dR} = a_{11} \tilde{x}_c(1, s) + a_{12} \tilde{x}_T(1, s)$$

$$\frac{d\tilde{x}_T(1, s)}{dR} + a_{21} \tilde{x}_c(1, s) + a_{22} \tilde{x}_T(1, s) = \frac{1}{3\text{Le}_p} [s\tilde{x}_T(1, s) - x_{T_i}(1)]$$

Even though it is possible to find a solution of this set of equations, we will limit our attention to the special case where the Lewis number, Le_f, is unity; this is a commonly accepted assumption for problems involving simultaneous diffusion and conduction. We can further simplify the problem by considering a specific disturbance, which has the form of a temperature pulse on the surface of the particle. With these assumptions, the mass and energy balances become identical. Now the solutions can by written either in terms of Bessel functions or in the form

$$\tilde{x}_c = \frac{1}{R} \left[c_1 \exp\left(\sqrt{s}R\right) + c_2 \exp\left(-\sqrt{s}R\right) \right]$$

$$\tilde{x}_T = \frac{1}{R} \left[c_3 \exp\left(\sqrt{s}R\right) + c_4 \exp\left(-\sqrt{s}R\right) \right]$$

7.10–38

Using the boundary conditions at $R = \infty$, we find that $c_1 = c_3 = 0$. The boundary condition involving the radial composition gradient leads to the expression

$$-c_2(1 + \sqrt{s}\,) \exp\left(-\sqrt{s}\,\right) = c_2 a_{11} \exp\left(-\sqrt{s}\,\right) + c_4 a_{12} \exp\left(-\sqrt{s}\,\right)$$

or

$$c_4 = -\frac{a_{11} + 1 + \sqrt{s}}{a_{12}} c_2$$

7.10–39

and the boundary condition involving the radial temperature gradient gives

$$-c_4(1 + \sqrt{s}\,) \exp\left(-\sqrt{s}\,\right) + c_2 a_{21} \exp\left(-\sqrt{s}\,\right) + c_4 a_{22} \exp\left(-\sqrt{s}\,\right)$$

$$= \frac{1}{3Le_p} c_4 s \exp\left(-\sqrt{s}\,\right) - \frac{1}{3Le_p} x_{T_i}$$

7.10–40

After eliminating c_4 from this equation, solving for c_2, and substituting into Eq. 7.10–38, we find that

$$\tilde{x}_c = \frac{-(a_{21} x_{T_i}/3RLe_p) \exp\left[-\sqrt{s}(R - 1)\right]}{(1/3Le_p)s^{3/2} + [1 + (1/3Le_p)(1 + a_{11})]s + (2 + a_{11} - a_{22})\sqrt{s} + (1 + a_{11} - a_{22})}$$

7.10–41

If we let

$$\sqrt{s} = \lambda$$

7.10–42

and restrict our attention to the dynamic behavior on the particle surface, $R = 1$, the result becomes

$$\tilde{x}_c = \frac{-(a_{21} x_{T_i}/3RLe_p)}{(1/3Le_p)\lambda^3 + [1 + (1/3Le_p)(1 + a_{11})]\lambda^2 + (2 + a_{11} - a_{22})\lambda + (1 + a_{11} - a_{22})}$$

7.10–43

Assuming that all the roots of the cubic equation in the denominator are distinct, we can factor the denominator and write

$$\tilde{x}_c = \frac{A_0}{(\lambda + \lambda_1)(\lambda + \lambda_2)(\lambda + \lambda_3)} = \frac{A_0}{(\sqrt{s} + \lambda_1)(\sqrt{s} + \lambda_2)(\sqrt{s} + \lambda_3)}$$

$$7.10\text{-}44$$

Using the Heaviside theorems and a table of transforms, we find that the form of the solution must be

$$x_c(\tau) = \sum_{i=1}^{3} A_i(\lambda_i)\left[\frac{1}{\sqrt{\pi\tau}} + \lambda_i e^{\lambda_i^2 \tau} \text{ erfc}\left(-\lambda_i\sqrt{\tau}\right)\right] \qquad 7.10\text{-}45$$

Providing that each λ_i is negative, or has a negative real part, this expression will approach zero as τ approaches infinity; that is, L'Hôpital's rule can be used to show that $\exp{(\lambda_i^2\tau)}$ erfc $(-\lambda_i\sqrt{\tau})$ becomes vanishingly small. Alternately, if any of the λ_i are positive, or have a positive real part, this term in the solution becomes unbounded. From our previous study of stability, we know that the roots of the characteristic equation

$$\frac{1}{3\text{Le}_p}\lambda^3 + \left[1 + \frac{1}{3\text{Le}_p}(1 + a_{11})\right]\lambda^2 + (2 + a_{11} - a_{22})\lambda + (1 + a_{11} - a_{22}) = 0$$

$$7.10\text{-}46$$

will not have negative real parts unless all the coefficients in the expression above are positive, or

$$1 + a_{11} - a_{22} > 0 \qquad 7.10\text{-}47$$

A second stability criterion can be obtained from the Routh array

$$2 + a_{11} - a_{22} > 0 \qquad 7.10\text{-}48$$

However, this is always satisfied when Eq. 7.10-47 is valid. Hence Eq. 7.10-47 is the stability criterion we were attempting to establish. Friedly and Peterson[7] showed that when this result is applied to the previous steady state solutions, the two extreme solutions were stable, whereas the intermediate solution was unstable.

Effect of Position of Measuring Element by Simulation

In the foregoing discussions we linearized the partial differential equations and boundary conditions describing the system about some steady state profile and then we either solved the linearized equations analytically or at least determined the nature of the solution in the neighborhood of this steady state. Although this approach does indicate the stability characteristics of the uncontrolled system and is often satisfactory for designing feedback controllers, it does not provide a global picture of the behavior of system. For example, if the system is unstable, we cannot use the linearized equations to determine the final state approached by the plant. In fact, for this kind of a problem, it is

[7] J. C. Friedly and E. E. Petersen, *Chem. Eng. Sci.*, **19**, 783 (1964).

necessary to solve the original nonlinear equations numerically. We often call this procedure *simulation*. Many of the advantages and disadvantages of this approach were discussed in Section 7.2.

As an illustration of the simulation approach, we consider an exothermic chemical reaction taking place in a heat exchanger. This kind of a unit seems economical in certain situations, for the heat released by the reaction is used to preheat the reactant stream. The system equations are

$$\rho_t \frac{\partial x_t}{\partial \theta} + \rho_t u_t \frac{\partial x_t}{\partial z} = -k_t x_t$$

$$\rho_t A_t C_p \frac{\partial T_t}{\partial \theta} + \rho_t u_t C_p \frac{\partial T_t}{\partial z} = U A_H (T_s - T_t) + (-\Delta H) A_t k_t x_t$$

$$\rho_s \frac{\partial x_s}{\partial \theta} - \rho_s u_s \frac{\partial x_s}{\partial z} = -k_s x_s$$

$$\rho_s A_s C_p \frac{\partial T_s}{\partial \theta} - \rho_s u_s A_s C_p \frac{\partial T_s}{\partial z} = U A_H (T_s - T_t) + (-\Delta H) A_s k_s x_s$$

7.10-49

where the subscript t represents the tube side of the exchanger and s represents the shell side. All the accumulation terms are equal to zero at steady state, but the equations are highly nonlinear; therefore the steady state composition and temperature profiles must be determined numerically. Some results for a particular set of process parameters[8] are shown in Figure 7.10-9.

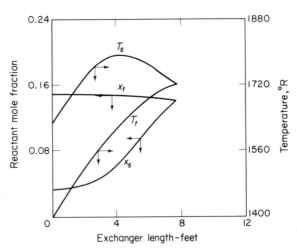

Figure 7.10-9. Steady state profiles. [Reproduced from J. M. Douglas, J. C. Orcutt, and P. W. Berthiaume, *Ind. Eng. Chem. Fundamentals*, **1**, 253 (1962), by permission of the ACS.]

[8] J. M. Douglas, J. C. Orcutt, and P. W. Berthiaume, *Ind. Eng. Chem. Fundamentals*, **1**, 253 (1962).

The stability of these steady state profiles could be studied in a number of ways. We could linearize the set of partial differential equations around the steady state solutions and attempt to determine whether or not one of the characteristic roots of the linearized equations with variable coefficients would have a positive real part. However, this approach normally requires an evaluation of the roots of a transcendental characteristic equation, which can be a difficult task. An alternate, and simpler, procedure would be to replace the spatial derivatives by a finite-difference approximation, in addition to linearizing the equations, so that we study the stability characteristics of a larger set of ordinary differential equations. For example, if we approximate each spatial derivative by four stirred-tank reactors, we would obtain 16 ordinary differential equations and 16 characteristic roots. Using the reactor sizes indicated on Figure 7.10–10 with this technique, and the parameters listed in the original paper, it can be shown that there are a pair of complex conjugate roots with positive real parts. Hence we expect the steady state solution of interest to be unstable.

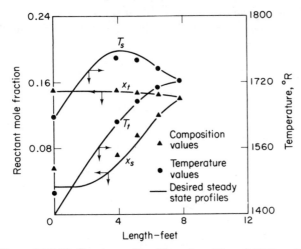

Figure 7.10-10. Reactor profiles. [Reproduced from J. M. Douglas, J. C. Orcutt, and P. W. Berthiaume, *Ind. Eng. Chem. Fundamentals*, **1**, 253 (1962), by permission of the ACS.]

Unfortunately, the stability analysis does not provide any indication of the final state that the system might approach. We expect that the system is bounded by zero and complete conversion, as well as the feed and adiabatic temperatures, so that an unstable steady state solution simply indicates that an arbitrarily small disturbance will force the system to move away from the initial region under consideration. To determine the final state, we must establish whether or not there are any other possible steady state solutions, whether there are any stable periodic solutions, and at least some of the possible paths from the unstable solution to these stable, final states. This task is quite difficult, for it requires that we obtain some numerical solutions of the original nonlinear, partial differential equations.

Until fairly recently it was not practical to attack problems of this complexity, even with a moderately large digital computer. Hence a common procedure for estimating the response of the system was to introduce the finite-difference approximation for the spatial derivative and then to solve the expanded set of nonlinear, ordinary differential equations with an analog computer. For the system parameters discussed above, and with the intervals for the difference approximation shown in Figure 7.10-10, the computer solutions indicated that the reactor would either blow out or blow up. In the blow-out case, there was essentially no reaction throughout the reactor and the temperature was uniform. At the blow-up steady state, all the conversion took place in the exchanger tubes, and the maximum temperature was much higher than the original steady state design condition. In fact, this temperature exceeded the safe operating level for the reactor material, so that a reactor failure was expected (a blow-up condition). One of these two final states was observed for a wide variety of disturbances entering the system; therefore it was assumed that these were the only possible final states. The blow-up condition is shown in Figure 7.10–11.

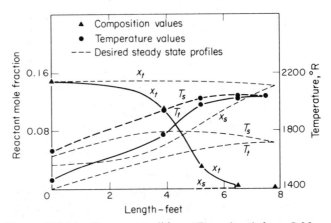

Figure 7.10-11. Blow-up condition. [Reproduced from J. M. Douglas, J. C. Orcutt, and P. W. Berthiaume, *Ind. Eng. Chem. Fundamentals*, **1**, 253 (1962), by permission of the ACS.]

Although it is helpful to know that our original steady state design can never be achieved in practice and to know what operating conditions will actually be observed instead of the desired one, this information still does not enable us to reach our original engineering goal. By analogy with our earlier work, we might hope that the installation of some kind of a feedback control system could be used to make the desired steady state design become stable rather than unstable. However, deciding whether it is better to measure one of the state variables somewhere along the reactor length rather than use a true feedback system and make the measurement at the shell outlet, or whether it is better to measure composition or temperature can only be established by making a number of case studies. In other words, we would like to be able to determine quickly the dynamic response of the nonlinear plant for various kinds of controllers with the measuring instruments located in a wide variety of positions. This process can be accomplished easily with the analog computer. In fact, the great advantage of the simulation approach is the ability to evaluate quickly the effects of modifying the original plant by adding controllers, holding tanks, piping changes, and so on. For the particular system under investigation, it can be shown that the effects of disturbances are minimized at the reactor outlet when the deviation of the tube outlet temperature from its desired steady state value is used as the error signal. The controller gain must be greater than a certain minimum value in order to make the process stable; but as the gain is increased, the system becomes very oscillatory and eventually becomes unstable again. Carlson[9] studied the same problem using a hybrid computer and found that the computation time could be reduced significantly with this approach.

Padé Approximation of a Dead Time

In the reactor simulation study presented above, a set of difference-differential equations rather than a set of partial differential equations was used to describe the system. This approximation, where the spatial derivatives are replaced by finite differences, was necessary to program the problem for an analog computer. A similar difficulty is encountered when attempts are made to use analog computers to simulate plants containing dead-time elements. The problem is inherent to analog machines because they operate continuously in time. Although recent developments of magnetic tape recorders and hybrid computers have made it possible to include dead times in a simulation, perhaps it is worthwhile to discuss briefly the Padé approximation of a dead time. This technique has been widely used in control studies and serves as an example of the kind of computational tricks employed by simulation experts.

We know that a dead time arises from the solution of a partial differential equation, and we have compared the response of the exact equation with the finite-difference approximation (see Section 5.1). The first-order, Padé approximation can be developed by writing the transfer function for a dead time as

$$H(s) = e^{-t_d s} = \frac{e^{-t_d s/2}}{e^{t_d s/2}} \qquad 7.10\text{--}50$$

then expanding the exponential functions in both the numerator and denomi-

[9] A. W. Carlson, Electronics Associates, Inc., *Application Bulletin No. 852034*, October 1968·

nator in a Taylor series, and, finally, truncating each series after the linear terms

$$H(s) = \frac{\tilde{y}}{\tilde{u}} = e^{-t_d s} \approx \frac{1 - (t_d/2)s}{1 + (t_d/2)s} \qquad 7.10\text{-}51$$

It is not difficult to show that this same expression can be obtained by taking the Laplace transform of the differential equation

$$\frac{t_d}{2}\frac{dy}{dt} + y = -\frac{t_d}{2}\frac{du}{dt} + u \qquad 7.10\text{-}52$$

and it is a simple task to develop an analog computer solution of this ordinary differential equation.

A comparison of the exact and approximate transfer functions given by Eq. 7.10–51 indicates that the gain curve is correct but that deviations start to occur in the phase curves at values of $t_d \omega$ greater than about 0.75. More accurate values of the phase lag can be predicted by using higher-order Padé approximations.[10] For example, a second-order approximation

$$H(s) = e^{-t_d s} \approx \frac{1 - (t_d s/2) + (t_d^2 s^2/12)}{1 + (t_d s/2) + (t_d^2 s^2/12)} \qquad 7.10\text{-}53$$

is valid up to values of $t_d \omega \approx 2.0$. After some manipulation, it can be shown that this form corresponds to a second-order, ordinary differential equation. Further improvements are possible by including additional terms. Similarly, other approximation techniques are available for dead-time systems.

Stability by Means of Liapunov Functions

In our analysis of the stability of lumped parameter processes, we noted that the direct method of Liapunov could sometimes be used to ascertain the stability of a plant without determining the characteristic roots or any other information about the behavior of the solution of the dynamic equations. The major limitations of this approach were the lack of any method for developing the required Liapunov functions and the fact that the predicted region of asymptotic stability was found to be very much smaller than the actual stable region. Some effort has been devoted to extending Liapunov's direct method to distributed parameter systems, and, as might be anticipated, the same difficulties are encountered. However, the method is still of some interest because it provides additional insight into the nature of partial differential equations from a different standpoint.

SECTION 7.11   SUMMARY

An introduction to some of the methods for designing control systems for processes with a single input and a single output has been presented. In the initial portion of the discussion, we described a plant where the optimum steady

---

[10] See p. 463 of the reference given in footnote 2 on p. 51.

state design was unstable. Then we showed some of the ways that various kinds of feedback controllers (i.e., proportional, derivative, and integral) could be used to make the unstable plant become stable. These examples demonstrated that the effect of adding a feedback controller to a process was to change the characteristic roots of the overall system. Hence it should be possible to use controllers to improve the dynamic response for both unstable and stable plants.

The way in which the feedback system alters the characteristic roots was studied for first-, second-, and third-order plants. The results indicate that perfect control can be achieved, at least conceptually, in the first two cases, but the last example illustrates that stable processes can become unstable if the controller gain becomes too large. In addition, the analysis revealed that an integral control mode increases the order of the system, and thereby tends to make it more unstable, even though it completely removes offset. For most plants, a derivative control mode gives a slower response but tends to stabilize the system. With this background, the similarities and differences of regulating and servomechanism control problems were discussed.

Some of the difficulties associated with implementing a control policy with actual hardware were described. Since this is a rapidly changing area of technology, only one simple control system was presented. Problems associated with automatic measurement, signal transmission, generating an error signal, implementing the desired control action, and altering valve settings to manipulate the control variables were considered.

Our previous work indicated that controllers could be used to make unstable systems become stable in some cases, whereas in other situations excessive controller gains could make stable plants become unstable. It is apparent, then, that stability considerations will play a major part in control system design. Hence a number of techniques that can be used to determine the stability of open-loop and closed-loop processes were introduced.

Finally, some of the most important procedures for synthesizing controllers for lumped parameter processes were presented. The application of Bode diagrams, the roots-locus method, and empirical approaches of Ziegler-Nichols and Cohen and Coon to the design problem were included in the discussion. Related problems, such as the treatment of nonideal controllers, loop tuning existing control systems, modifications for nonlinear processes, and the selection and location of the measuring instrument, were also described.

Design procedures for feedforward control systems were developed next. Originally only linear plants were considered, but the required extensions for nonlinear processes were noted. Initially it appears as if a feedforward controller always provides a perfect control, but a more careful investigation of the assumptions implicit in the synthesis technique reveals that perfect control can never be realized. Thus feedforward systems should always be augmented by a feedback controller.

In the last portion of the chapter, some of the existing work on the feed-

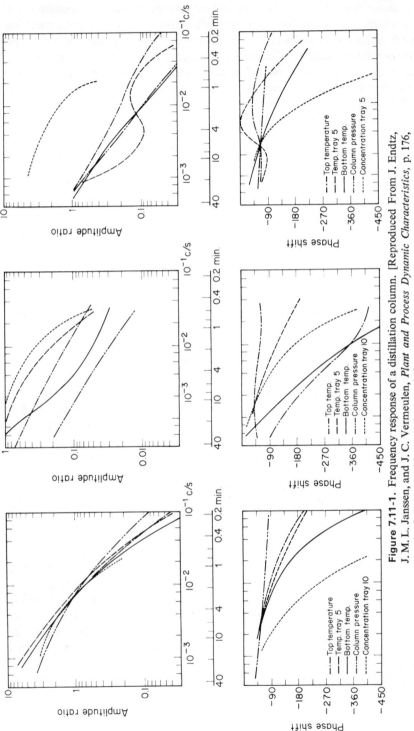

Figure 7.11-1. Frequency response of a distillation column. [Reproduced From J. Endtz, J. M. L. Janssen, and J. C. Vermeulen, *Plant and Process Dynamic Characteristics*, p. 176, Academic Press, N. Y., 1957, by permission.]

back control of distributed parameter processes was reviewed. The basis of the Cohen and Coon empirical procedure was derived. This problem illustrates that the characteristic equations of most distributed parameter plants have an infinite number of roots, which means that control-system synthesis techniques will be more difficult to apply for this type of process. Some stability problems were considered, and these helped to indicate the kind of characteristic equations that may be encountered.

EXERCISES

1. (A) The response of the outlet temperature of a steam-heated exchanger to a step change in the steam pressure was presented in Exercise 2, p. 293, Volume 1. Calculate the gains you would use for various kinds of controllers for this system.

2. (B*) Endtz, Jannsen, and Vermeulen[1] measured the frequency response of an industrial furnace and found that the transfer function relating the temperature of the effluent stream to the pressure supplied to a pneumatic motor valve on the fuel line could be represented by

$$\frac{\tilde{T}}{\tilde{P}} = \frac{40e^{-20s}}{(900s + 1)(25s + 1)} \quad (°C/psi)$$

Also, the response of the outlet temperature to feed temperature changes was

$$\frac{\tilde{T}}{\tilde{T}_f} = \frac{e^{-600s}}{60s + 1}$$

where the time constants are given in seconds.

If a fast-acting temperature-measuring device is available, which has a gain 0.2 (psi/°C), select the gains for various kinds of pneumatic controllers that use air pressure for the input and output signals. Then calculate the response of the closed-loop systems to a 20°C step change in the feed temperature.[2]

3. (B) Consider the reactor system discussed in Section 7.1 and evaluate the possibility of stabilizing the unit by manipulating the feed rate rather than the flow rate of heating fluid. Study the effect of adding both proportional and derivative controllers that are activated by both composition and temperature-error signals.

4. (B*) A model for a plate, gas absorption unit containing only two trays was described in Exercises 9, p. 106, and 20, p. 215, in Volume 1. We studied some additional features of this plant in Exercises 21 and 22, p. 216, Volume 1. Now suppose that we want to design several feedback control systems for the unit in order to demonstrate the principles of control theory in the laboratory.

First, calculate the response of the liquid composition leaving the bottom of the column to a step change in the liquid flow rate. Use the linearized model that includes the tray hydraulics for this calculation. Approximate this step response by the

[1] J. Endtz, J. M. L. Janssen, and J. C. Vermeulen, *Plant and Process Dynamic Characteristics*, p. 176, Butterworths, London, 1957.

[2] Taken from D. D. Perlmutter, *Introduction to Chemical Process Control*, p. 142, Wiley, N.Y., 1965.

reaction curve method and apply Cohen and Coon's rules to estimate the controller gains.

Next, prepare a Bode plot from the system transfer function. Determine the ultimate gain and the ultimate period from this diagram, and apply the Ziegler-Nichols rules to calculate the controller gains.

To obtain more accurate values for the gains, use the gain margin and phase margin criteria to design various kinds of control systems.

Calculate the response of the closed-loop plant with just a proportional controller to a step change of 10 percent in the inlet vapor composition. Also, prepare a Bode plot relating the outlet liquid composition to the inlet vapor composition for the closed-loop plant.

Discuss how you would go about implementing your control system in the laboratory.

5. (C†) Perhaps the next step to test how well you have mastered control theory would be to repeat the exercise described above for a column containing three trays.

6. (B†) Dassau and Wolfgang[3] used an analog computer to simulate the dynamic behavior of a proposed plant while it was still in the design stage. Their calculations revealed that the desired operating condition would be unstable and that conventional control systems would be too slow ever to achieve stable operation. Hence they redesigned the plant so that unstable equilibrium points would not be encountered.

Under what conditions would you attempt to convince management to redesign a plant, rather than look for control systems that would give a satisfactory and stable operation?

As part of the investigation, they used a radioactive tracer technique to measure experimentally the residence time distribution of a pilot plant reactor. They found that a second-order dynamic model gave an adequate description of the data. Since they knew from previous work that their reaction could be described by a first-order rate equation, they compared the transfer function for two isothermal CSTRs in series with the transform of their residence-time distribution curve and selected the value of the rate constant so that the two curves matched. By repeating this procedure at several temperatures, they were able to establish the constants in the Arrhenius equation.

Suppose that the first bed in the multibed commercial reactor is geometrically similar to the pilot plant unit. Moreover, assume that this bed would exhibit the same residence time distribution, rather than the perfect mixing characteristics considered by the authors. How would this modification of the system equations change the stability analysis presented in the paper? Similarly, how would it modify the problem of designing a satisfactory control system?

7. (B†) Zoss and Wilson[4] describe the application of the Ziegler-Nichols rules to select the gains for a feedback control system installed on a steam-heated exchanger. They assume that the effluent temperature of the water leaving the exchanger, T, can be related to the valve stem position, x, on the steam side of the ex-

[3] W. J. Dassau and G. H. Wolfgang, *Chem. Eng. Prog.*, **59**, No. 4, 43 (1963) and *Hydrocarbon Process Petrol. Refiner*, **43**, No. 12, 76 (1964).

[4] L. M. Zoss and H. S. Wilson, *ISA Journal*, **9**, No. 12, 59 (1962).

changer by the simple second-order transfer function

$$\frac{\Delta \tilde{T}}{\Delta \tilde{x}} = \frac{224(°F/in.)}{(1 + 0.017s)(1 + 0.432s)}$$

where the time constants are measured in minutes. A derivation of this simple model and a procedure for estimating the constants are presented in the paper. The outlet water temperature is measured using a sensing element and a transmitter. The transfer function for this block is assumed to be

$$\frac{\Delta \tilde{P}_T}{\Delta \tilde{T}} = \frac{0.12(psi/°F)}{0.024s + 1}$$

Also, the transfer function describing the behavior of the valve top and positioner is given as

$$\frac{\Delta \tilde{x}}{\Delta \tilde{P}_c} = \frac{0.047(in./psi)}{0.083s + 1}$$

Use the gain and phase margin criteria, as well as the Ziegler-Nichols rules, to determine the gain settings in a controller activated by a pressure signal from the temperature-sensing element, $\Delta \tilde{P}_T$, and sending a pressure signal $\Delta \tilde{P}_c$ to the valve motor.

8. (B†) Endtz, Jannsen, and Vermeulen[5] made a number of frequency response measurements on a pilot-plant scale distillation column. Some of their results are shown in Figure 7.8-11.

(a) If you wanted to maintain the column pressure constant, which variable would you select to manipulate?

(b) If you wanted to control the temperature on tray 5, which variable would you select to manipulate? Determine the values of the controller gains.

(c) Design a control system for maintaining the composition on tray 10 approximately constant.

(d) How would you decide which of the control systems described above would be the most desirable to implement?

9. (B) In Example 7.4-2, Routh's criterion was used to evaluate the stability of a closed-loop system described by the characteristic equation

$$100s^3 + 215s^2 + 30.5s + 1 + K_c[(1 + 0.667s) + (0.333/s)] = 0$$

Sketch the Mikhailov curves for cases where $K_c = 0$, 100, and 1000, and use these diagrams to assess the stability of the plant.

10. (B) Develop a Bode diagram for the battery of reactors described in Example 7.5-9 when the feedback loop around the battery is closed. Compare the result to the Bode diagram for the five reactors when a feedback control system is installed around the first three.

11. (B†) A simulation study on an analog computer was used by Parker and Prados[6] to design a control system for the production of ethylene glycol in a contin-

---

[5] See footnote 1, p. 169.

[6] W. A. Parker and J. W. Prados, *Chem. Eng. Prog.*, **60**, No. 6, 74 (1964).

uous-stirred-tank reactor. The reactions, which take place in the liquid phase and which are catalyzed by sulfuric acid, are

$$EO + H_2O \rightarrow EG$$

$$EO + EG \rightarrow DEG$$

$$EO + DEG \rightarrow TEG$$

$$EO + TEG \rightarrow FEG$$

where EO = ethylene oxide, EG = ethylene glycol, DEG = diethylene glycol, TEG = triethylene glycol, and FEG = tetraethylene glycol. A trial-and-error procedure was used to show that the rate of each reaction was proportional to the product of the concentrations of the reacting species and that published data could be predicted using the rate constants

$$k_1 = N\{\exp[30.163 - (10,583/T)]\} \text{ (liter)(mole)}^{-1}(\min)^{-1}$$

and

$$k_1 : k_2 : k_3 : k_4 = 1.0 : 2.1 : 2.2 : 1.9$$

where N = normality of H_2SO_4 catalyst. They studied a system where the reactor volume was 100 ft^3, the feed was 200 lb/min of 9 percent EO solution at 20°C, the catalyst concentration was 0.5 percent, the overall heat-transfer coefficient was 233 for a coil having 111 ft^2 of heat-transfer surface, and the flow rate of coolant was 600 lb/min. Their control system adjusted the inlet temperature of the cooling stream according to measured deviations in the reactor temperature. This step was accomplished by mixing steam or cold water with the stream leaving the cooling coil. At normal operating conditions, the reactor temperature was 50°C.

Derive the material and energy balances needed to describe the dynamic behavior of the reactor. Determine a steady state solution to the equations and linearize them about this equilibrium point. Then design a control system that would alter the flow rate of the heat-transfer fluid in such a way that the reactor temperature would remain constant.

Under normal circumstances, we would prefer to have a control system that would maintain the composition of ethylene glycol constant. Can you think of a way to achieve this objective?

12. (B†) The effect of recycle loops on the stability of a plant was studied by Gilliland, Gould, and Boyle.[7] Their analysis indicates that the presence of the recycle loop increases the sensitivity of the plant to disturbances, it makes the plant less stable, it makes the overall response of the plant much slower than would be anticipated from the time constants of the individual units, and that these detrimental features become amplified as the fraction recycled increases.

To gain some experience with plants of this type, consider a first-order, irreversible reaction in a nonisothermal CSTR. Suppose that the reactor effluent enters a distillation column where it is separated into a pure product, which goes overhead, and pure reactants, which are recycled. Assume that both compositions are held constant, irrespective of any changes in the feed rate or feed composition to the

---

[7] E. R. Gilliland, L. A. Gould, and T. J. Boyle, *Proceedings of the Joint Automatic Control Conference*, p. 140, Stanford University, Stanford, Calif, June 1964.

column, by manipulating the reflux and boil up rates (and therefore the distillate and bottoms rates). In addition, assume that the reactor effluent temperature has no effect on the column operation and that the temperature of the recycle stream is identical to that of the feed stream. For simplicity, consider a very rough approximation of the column dynamics, where a single, column-time constant multiplied by an accumulation term is introduced into the linearized material balance relating the amount of reactant leaving the bottom of the column in the recycle stream to the reactant fed to the column.

Derive the dynamic equations describing the plant and linearize them around the steady state operating condition. Develop a stability criterion for the recycle loop. If the reaction is exothermic and the reactor is stable, can the plant ever be unstable? What about if the reaction is endothermic?

Discuss the effect that the recycle loop has on plant start-up. To simplify the problem, consider an isothermal reactor and describe how you would attempt to bring the plant to steady state conditions.

13. Design a feedforward control system for the furnace described in Exercise 2 above that would manipulate the valve in the fuel line to compensate for measured changes in the feed temperature. Discuss your results.

14. (B*) Another interesting laboratory experiment that would illustrate the principles of control theory would be to install a feedforward controller on the two-plate, gas absorption column described in Exercise 4 above. Design a control system that would manipulate the inlet liquid flow rate to compensate for measured changes in the inlet vapor composition. Discuss in detail the problems you might encounter during the implementation of the controller.

15. (C) Design a feedback control system for the cocurrent heat-exchanger problem described in Exercise 9, p. 295, Volume 1.

16. (C*) A simple plant with a reactor and a separator was described in Exercise 7, p. 105, Volume 1. Design a control system for the plant that would maintain the product composition of the rich phase that leaves the separator approximately constant by manipulating the recycle flow rate. Then estimate the response of the closed-loop plant to a 10 percent step change in the feed composition. Consider the system parameters listed in the original paper.

17. (B*) A transfer function for a furnace that was based on experimental frequency response measurements was presented in Exercise 2 above. The authors note that data can equally well be described by the equation

$$\frac{\tilde{T}}{\tilde{P}} = \frac{40(°C/psi)}{(900s + 1)(13s + 1)^3}$$

How does this modification in the process equation alter the feedback control system discussed in Exercise 2? How does it affect the feedforward control system discussed in Exercise 13 above?

Multivariable
Control
Systems

8

In the preceding chapter we developed control-system design procedures for plants having a single output and a single manipulative input. Although some chemical processes fall into this category, they constitute a small percentage of the total number of systems of interest. In fact, almost all chemical processing units have multiple state variables and a variety of inputs that can be used as control variables. We might try to apply our previous theory to these multivariable plants by selecting pairs of inputs and outputs and designing single-loop control systems. However, doing so raises questions: Which input should we couple with which output? Will the single-loop systems interact with each other? Is there some better way to approach the problem? In this chapter we will try to answer these questions.

First, we describe the methods that can be used to extend classical servomechanism theory to multivariable processes. Next, a more general discussion of the ideas of controllability and observability is presented, and this is followed by a study of the feedforward control of multivariable plants. Finally, some of the challenges associated with the design of control systems for a complete chemical plant are noted.

SECTION 8.1 MULTIVARIABLE FEEDBACK CONTROL

The advantages of feedback control were described in the previous chapter for plants having a single input and a single output. One of the uncertainties associated with the design of these simple closed-loop systems was how to select an appropriate output to maintain constant and how to choose the manipulative input that would correspond to the best closed-loop performance. We noted that a case-study approach could be used to evaluate various combinations of input and output pairs, but now we want to consider the possibility that a better procedure may exist.

In particular, in our study of the battery of stirred-tank reactors in Example 7.8–2, we found that it might be advantageous to locate one measuring device as close as possible to the expected source of disturbances and another at the output from the plant. It does not take much imagination to extend this idea of measuring several variables to an approach where we also manipulate several of the inputs simultaneously. Whenever we measure a number of outputs and use a combination of error signals to alter a number of control variables, we say that we have a *multivariable feedback control* system.

Several procedures for synthesizing multivariable controllers are presented in this section. Initially we consider cascade control systems, for they were mentioned in Example 7.8–2. The problem of the interactions between single control loops is treated next, and then we discuss a method that can be used to select pairs of input and output variables so that the interaction is minimized. Finally, we present techniques for synthesizing noninteracting control systems, including both the use of three-mode controllers and the development of modal analysis. It should be recognized that a noninteracting controller is not necessarily the best possible system, and therefore a comprehensive treatment of optimal multivariable feedback control is given in the next chapter.

Cascade Control

Our basic approach in feedback control theory is to measure the output from the plant, compare this measured value to the desired output, and use the error signal generated in this way to activate the controller. However, our brief study in Section 7.8 of the effect of the measured variable revealed that we could improve the performance of the plant by measuring disturbances as close as possible to their source. Unfortunately, this procedure destroys some of the feedback nature of the controller, so that no compensation will be provided for any disturbances entering the system after the measuring point. Obviously it would be desirable to find some way of combining the best features of these two kinds of control configurations. One way of accomplishing this merger is to use the arrangement shown in Figure 8.1–1. In other words, we implement the control system shown in Figure 7.8–3, but the setpoint for this

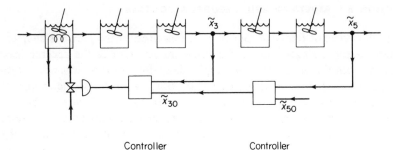

Controller   Controller

Figure 8.1-1. Combined control systems.

controller is now taken to be the output from another controller, which acts on the error signal at the output from the plant.

With this combination of the two controllers, the one on the left, which is activated by the error signal ($\tilde{x}_{30} - \tilde{x}_3$), provides rapid corrections for disturbances in the feed stream, or disturbances which enter any of the first three tanks. Also, compensation is made for disturbances affecting the operation of either of the last two tanks, because the error signal at the plant output, ($\tilde{x}_{50} - \tilde{x}_5$), changes the setpoint of the first controller. This kind of an arrangement, where we use the output of one controller to change the setpoint of another, is commonly encountered in industry and is called *cascade control*.

A block diagram of a cascade control system is given in Figure 8.1-2. We see that there are two measuring elements and two controllers, but only one input variable that is manipulated. The controller in the inner loop is called the secondary, or slave, controller, whereas that in the outer loop is the primary,

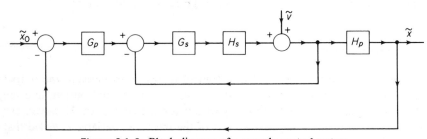

Figure 8.1-2. Block diagram of a cascade control system.

or master, controller. The single process variable we are attempting to maintain constant is denoted as $\tilde{x}$ and the desired value of this output is $\tilde{x}_0$, which may be equal to zero. The main disturbance affecting the plant is denoted as $\tilde{v}$, and with the proper design it enters the inner loop. The blocks H_s and H_p represent the transfer functions of various portions of the plant, although, for simplicity, we are following our normal procedure of neglecting the dynamics of the elements in the feedback loop.

Some other illustrations of cascade controllers are presented in Figures 8.1-3 through 8.1-5. In the first case, the secondary control loop compensates for disturbances in the supply pressure of the fuel almost immediately, while the primary loop provides correction for less frequent fluctuations in the feed temperature of the process fluid. The cascade combination will be faster than a single controller, which adjusts the fuel flow rate as some function of the measured deviation of the effluent temperature of the process fluid. Similarly, the secondary loops in the other two figures make it possible to achieve rapid corrections for fluctuations in the inlet coolant temperature to each process. Compensation for less drastic disturbances in the inlet composition or temperature of the process fluid entering the tank reactor, or the inlet temperature to the shell side of the exchanger, is supplied by the primary loops. Improvements can be obtained via the cascade arrangement, for it is not necessary to wait for

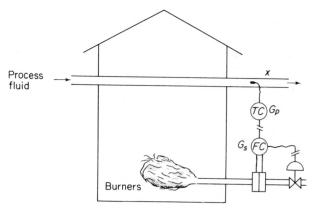

Figure 8.1-3. Cascade control of furnace.

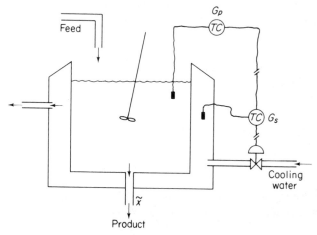

Figure 8.1-4. Cascade control of a jacketed reactor.

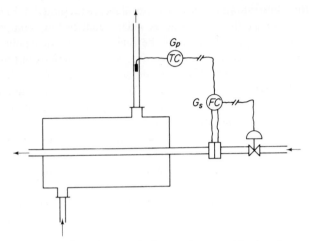

Figure 8.1-5. Cascade control of a countercurrent heat exchanger.

the main disturbance—the inlet coolant temperature—to affect the measured value of the outlet process stream before we take corrective action. In other words, cascade control has the advantage of providing improved control because it reduces the effect of a disturbance near its source. Also, as we will show below, the cascade arrangement increases the stability of the system; that is, the presence of the inner loop shifts the crossover frequency to higher values so that higher-loop gains can be used.

From the preceding discussion we see that we want to design the cascade system so that the major disturbances enter the inner loop and so that this loop is much faster (conventional practice indicates 3 to 5 times) than the outer loop. In order to establish the quantitative design conditions, we must develop the equations describing the cascade controller. For the purpose of comparison, we note that modifying Figure 8.1–2 to correspond to a regulating, single-loop control system with $\tilde{x}_0 = 0$ leads to the expression

$$\frac{\tilde{x}}{\tilde{v}} = \frac{H_p}{1 + G_p H_p H_s} \qquad \text{8.1–1}$$

For the cascade system, we have

$$\tilde{x} = H_p \tilde{v} + H_p \tilde{y}_1 \qquad \text{8.1–2}$$

where $\tilde{y}_1$ is the signal leaving the block H_s in Figure 8.1–2. However, we also find that

$$\tilde{y}_1 = H_s G_s(\tilde{y}_2 - \tilde{y}_1 - \tilde{v}) \qquad \text{8.1–3}$$

where $\tilde{y}_2$ is the signal leaving the primary controller, G_p; that is, it is the set point of the secondary controller. Finally, we note that

$$\tilde{y}_2 = G_p(\tilde{x}_0 - \tilde{x}) \qquad \text{8.1–4}$$

Letting $\tilde{x}_0 = 0$, solving Eq. 8.1-3 for $\tilde{y}_1$ in terms of $\tilde{y}_2$ and $\tilde{v}$, and then using this result and Eq. 8.1-4 to eliminate $\tilde{y}_1$ and $\tilde{y}_2$ from Eq. 8.1-2, we obtain

$$\frac{\tilde{x}}{\tilde{v}} = \frac{H_p}{1 + G_p H_p H_s G_s + G_s H_s} \qquad 8.1\text{-}5$$

Comparing this result for the cascade controller with the single-loop system given by Eq. 8.1-1, we find that the transfer function of the secondary controller appears in two terms in the denominator. This means that we can significantly decrease the effect of disturbances on the plant.

The stability of the cascade control system is determined by the denominator of the transfer function, and the limit of stability will be obtained when

$$1 + G_p H_p H_s G_s + G_s H_s = 0 \qquad 8.1\text{-}6$$

Rearranging this result so that it is equivalent to the form we used to develop the root-locus rules, we obtain

$$G_p \left(\frac{G_s H_s}{1 + G_s H_s} \right) H_p = -1 \qquad 8.1\text{-}7$$

The stability limit occurs at the frequency for which the phase angle of the function $G_p[G_s H_s/(1 + G_s H_s)]H_p$ is equal to -180 deg. Providing that the inner loop is much faster than the outer loop, the crossover frequency for this secondary loop will occur at a much higher value than that corresponding to the primary loop. Also, the presence of the inner loop causes the crossover frequency of the overall system to increase, so that higher gains can be used.

The actual settings for the controller are established by first applying the methods described in the previous chapter to find the controller gain for the inner loop; generally only a proportional controller is used as the secondary controller, although an integral mode sometimes is added if the loop gain is small. Next, the crossover frequency for the overall system is determined from a Bode plot, and again we use the simple criteria given in Chapter 7 to find the controller settings. In other words, the design is accomplished by first considering the inner loop and then imbedding this result in the outer loop. It is interesting to note that stable operation of the total system can often be achieved even when the inner loop is unstable, although this approach is avoided in practice in case one of the elements of the outer loop fails or the primary controller is temporarily disconnected.

Of course, it would be possible to extend the ideas of cascade control to complicated multivariable plants. Similarly, as an alternative approach, we could use the output from one or more controllers to change the gains of the various modes of another controller. The possibilities are limited only by the designer's imagination. However, it should be recognized that the potential advantages of this type of multivariable control system have never been explored. Instead, common practice has been to attach several cascade systems to a unit and hope that the interactions between the systems do not lead to serious difficulties.

Example 8.1–1 Cascade control of a series of isothermal CSTRs

As an illustration of the quantitative technique for designing a cascade control system, consider the series of isothermal CSTRs shown in Figure 8.1–1. We want to maintain the composition of the fifth reactor constant despite fluctuations in the feed composition, and we attempt to accomplish this by manipulating the temperature, or rate constant, in the first vessel. Single-loop control systems for this plant were discussed in Example 7.8–2, and now we want to compare these results with the cascade control system shown above.

The system equations are developed in detail in Example 7.8–2. The transfer functions of interest are

$$\tilde{x}_3 = \frac{\alpha_2^2 \beta}{(s + \alpha_1)^3} \tilde{u} + \frac{\alpha_2^3}{(s + \alpha_1)^3} \tilde{v} \qquad\qquad 8.1–8$$

and

$$\tilde{x}_5 = \frac{\alpha_2^2}{(s + \alpha_1)^2} \tilde{x}_3 \qquad\qquad 8.1–9$$

If we select a proportional controller for the inner loop and remember that the system gain is negative (i.e., $\beta < 0$), we can write

$$\tilde{u} = -K_{c1}(SP - \tilde{x}_3) \qquad\qquad 8.1–10$$

The set-point signal for this secondary controller, SP, is the output of the primary controller. If we again choose a proportional controller for this outer loop, we have

$$SP = +K_{c2}(\tilde{x}_{50} - \tilde{x}_5) \qquad\qquad 8.1–11$$

By combining all these expressions, we see that the transfer function for the cascade system is

$$\frac{\tilde{x}_5}{\tilde{v}} = \frac{\alpha_2^5/(s + \alpha_1)^5}{1 + K_{c1}[(-\beta)\alpha_2^2/(s + \alpha_1)^3] + K_{c1}K_{c2}[(-\beta)\alpha_2^4/(s + \alpha_1)^5]} \qquad 8.1–12$$

In order to design the cascade controller, we first consider just the inner loop. However, the control transfer function given by Eq. 8.1–8 is exactly the same as that discussed in Example 7.8–2; therefore we can use the results obtained in that problem. Once we have fixed the value of the controller gain for the inner loop at $K_{c1} = 77.4$, we can replace the inner control loop by its equivalent closed-loop transfer function. As mentioned earlier, the simplest procedure for evaluating a closed-loop response from information about the open-loop plant is to prepare a Nichols chart. This is shown in Figure 8.1–6 for the case where we use the straight-line approximations for the transfer function describing three CSTRs in series. With the information on the Nichols chart, it is a simple matter to prepare a Bode diagram for the closed-loop response of the secondary control system. The curve of interest is shown on Figure 8.1–7.

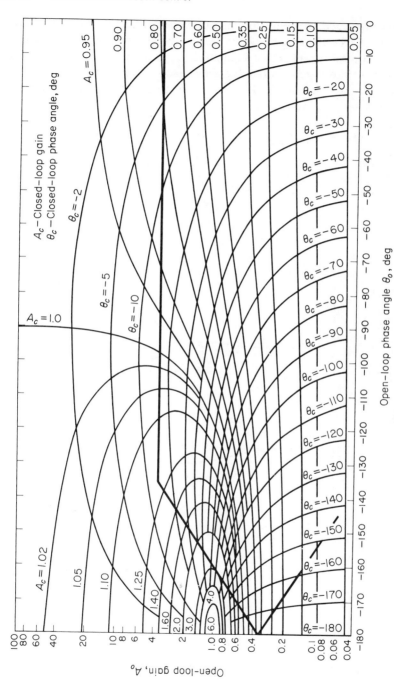

Figure 8.1-6. Nichols chart for three CSTRs in series.

When we add the result for the inner control loop to the Bode diagram for two CSTRs in series, we obtain the transfer function needed to design the primary control system. From Figure 8.1-7, we see that when the phase lag of the combined systems is 180 deg, the crossover frequency is $\omega_{co} = 8.1$, and the gain at this frequency is $|G| = 1.13$. Using a gain margin criterion to select the controller gain, we find that

$$K_{c2} = \frac{1}{1.75(1.13)} = 0.505 \quad \text{or} \quad K_{c2} = \frac{1}{2(1.13)} = 0.442$$

Alternately, the phase margin approach leads to the values

$$K_{c2} = \frac{1}{(G|_{-135°})} = 1.19 \quad \text{or} \quad K_{c2} = \frac{1}{(G|_{-120°})} = 1.10$$

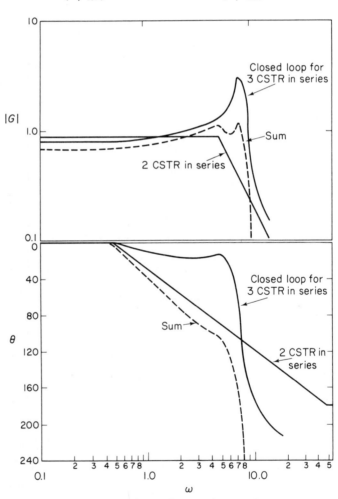

Figure 8.1-7. Bode plot for cascade control system.

It should be noted that the Bode diagram is quite steep in the neighborhood of the crossover frequency; therefore a careful design of the cascade control system would require plots of the exact transfer functions, rather than the asymptotic approximations. For the purpose of illustrating the design procedure, we will neglect any error introduced in this way and will select the gain of the primary controller as $K_{c1} = 0.5$. If we evaluate the offset of the cascade controller, again using the plant parameters presented in Example 7.8–2, we find that

$$
x_5 = \frac{\alpha_2^5 v}{\alpha_1^5 + K_{c1}(-\beta)\alpha_1^2\alpha_2^2 + K_{c1}K_{c2}(-\beta)\alpha_2^4}
$$

$$
= \frac{(5)^5(0.15)}{(5.33)^5 + (0.313)(5.33)^2(5)^2(77.4) + (0.313)(5)^4(77.4)(0.5)}
$$

$$
= 0.01642
$$

whereas in our previous study we obtained offset values of 0.0456 and 0.0230. Hence we expect the cascade controller to have a better performance than either of the other systems.

Pseudo-Cascade Control Systems

There have been a number of control systems discussed in the literature that have been called cascade controllers but where the primary controller is just a measuring instrument. In other words, the measured value of one process variable becomes the set point, within a proportionality factor, for a control system. Since there is no flexibility left in the design of the primary controller in this case, it has been suggested that these systems be called *pseudo-cascade controllers* in order to distinguish them from their more general counterparts.

An industrially important example of a pseudo-cascade system is a *ratio controller*. This type of instrument is particularly useful when two streams are being mixed to give some desired composition—for example, in a blending operation, the mixing of fuel and air supplied to the burners in a furnace, or the supply of two reacting components to a reactor. With the ratio controller, we can make certain that the proper proportion of the two components is always

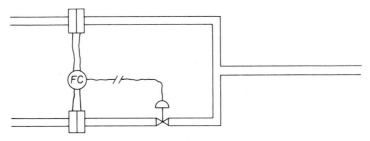

Figure 8.1-8. Ratio control system.

available even if the total flow rate varies somewhat due to disturbances in the individual flow rates. A better performance is usually obtained in this way. A schematic drawing of a ratio control system is presented in Figure 8.1-8. The design of the control loop seldom offers any difficulty.

Another example of a pseudo-cascade control system is given in Figure 8.1-9. A bulb containing a liquid mixture of the desired composition is placed on the tray of a distillation column. The vapor pressure in the bulb is used as the set point for a control system, which measures the actual vapor pressure on the same tray and adjusts the steam rate to the reboiler as some function of this difference in vapor pressure. Again the design of the control system generally is straightforward.

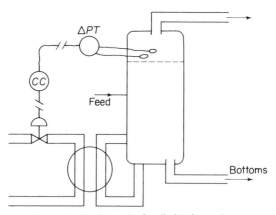

Figure 8.1-9. Control of a distillation column.

Interaction of Single-Loop Control Systems

Perhaps the most obvious way to generalize the configuration for the cascade controller just discussed would be to install two separate single-loop control systems. In other words, we merely select two pairs of input and output variables and design a control system for each pair. The additional flexibility introduced by adding a second, manipulative input might make this kind of a controller have a better performance than the cascaded loops. However, we find that the two control loops interact with each other if we follow this approach, and in some cases this interaction can be disastrous.

In order to demonstrate the interaction, let us consider the control of a nonisothermal stirred-tank reactor. The system equations were developed in Example 3.3-1 and were discussed in Section 7.1.

$$\frac{dA}{dt} = \frac{q}{V}(A_f - A) - kA$$

$$\frac{dT}{dt} = \frac{q}{V}(T_f - T) + \frac{UA_H Kq_H}{C_p \rho (1 + Kq_H)}(T_H - T) + \frac{(-\Delta H)}{C_p \rho} kA \qquad 8.1\text{-}13$$

Linearizing these relationships, we obtain

$$\frac{dx_1}{dt} = a_{11}x_1 + a_{12}x_2 + b_{11}u_1 + c_{11}x_{1f}$$

$$\frac{dx_2}{dt} = a_{21}x_1 + a_{22}x_2 + b_{21}u_1 + b_{22}u_2 + c_{22}x_{2f} + c_{23}x_{2H} \qquad 8.1\text{-}14$$

One possible way of compensating for disturbances in feed composition x_{1f}, feed temperature x_{2f}, or the inlet temperature of heating fluid x_{2H}, even if we do not measure these disturbances, is to measure the deviation in the outlet composition from its desired steady state value x_1 and use this error signal to change the reactor feed rate u_1 according to some control action G_1 (see Figure 8.1-10). Simultaneously, we might attempt to measure the deviation of the

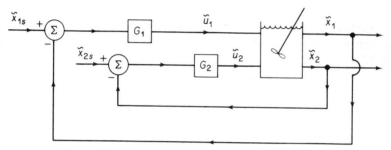

Figure 8.1-10. Two single loop controllers on a CSTR.

reactor temperature from its design value x_2 and use this error signal to change the flow rate of fluid through the heating coil u_2 according to some other control function G_2. In other words, we use two single-loop control systems of the type described in the previous chapter, with the hope that the two-loop controller might give a better performance than either of the single-loop systems. Of course, our choice of which error signal we use to change one or the other of the inputs is arbitrary, just as our selection of the single output and single control variable for a single-loop controller would be arbitrary. Nevertheless, since we sometimes obtained interesting and useful results for the single-loop case, we will plunge ahead with our analysis, and later, if necessary, we can evaluate the performance of the other configurations.

Writing the equations in matrix form, taking their Laplace transforms, and solving for the transform of the state variables, we obtain

$$\tilde{\mathbf{x}} = (s\mathbf{I} - \mathbf{A})^{-1}\mathbf{B}\tilde{\mathbf{u}} + (s\mathbf{I} - \mathbf{A})^{-1}\mathbf{C}\tilde{\mathbf{v}} \qquad 8.1\text{-}15$$

Now if we let

$$\mathbf{H}(s) = (s\mathbf{I} - \mathbf{A})^{-1}\mathbf{B} \qquad \tilde{\mathbf{w}} = (s\mathbf{I} - \mathbf{A})^{-1}\mathbf{C}\tilde{\mathbf{v}} \qquad 8.1\text{-}16$$

The equations become

$$\begin{pmatrix} \tilde{x}_1 \\ \tilde{x}_2 \end{pmatrix} = \begin{pmatrix} H_{11} & H_{12} \\ H_{21} & H_{22} \end{pmatrix} \begin{pmatrix} \tilde{u}_1 \\ \tilde{u}_2 \end{pmatrix} + \begin{pmatrix} \tilde{w}_1 \\ \tilde{w}_2 \end{pmatrix} \qquad 8.1\text{-}17$$

Considering a case where we are allowed to change the controller set points—that is, $\tilde{x}_{1s} = \tilde{x}_{10}$ and $\tilde{x}_{2s} = \tilde{x}_{20}$—the control equations become

$$\tilde{u}_1 = G_1(s)(\tilde{x}_{10} - \tilde{x}_1) \qquad \tilde{u}_2 = G_2(s)(\tilde{x}_{20} - \tilde{x}_2) \qquad 8.1\text{-}18$$

where $G_1(s)$ and $G_2(s)$ are the as yet unspecified transfer functions of the controllers. Substituting these expressions into Eq. 8.1-17

$$\tilde{x}_1 = H_{11}\tilde{u}_1 + H_{12}\tilde{u}_1 + \tilde{w}_1 \qquad \tilde{x}_2 = H_{21}\tilde{u}_1 + H_{22}\tilde{u}_2 + \tilde{w}_2 \qquad 8.1\text{-}19$$

and solving the first equation for $\tilde{x}_1$ and the second equation for $\tilde{x}_2$, we find that

$$\tilde{x}_1 = \frac{H_{11}G_1\tilde{x}_{10} + H_{12}G_2\tilde{x}_{20} - H_{12}G_2\tilde{x}_2 + \tilde{w}_1}{1 + H_{11}G_1} \qquad 8.1\text{-}20$$

$$\tilde{x}_2 = \frac{H_{21}G_1\tilde{x}_{10} + H_{22}G_2\tilde{x}_{20} - H_{21}G_1\tilde{x}_1 + \tilde{w}_2}{1 + H_{22}G_2} \qquad 8.1\text{-}21$$

A schematic diagram of the reactor and the control systems, showing the transfer functions of the various components in the plant, is presented in Figure 8.1-11.

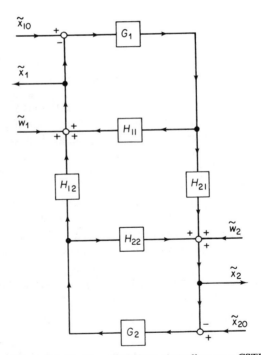

Figure 8.1-11. Two single loop controllers on a CSTR.

We can see immediately from either Eqs. 8.1–20 and 8.1–21 or from Figure 8.1–11 that any disturbance entering the system, or any change in the controller set points, will activate both controllers and cause changes in both outputs. In physical terms this means that if for some reason we measure a deviation in one of the outputs, say composition, but that the temperature is at its desired value, the first controller will change the feed rate in an attempt to compensate for the composition upset. However, this change in the feed rate also causes the reactor temperature to vary, so that the temperature is driven away from the desired value. Then the second control loop attempts to compensate for this variation, but at the same time it also causes a change in the reactor composition. Similarly, suppose that the reactor is operating at the desired conditions, but for some reason we want to change the reactor composition slightly by changing $\tilde{x}_{10}$. Initially this will tend to make $\tilde{x}_1$ change in an appropriate manner; but since it also causes a change in the reactor temperature, which activates the second controller and causes the heating flow rate to vary, and this, in turn, produces an additional change in the composition, etc., we might not obtain a very desirable response.

Serious stability problems can also be encountered with the multiple-loop system. From our experience with single-loop controllers, we would expect that the stability of the individual loops depends on the denominators of Eqs. 8.1–20 and 8.1–21. In other words, if all the roots of the functions

$$1 + H_{11}(s)G_1(s) \qquad 1 + H_{22}(s)G_2(s) \qquad\qquad 8.1\text{–}22$$

had negative real parts, we would expect the two, individual, control loops to be stable. However, because of the interaction between the loops, the dynamic behavior of one loop depends on the dynamics in the other loop; that is, Eq. 8.1–20 for $\tilde{x}_1$ is a function of $\tilde{x}_2$. Thus, it becomes necessary to solve the simultaneous equations, Eqs. 8.1–20 and 8.1–21, explicitly for $\tilde{x}_1$ and $\tilde{x}_2$. When this is done for a case where $\tilde{x}_{10} = \tilde{x}_{20} = 0$, we obtain

$$\tilde{x}_1 = \frac{(1 + H_{22}G_2)\tilde{w}_1 - H_{12}G_2\tilde{w}_2}{(1 + H_{11}G_1)(1 + H_{22}G_2) - H_{12}H_{21}G_1G_2} \qquad\qquad 8.1\text{–}23$$

$$\tilde{x}_2 = \frac{-H_{21}G_1\tilde{w}_1 + (1 + H_{11}G_1)\tilde{w}_2}{(1 + H_{11}G_1)(1 + H_{22}G_2) - H_{12}H_{21}G_1G_2} \qquad\qquad 8.1\text{–}24$$

Now it is apparent that in order to obtain a stable system, the three functions

$$1 + H_{11}G_1, \quad 1 + H_{22}G_2, \quad (1 + H_{11}G_1)(1 + H_{22}G_2) - H_{12}H_{21}G_1G_2$$

$$8.1\text{–}25$$

must all have roots with negative real parts. Normally we require that all three of these criteria are satisfied, rather than just the last one, so that if one of the control systems fails for some reason, the other system will still be stable.

Example 8.1–2 Feedback control of a distillation column

For the control of a distillation column, we know that we want to maintain the outlet compositions from the top and bottom trays approximately constant. However, compositions are difficult to measure, and therefore it is common practice to attempt to maintain the vapor pressure and temperature at two points within the column at fixed values with the hope that this procedure will lead to satisfactory operation. Hence one possible control scheme would be to manipulate the flow rate of cooling water to the condenser to compensate for measured deviations of the column pressure and simultaneously to change the steam rate to the reboiler to correct for measured deviations of the column temperature at some point. Alternately, we could switch the measured and manipulative variables in the scheme above (see Figure 8.1–12). Similarly, other configurations could be developed using the reflux rate and/or the feed rate as control variables.

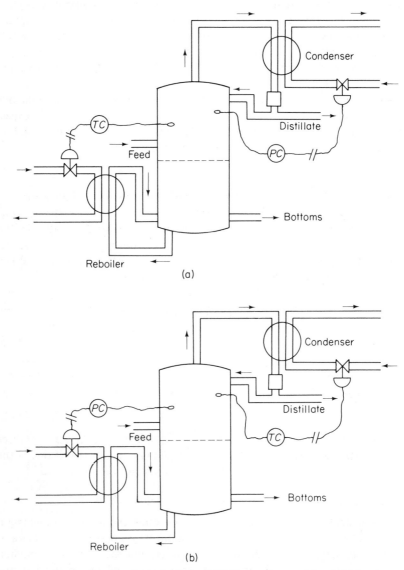

Figure 8.1-12. Control of a distillation column: (a) operable scheme; (b) inoperable scheme.

Since all these control systems involve two interacting loops, we want to make certain that each loop will perform in a satisfactory manner if the other loop is shut off, or fails, and that both loops taken together will give a desirable response. If we consider the control system sketched in Figure 8.1–12b, we find that we obtain very different behavior, depending on whether or not the pressure control loop is closed. For example, when this loop is open, we expect that an increase in the cooling water flow rate will lead to a lower temperature in the column. On the contrary, when the pressure control loop is closed, the temperature variations are directly related to compositions. For this case, an increase in the cooling-water flow rate generally produces more top product, and correspondingly less bottom product, so that heavier compositions and higher temperatures are observed at the top of the column. Thus, we find that the sign of the temperature response reverses when we close the pressure control loop, and we expect the control system to make the column inoperable under certain circumstances. This means that it would be much better to use the control scheme shown in Figure 8.1–12a, although it is still necessary to determine the four system transfer functions (i.e., how the column pressure and temperature at the points of interest change with steam flow to the reboiler and cooling water flow to the condenser) and then use the preceding criteria to verify that each of the single loops, as well as the two-loop system, is stable.

Selecting Control Variables to Achieve Noninteraction

Once we recognize that the interaction between two single-loop control systems can lead to serious problems, we would like to know how to select pairs of input and output variables in order to minimize the amount of interaction. Fortunately Bristol[1] has developed a procedure for this purpose. It is a particularly simple method to apply because it is based only on the steady state system equations.

To develop Bristol's technique, we consider a multivariable process that is described by the linearized equations

$$\dot{\mathbf{x}} = \mathbf{A}\mathbf{x} + \mathbf{B}\mathbf{u} \qquad 8.1\text{–}26$$

Initially we will consider a case where there are the same number of control variables as state variables, and later we will attempt to generalize the results. At steady state conditions, the accumulation terms are equal to zero, and the foregoing reduces to

$$\mathbf{x} = \mathbf{A}^{-1}\mathbf{B}\mathbf{u} \qquad 8.1\text{–}27$$

Exactly the same result is obtained if we consider the transformed dynamic equations

$$\tilde{\mathbf{x}} = (s\mathbf{I} - \mathbf{A})^{-1}\mathbf{B}\tilde{\mathbf{u}} \qquad 8.1\text{–}28$$

$$\tilde{\mathbf{x}} = \mathbf{H}(s)\tilde{\mathbf{u}} \qquad 8.1\text{–}29$$

[1] E. H. Bristol, *IEEE Transactions on Automatic Control,* **AC-11,** No. 1, 133 (1966).

and apply the final value theorem to every element in the transfer matrix. To keep the notation to a minimum, we let

$$-\mathbf{A}^{-1}\mathbf{B} = \mathbf{H} = \begin{pmatrix} H_{11} & H_{12} \\ H_{21} & H_{22} \end{pmatrix} \qquad 8.1\text{-}30$$

where the elements of $\mathbf{H}$ are constants, which represent the steady state gains of the process.

We are attempting to select pairs of control and state variables so that the interactions between any single-loop systems we install will be minimized. As a measure of the amount of interaction, Bristol suggested that we consider the ratio of the steady state gain between an input and output for the open-loop plant to the steady state gain between the same input and output when all of the other input-output pairs are under closed-looped control. Thus the values of the numerator of this measure are just the elements of the matrix given by Eq. 8.1-30, H_{ij}. To determine the factors for the denominator, we must evaluate the effect of closed-loop control on the process. For a process unit with two inputs and two outputs, we can draw a schematic of two single-loop control systems as shown in Figure 8.1-11. By rearranging the elements in the diagram, we can develop a relationship between an input-output pair when the second feedback loop is closed. Figure 8.1-13 shows the schematic for this case. Using our previous methods for treating block diagrams, it is easy to prove that the transfer function relating $\tilde{x}_1$ to $\tilde{u}_1$ becomes

$$\frac{\tilde{x}_1}{\tilde{u}_1} = H_{11} - H_{12}H_{21}\left(\frac{G_2}{1 + G_2 H_{22}}\right) \qquad 8.1\text{-}31$$

This result illustrates that the design of any controller relating $\tilde{x}_1$ to $\tilde{u}_1$ will depend on the setting of the second controller G_2.

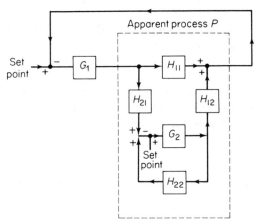

Figure 8.1-13. Alternate diagram for two single loop control systems.

In order to proceed, we need to introduce some simplifications. At steady state conditions, the terms on the right-hand side of the equation are constant and independent of the Laplace parameter s. Furthermore, we assume that the second control loop achieves perfect control, which means that the controller gain becomes infinite. According to Eq. 8.1–31, as G_2 approaches infinity, the transfer function approaches

$$\frac{\tilde{x}_1}{\tilde{u}_1} = \frac{H_{11}H_{22} - H_{12}H_{21}}{H_{22}} \qquad 8.1\text{--}32$$

However, we note that this result is simply the reciprocal of the first element of the inverse of the H matrix, given by Eq. 8.1–30. Our measure of interaction then becomes H_{11} divided by this value or

$$M_{11} = \frac{H_{11}H_{22}}{H_{11}H_{22} - H_{12}H_{21}} \qquad 8.1\text{--}33$$

If we repeat the procedure described above for all possible input-output pairs and arrange the values of the measure in a matrix, we find that

$$M_{ij} = H_{ij}(\mathbf{H}^{-1})_{ji} \qquad 8.1\text{--}34$$

In other words, the elements of $\mathbf{M}$ are the product of an element in $\mathbf{H}$ and the corresponding element in the transpose of $\mathbf{H}^{-1}$. For our simple 2×2 matrix, this becomes

$$\mathbf{M} = \frac{1}{H_{11}H_{22} - H_{12}H_{21}} \begin{pmatrix} H_{11}H_{22} & -H_{12}H_{21} \\ -H_{12}H_{21} & H_{11}H_{22} \end{pmatrix} \qquad 8.1\text{--}35$$

Now, suppose that we consider a few simple examples, where the problem of selecting which control variable to couple with which state variable is trivial, and evaluate our measure of the interaction, $\mathbf{M}$. If $\mathbf{H}$ is diagonal

$$\mathbf{H} = \begin{pmatrix} a_{11} & 0 \\ 0 & a_{22} \end{pmatrix} \qquad 8.1\text{--}36$$

we find that

$$\mathbf{M} = \begin{pmatrix} 1 & 0 \\ 0 & 1 \end{pmatrix} \qquad 8.1\text{--}37$$

Similarly, if

$$\mathbf{H} = \begin{pmatrix} 0 & a_{12} \\ a_{21} & 0 \end{pmatrix} \qquad 8.1\text{--}38$$

we get

$$\mathbf{M} = \begin{pmatrix} 0 & 1 \\ 1 & 0 \end{pmatrix} \qquad 8.1\text{--}39$$

Also, if

$$\mathbf{H} = \begin{pmatrix} a_{11} & 0 \\ a_{21} & a_{22} \end{pmatrix} \qquad\qquad 8.1\text{-}40$$

we obtain

$$\mathbf{M} = \begin{pmatrix} 1 & 0 \\ 0 & 1 \end{pmatrix} \qquad\qquad 8.1\text{-}41$$

After considering a number of illustrations of this type, Bristol came to the conclusion that the state and control variable pairs, x_i and u_j, should be chosen such that their measures of interaction, M_{ij}, are positive and as close as possible to unity. He applied this criterion to some process problems and found that it worked well.

In addition, he noted that there were a number of inherent properties of the measure:

1. The sum of the elements in any row or column is equal to unity.

2. A similarity transformation does not change the measure.

3. Changing the order of the rows or columns in $\mathbf{H}$ results in the same change in the order of $\mathbf{M}$.

4. Values of the measure much larger than unity imply that the gain matrix is nearly singular.

5. The values of the submeasure of a subprocess that is effectively isolated from the remainder of the plant is the same as the measure of the subprocess.

6. The quantities used to define the measure are also encountered when attempts are made to determine the effect a small change in a process parameter has on the performance of a control system; that is, the measure is an approximate measure of its own sensitivity.

7. The transfer function relating x_i to u_j with all other loops closed will be nonminimum phase or unstable if M_{ij} is negative.

8. A plant with two inputs and two outputs controlled by two negative feedback controllers will be stable if the control loops are selected so that their measures are positive. Additional stability criteria can be developed for cases where one measure is negative and for third-order plants.

The design rule given above makes it possible to select sets of single-loop controllers so that the interaction between these single-loop systems is minimized. For many plants this means that the problem of choosing appropriate values for the controller gains, or tuning the controllers, will be simplified. However, the procedure does not guarantee that there will not be any interaction or that all problems associated with interaction can be avoided. Nevertheless, the method is simple to apply and it does provide a starting point for a more detailed study. This can be a great advantage, particularly for units described by a large number of equations.

The technique, as presented, requires that $\mathbf{H}$ is a square matrix, which means that there are the same number of control variables as state variables. Whenever the number of state variables exceeds the number of control variables, generally it is not possible to force the steady state plant to an arbitrary final condition. This is in contrast to the dynamic system, as we will show in the next section, providing that it is completely controllable. The apparent discrepancy between the steady state and dynamic analyses is not treated in the original paper. However, intuitively it seems reasonable to arbitrarily choose a number of state variables equal to the number of control variables and then to evaluate the measure. By repeating this procedure for all the possible combinations and selecting the situation where the values of the measure are closest to unity, we might expect to minimize the interaction. Cases where the number of control variables exceeds the number of state variables are simple to treat, because it is now possible to consider a number of square submatrices that include all the state variables. After calculating the measures corresponding to all the possible cases, we can select the control systems based on the normal design rule.

Bristol suggests that another advantage of calculating the measure is that it emphasizes the important interactions in the plant. Therefore, in any attempt to simplify the dynamic model describing a process, we would want to preserve these interactions. However, until the approach has been tested for a wide variety of chemical processes and proven to give valid results, the control systems suggested by the design rule should be considered as likely possibilities or first estimates rather than theoretically specified systems. It would seem as if there might be plants where a significant error would be introduced when the single-loop controllers were chosen only on the basis of steady state considerations.

Example 8.1–3 Single-loop controllers for a stirred-tank reactor

Consider the linearized equations for a continuous-stirred-tank reactor with a cooling coil

$$\dot{x}_1 = a_{11}x_1 + a_{21}x_2 + b_{11}u_1$$
$$\dot{x}_2 = a_{21}x_1 + a_{22}x_2 + b_{21}u_1 + b_{22}u_2$$

$$8.1\text{–}42$$

and the system parameters

$$a_{11} = -0.424, \quad a_{12} = -0.1137, \quad a_{21} = +0.414, \quad a_{22} = +0.0987$$
$$b_{11} = 9.76 \times 10^{-4}, \quad b_{21} = -6.32 \times 10^{-4}, \quad b_{22} = -3.44 \times 10^{-4}$$

Select the pair of single-loop control systems for the reactor so that we obtain the minimum amount of interaction.

Solution

From Eq. 8.1–30 we find that

$$\mathbf{H} = \mathbf{A}^{-1}\mathbf{B} = 10^{-2}\begin{pmatrix} 0.4686 & -0.751 \\ -2.608 & 2.804 \end{pmatrix}$$

Then

$$\mathbf{H}^{-1} = 10^2 \begin{pmatrix} -4.3476 & -1.1645 \\ -4.0443 & -0.7266 \end{pmatrix}$$

and

$$(\mathbf{H}^{-1})^T = 10^2 \begin{pmatrix} -4.3476 & -4.0443 \\ -1.1645 & -0.7266 \end{pmatrix}$$

Finally, from the definition of the measure, $M_{ij} = H_{ij}(\mathbf{H}^{-1})_{ji}$, we obtain

$$\mathbf{M} = \begin{pmatrix} -2.03 & 3.04 \\ 3.04 & -2.03 \end{pmatrix}$$

This result indicates that if we plan to install two single-loop feedback controllers, we should use the coolant flow rate to control composition and manipulate the feed rate to compensate for variations in the reactor temperature. Other configurations are likely to lead to stability problems and poor control. Two examples of optimal controllers for this reactor, both of which include interaction, are described in Chapter 9.

It should be noted that for most problems it is possible to find a finite range of controller gains that will satisfy the stability criteria discussed in Eq. 8.1–25, even if the controllers are connected so that the measure is negative; Bristol's analysis assumes infinite values for the controller gains in all the loops not under consideration. Also, it is possible to show that for a CSTR with a heating coil, the best control is achieved when composition variations are corrected by manipulating the feed rate and temperature errors are compensated by changing the flow rate of heating fluid, in contrast to the example above. The control system with heat addition agrees better with our intuition, because the flow rate of heating fluid appears only in the energy equation and therefore offers a more direct control of temperature. The case with heat addition also is the only situation where an optimum economic design can be obtained (see Chapter 2).

Design of Noninteracting Control Systems

The problem of interaction of single-loop control systems on a multivariable plant also implies that the simple criteria we used to establish the controller gains—that is, a $\frac{1}{4}$ decay ratio, a gain margin of 1.7, a phase margin of $35°$, etc.—can no longer be applied directly, for any attempt to design or tune one control loop will affect the dynamic characteristics of the other loop. Therefore, unless we can find some way of modifying our approach, it will be necessary to abandon all our experience with single-loop systems. Obviously we would prefer not to take such a drastic step. Instead we would like to develop a method whereby we can treat the single-loop systems in such a way that they do not interact.

We first recognize that the difficulty with interaction arises because when we change the setting of any one of the control variables, all the system outputs are affected. However, suppose that we simultaneously change all the other control variables and do this in such a way that we exactly cancel the effect of the first control variable on all the outputs except that used to generate the original error signal. In this way a measured output deviation will cause all of the control variables to change, but only one output will vary, so that it will appear to be a single-loop control system. We can apply this technique for as many error signals as we please and eventually windup with a multivariable controller that looks like a large number of single-loop systems, where each of the single loops are independent of the others—that is, the control systems do not interact.

The quantitative conditions necessary to obtain a noninteracting controller can be developed by considering the equations for the transfer functions of the plant

$$\tilde{\mathbf{x}} = \mathbf{H}(s)\tilde{\mathbf{u}} + \tilde{\mathbf{w}} \qquad 8.1\text{-}43$$

and letting the feedback control action have the form

$$\tilde{\mathbf{u}} = \mathbf{G}(s)(\tilde{\mathbf{x}}_0 - \tilde{\mathbf{x}}) = \mathbf{G}(s)\tilde{\mathbf{e}} \qquad 8.1\text{-}44$$

where $\mathbf{G}(s)$ is a matrix of as yet unspecified transfer functions and $\tilde{\mathbf{e}}$ represents the vector of error signals which we measure. Substituting, we obtain (see Figure 8.1–14)

$$\tilde{\mathbf{x}} = \mathbf{H}(s)\mathbf{G}(s)(\tilde{\mathbf{x}}_0 - \tilde{\mathbf{x}}) + \tilde{\mathbf{w}} = \mathbf{H}(s)\mathbf{G}(s)\tilde{\mathbf{e}} + \tilde{\mathbf{w}} \qquad 8.1\text{-}45$$

so that in order to have the error signal corresponding to one particular output

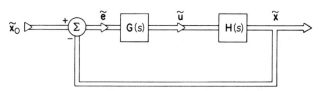

Figure 8.1-14. Multivariable control system using vector–matrix representations.

affect only that same output, we must make **HG** a diagonal matrix. Hence the criterion for a noninteracting control system is that

$$\mathbf{H}(s)\mathbf{G}(s) = \mathbf{K}(s) \qquad 8.1\text{-}46$$

where $\mathbf{K}(s)$ is a diagonal matrix of transfer functions. By setting the diagonal elements of $\mathbf{K}(s)$ equal to our normal three-mode or other simpler controllers and using the procedures described in Chapter 7 to find the appropriate controller gains, since the controllers no longer interact, we can solve Eq. 8.1–46 to obtain the desired values of $\mathbf{G}(s)$.

$$\mathbf{G}(s) = \mathbf{H}^{-1}(s)\mathbf{K}(s) \qquad 8.1\text{-}47$$

From Eq. 8.1–45 we find that the system response is given by

$$\tilde{x} = (I + HG)^{-1}HGx_0 + (I + HG)^{-1}\tilde{w} \qquad 8.1\text{-}48$$

or

$$\tilde{x} = (I + K)^{-1}K\tilde{x}_0 + (I + K)^{-1}\tilde{w} \qquad 8.1\text{-}49$$

Example 8.3–2 Noninteracting control of a CSTR

The linearized equations describing a nonisothermal CSTR can be written as (see Example 3.3–1)

$$\dot{x}_1 = a_{11}x_1 + a_{21}x_2 + b_{11}u_1 + c_{11}x_{1f}$$
$$\dot{x}_2 = a_{21}x_1 + a_{22}x_2 + b_{21}u_1 + b_{22}u_2 + c_{22}x_{2f} + c_{23}x_{2H} \qquad 8.1\text{-}50$$

or

$$\dot{x} = Ax + Bu + Cv \qquad 8.1\text{-}51$$

Taking Laplace transforms and solving for $\tilde{x}$, we obtain

$$\tilde{x} = (sI - A)^{-1}B\tilde{u} + (sI - A)^{-1}C\tilde{v} = H\tilde{u} + \tilde{w} \qquad 8.1\text{-}52$$

where

$$(sI - A)^{-1} = \frac{1}{[s^2 - (a_{11} + a_{22})s + a_{11}a_{22} - a_{12}a_{21}]}\begin{pmatrix} s - a_{22} & a_{12} \\ a_{21} & s - a_{11} \end{pmatrix}$$

$$8.1\text{-}53$$

$$H(s) = (sI - A)^{-1}B = \left[\frac{1}{s^2 - (a_{11} + a_{22})s + a_{11}a_{22} - a_{12}a_{21}}\right]$$
$$\times \begin{pmatrix} b_{11}(s - a_{22}) + b_{21}a_{12} & a_{12}b_{22} \\ b_{11}a_{21} + b_{21}(s - a_{11}) & (s - a_{11})b_{22} \end{pmatrix} \qquad 8.1\text{-}54$$

and

$$\tilde{w} = (sI - A)^{-1}C\tilde{v} = \left[\frac{1}{s^2 - (a_{11} + a_{22})s + (a_{11}a_{22} - a_{12}a_{21})}\right]$$
$$\times \begin{pmatrix} (s - a_{22})c_{11}x_{1f} + a_{12}(c_{22}x_{2f} + c_{23}x_{2H}) \\ a_{21}c_{11}x_{1f} + (s - a_{11})(c_{22}x_{2f} + c_{23}x_{2H}) \end{pmatrix} \qquad 8.1\text{-}55$$

For simplicity, we consider a noninteracting, proportional, feedback control system, so that

$$K(s) = \begin{pmatrix} k_{11} & 0 \\ 0 & k_{22} \end{pmatrix} \qquad 8.1\text{-}56$$

Then our feedback control matrix $\mathbf{G}(s)$ becomes

$$\mathbf{G}(s) = \mathbf{H}^{-1}(s)\mathbf{K}(s)$$

$$= \frac{1}{b_{11}b_{22}} \begin{pmatrix} b_{22}(s - a_{11})k_{11} & -a_{12}b_{22}k_{22} \\ -b_{11}a_{21}k_{11} - b_{21}(s - a_{11})k_{11} & b_{11}(s - a_{22})k_{22} + b_{21}a_{12}k_{22} \end{pmatrix}$$

$$= \begin{pmatrix} G_{11} & G_{12} \\ G_{21} & G_{12} \end{pmatrix} \qquad\qquad 8.1\text{-}57$$

and the control policy that we must implement is

$$\begin{pmatrix} u_1 \\ u_2 \end{pmatrix} = \begin{pmatrix} G_{11} & G_{12} \\ G_{21} & G_{22} \end{pmatrix} \begin{pmatrix} \tilde{x}_{10} - \tilde{x}_1 \\ \tilde{x}_{20} - \tilde{x}_2 \end{pmatrix} \qquad\qquad 8.1\text{-}58$$

which resembles a multivariable proportional plus derivative controller. For more complex control policies, we replace k_{11} and k_{22} in Eq. 8.1-57 by the appropriate transfer functions.

Modal Analysis

In the foregoing discussion of noninteracting control systems, we simply devised a way of applying the elementary techniques of the design of single-loop control systems to multivariable processes. We implicitly assumed that we knew which variables should be manipulated and which we should measure and attempt to maintain constant, or else we recognized that it would be necessary to evaluate the plant response for various combinations of control and controlled variables in order to see which configuration gives the best response. However, there is an alternate approach that leads to a noninteracting control system and that also provides an indication of which variables should be measured and which should be controlled. This technique is called *modal analysis*,[2,3] because it is based on the postulate that the dynamic response of a plant is primarily determined by the slowest modes—that is, the transient modes associated with the smallest eigenvalues (see Appendix A)—and that by using a control system to increase these eigenvalues, we can speed up the response of the plant. In addition, the method suggests that we should approximate complicated, high-order systems by simpler systems having exactly the same modes. However, it should be recognized that the disturbances entering the plant excite the different modes in different ways; therefore even if we provide a considerable amount of damping in the normally slow modes, we still might not obtain a very satisfactory response if the disturbances have their greatest effect on the faster modes. In other words, the method has certain limitations, which is also true for any other approach.

We consider the linearized system equations

$$\dot{\mathbf{x}} = \mathbf{A}\mathbf{x} + \mathbf{B}\mathbf{u} \qquad\qquad 8.1\text{-}59$$

[2] J. K. Ellis, and G. W. T. White, *Control*, April, May, June 1965.

[3] H. H. Rosenbrock, *Chem. Eng. Prog.*, **58**, 43 (1962).

and assume that the outputs that we measure are given by the expression

$$\mathbf{z} = \mathbf{Tx} \qquad 8.1-60$$

We use this notation to indicate that normally the number of outputs we can measure is less than the total number of state variables, as, for example, in a distillation column where we can only measure the compositions on the top and bottom trays. For realistic process problems, $\mathbf{A}$ will be an $n \times n$ matrix, $\mathbf{B}$ will be an $n \times m$ matrix, and T will be an $l \times n$ matrix, where m and l are less than n. Of course, the actual number of measured and control variables is determined by economic considerations. Initially we will consider the simplest case where $n = m = l$, and later we will extend the results to the more realistic plant.

Simple Plant

The problem of designing a control system is to choose a relationship between the control variables $\mathbf{u}$ and the measured or output variables $\mathbf{z}$ so that the controlled process has a more favorable dynamic response than the uncontrolled plant. In the modal analysis method, we choose this to be a proportional control matrix $\mathbf{K}$, where $\mathbf{K}$ is diagonal

$$\mathbf{u} = \mathbf{Kz} = \mathbf{KTx} \qquad 8.1-61$$

Substituting this expression into the system equations, we obtain

$$\dot{\mathbf{x}} = (\mathbf{A} + \mathbf{BKT})\mathbf{x} \qquad 8.1-62$$

At this point we introduce some of the results from linear algebra that we developed in Appendix A. One expression of interest for plants having distinct eigenvalues is

$$\mathbf{C\Lambda} = \mathbf{AC} \qquad 8.1-63$$

where $\mathbf{\Lambda}$ is a diagonal matrix of eigenvalues and $\mathbf{C}$ is the matrix of right-hand eigenvectors, which for our present purposes we will assume have been normalized. Similarly, we found that

$$\mathbf{\Lambda M} = \mathbf{MA} \qquad 8.1-64$$

where $\mathbf{M}$ is the matrix of left-hand eigenvectors; again we assume that they have been normalized. Finally, we also proved that

$$\mathbf{MC} = \mathbf{I} \qquad 8.1-65$$

Using these results, it is a simple matter to show that

$$\mathbf{MAC} = \mathbf{\Lambda} \qquad 8.1-66$$

and

$$\mathbf{A} = \mathbf{C\Lambda M} \qquad 8.1-67$$

Now, if, we substitute this last expression into Eq. 8.1-62,

$$\dot{x} = (C\Lambda M + BKT)x \qquad \qquad 8.1\text{-}68$$

arbitrarily choose the control matrix B to be the matrix of right-hand eigenvectors, C (see Appendix A),

$$B = C \qquad \qquad 8.1\text{-}69$$

and choose the output matrix T to be the matrix of left-hand eigenvectors, M,

$$T = M \qquad \qquad 8.1\text{-}70$$

the system equations become

$$\dot{x} = C(\Lambda + K)Mx \qquad \qquad 8.1\text{-}71$$

This is the result we have been looking for. It indicates that the controlled plant has the same eigenvectors as the uncontrolled plant, but the eigenvalues can be increased by the proportional gain matrix K. In other words, if all the eigenvalues λ_i are negative and real, we can make these as large as we please, $\lambda_i + k_i$, by adding an appropriate amount of negative feedback control. Also, this shifting of the eigenvalues can be accomplished without interaction, for Λ and K are both diagonal matrices. The system response for this case is

$$x = \sum_{i=1}^{n} \alpha_i c_i \exp\left[(\lambda_i + k_i)t\right] \qquad \qquad 8.1\text{-}72$$

where c_i is a right-hand eigenvector and α_i is the integration constant corresponding to the ith mode and is determined by the initial conditions.

One of the most interesting features of the analysis is that it tells us how we should choose our control variables and our outputs, Eqs. 8.1-69 and 8.1-70. Thus, instead of measuring the actual state variables, we might be better off measuring a linear combination of these variables

$$z = Tx = Mx \qquad \qquad 8.1\text{-}73$$

because these outputs are more closely related to the modes of the system. This procedure can always be accomplished if we can measure all the state variables. Because of practical considerations, where we choose control variables to be those that can easily be manipulated by turning a valve, normally it will not be possible to choose the control matrix $B = C$. In fact, for most plants, the matrix B is fixed. However, we can obtain the same final result by replacing K in Eq. 8.1-61 by a more general matrix G, which contains off-diagonal elements, and eventually requiring that

$$BGT = CKM \qquad \qquad 8.1\text{-}74$$

where we again choose K to be diagonal. Hence for a case where we make $T = M$, the unknown control matrix G becomes

$$G = B^{-1}CK \qquad \qquad 8.1\text{-}75$$

Limited Number of Control Variables

In order to illustrate the way in which the analysis must be modified when the number of control variables and outputs is not equal to the number of state variables (which is the most common case), we will consider a plant where the number of control variables is m and $m < n$. For simplicity, we still assume that the number of outputs is equal to the number of state variables and that we can choose the $n \times m$ control matrix $\mathbf{B}$ and the $n \times n$ output matrix $\mathbf{T}$ in any way that we please. If these assumptions are not valid, additional modifications will be necessary.

First we arrange the eigenvalues in order of their increasing absolute magnitude and then we choose the columns of $\mathbf{B}$ to be the m eigenvectors $\mathbf{c}_j$ corresponding to the m smallest eigenvalues λ_j. We again choose the relationship between the control variables and the outputs to be a diagonal matrix $\mathbf{K}$ and select the output matrix $\mathbf{T}$ to be equal to the matrix $\mathbf{M}$. Thus our system equation becomes

$$\dot{\mathbf{x}} = (\mathbf{A} + \mathbf{BKM})\mathbf{x} \qquad\qquad 8.1\text{--}76$$

but as we shall demonstrate, $n - m$ of the measured values of the outputs and controller gains are redundant.

If we augment the $n \times m$ $\mathbf{B}$ matrix with $n - m$ columns of zeros and partition the matrix product $\mathbf{BK}$, we obtain

$$\mathbf{BK} = (\mathbf{c}_1 \mathbf{c}_2, \cdots, \mathbf{c}_j 00, \cdots, 0) \begin{pmatrix} k_{11} & 0 & \cdots & 0 \\ 0 & k_{22} & \cdots & 0 \\ \vdots & \vdots & & \vdots \\ 0 & 0 & \cdots & k_{nn} \end{pmatrix} \qquad 8.1\text{--}77$$

$$\mathbf{BK} = (\mathbf{C}_m \,\vdots\, \mathbf{0}) \left(\begin{array}{c|c} \mathbf{K}_m & \mathbf{0} \\ \hline \mathbf{0} & \mathbf{K}_{n-m} \end{array} \right) \qquad\qquad 8.1\text{--}78$$

where $\mathbf{C}_m$ is an $m \times m$ matrix whose columns are the m eigenvectors corresponding to the m smallest eigenvalues, $\mathbf{K}_m$ is an $m \times m$ diagonal matrix, and $\mathbf{K}_{n-m}$ is an $(n - m) \times (n - m)$ diagonal matrix. Expanding the last expression, we find that

$$\mathbf{BK} = (\mathbf{C}_m \mathbf{K}_m \,\vdots\, \mathbf{0}) \qquad\qquad 8.1\text{--}79$$

which shows that the last $n - m$ measurements and values of $\mathbf{K}$ are neglected. Similarly, it is possible to show that

$$\mathbf{C}\boldsymbol{\Lambda} = (\mathbf{C}_m \,\vdots\, \mathbf{C}_{n-m}) \left(\begin{array}{c|c} \boldsymbol{\Lambda}_m & \mathbf{0} \\ \hline \mathbf{0} & \boldsymbol{\Lambda}_{n-m} \end{array} \right) \qquad\qquad 8.1\text{--}80$$

where the definitions of the elements of the partitioned matrices correspond to those given before. The expansion of this product gives the expression

$$\mathbf{C\Lambda} = (\mathbf{C}_m \mathbf{\Lambda}_m \mid \mathbf{C}_{n-m} \mathbf{\Lambda}_{n-m}) \qquad 8.1\text{--}81$$

The same procedure can be used to partition the matrix of left-hand eigenvectors

$$\mathbf{M} = \begin{pmatrix} M_m \\ M_{n-m} \end{pmatrix} \qquad 8.1\text{--}82$$

where M_m is an $m \times n$ matrix whose rows are the left-hand eigenvectors corresponding to the m smallest eigenvalues and M_{n-m} is an $(n-m) \times n$ matrix whose rows include the remaining left-hand eigenvectors. Substituting all these expressions into the system equations, we obtain

$$\dot{\mathbf{x}} = [\mathbf{C}_m(\mathbf{\Lambda}_m + \mathbf{K}_m)M_m + \mathbf{C}_{n-m}\mathbf{\Lambda}_{n-m}M_{n-m}]\mathbf{x} \qquad 8.1\text{--}83$$

This result shows that the m smallest eigenvalues can be made as large as desirable by using an appropriate amount of negative feedback control, $\mathbf{K}_m$. All the remaining eigenvalues and the system eigenvectors are not affected by this procedure, and the control is accomplished without interaction.

A similar type of analysis can be used to extend the results to cases where the number of outputs is less than the number of state variables. Also, systems having one or more pairs of complex conjugate roots can be handled.

SECTION 8.2 CONTROLLABILITY AND OBSERVABILITY

Up to this point we have implicitly assumed that it is possible to use some kind of a control system to alter the dynamic characteristics of a plant in such a way that we obtain a desirable, dynamic response. However, this assumption may not be valid for very large and complicated processes, even in cases where a large number of control variables (although not necessarily the right ones) may be available. Thus it becomes important to introduce a precise definition of controllability.

A related problem is concerned with how many and which of the output variables must be measured in order to be certain that all the system dynamics become apparent. In other words, we need to provide a quantitative definition of the concept of observability of the process dynamics. Both concepts will be discussed in some detail in this section.

Definition of Controllability

Although it is possible to define the controllability of a plant in a great number of ways,[1,2] the general definition we choose to adopt is

> A system is said to be completely controllable if it is possible to adjust its control variables $u_1(t), u_2(t), \cdots, u_m(t)$ in such a way that it

[1] R. E. Kalman, *J. Control (SIAM)*, Ser. A, **1**, No. 2, 152 (1963).

[2] E. G. Gilbert, *J. Control (SIAM)*, Ser. A, **1**, No. 2, 128 (1963).

will force the system to go from an arbitrary state $x_1(0)$, $x_2(0)$, $\cdots$, $x_n(0)$ to some other, arbitrary final state $x_1(\theta)$, $x_2(\theta)$, $\cdots$, $x_n(\theta)$ in a finite time θ.

In other words, we say that the plant is completely controllable if we can make it do whatever we please. Perhaps this definition is too restrictive in the sense that we are asking too much of the plant. But if we are able to show that the system equations satisfy this definition, certainly there can be no intrinsic limitation for the design of a control system for the plant. Of course, the fact that we can satisfy the definition of controllability does not provide any information about how to design an acceptable controller. Also, if the system turns out not to be controllable, it does not necessarily mean that the plant can never be operated in a satisfactory manner. Providing that a control system will maintain the important outputs in an acceptable region, the fact that we cannot make the plant do whatever we wish is immaterial. Alternately, it is often possible to increase the number of control variables and make the plant completely controllable.

In order to illustrate the way in which the definition is used to ascertain the controllability of a plant, we will examine the behavior of a first-order system. From our previous experience with this kind of a process, we expect it to be controllable. Therefore it should serve to illustrate the quantitative implications of the definition in the simplest possible manner. As a specific example, we consider the behavior of an isothermal CSTR.

Example 8.2-1 Controllability of a first-order system

The dynamic equations describing a first-order reaction in an isothermal CSTR are

$$V\frac{dA}{dt} = q(A_f - A) - kVA \qquad\qquad 8.2\text{-}1$$

Letting the control variable be the reactor temperature or, actually, the rate constant k, introducing the quantities

$$x = A - A_s, \quad u = k - k_s, \quad a = -\left(\frac{q}{V} + k_s\right), \quad b = -A_s \qquad 8.2\text{-}2$$

and linearizing the equation, we obtain

$$\frac{dx}{dt} = ax + bu \qquad\qquad 8.2\text{-}3$$

The solution of this equation is just

$$x(\theta) = e^{a\theta}x_0 + e^{a\theta}\int_0^\theta e^{-at}bu(t)\,dt \qquad\qquad 8.2\text{-}4$$

From our definition of complete controllability, we want to see if we can find some control function $u(t)$ that will make the plant proceed from an arbitrary initial condition x_0 to an arbitrary final state $x(\theta)$ in a finite time θ. This can indeed be accomplished if we can find a control function $u(t)$ in the time interval $0 \leq t \leq \theta$ that will make the term

$$\int_0^\theta e^{-at} bu(t)\, dt \qquad\qquad 8.2\text{-}5$$

take on any particular value we happen to choose. If we consider the simplest possible control function—namely, that $u(t)$ is equal to a constant u_c—the preceding term has the value

$$\frac{b}{a} u_c (1 - e^{-a\theta}) \qquad\qquad 8.2\text{-}6$$

Providing that we put no restrictions on the magnitude or the sign of u_c, this term can take on any desired value. Hence the first-order linear system is completely controllable.

It is important to note that we could have broken the time interval $0 \leq t \leq \theta$ up into two pieces, $0 \leq t \leq t_1$ and $t_1 \leq t \leq \theta$, and chosen constant control settings, u_{c1} and u_{c2}, for these intervals. Then the expression given by Eq. 8.2-5 becomes

$$\frac{b}{a} u_{c1}(1 - e^{-at_1}) + \frac{b}{a} u_{c2}(e^{-at_1} - e^{-a\theta}) \qquad\qquad 8.2\text{-}7$$

We now consider the possibility of whether or not this second control policy can be used to transfer the system from exactly the same initial condition to the same final point in the same time θ. Alternately, we would like to know if the proper choice of u_{c1}, u_{c2}, and t_1 will make the expression given by Eq. 8.2-7 identical to that in Eq. 8.2-6 for arbitrary choices of u_c and θ. The answer to this question is yes; in fact, there are an infinite number of ways in which the values can be selected. Similarly, it is possible to show that other control functions can also lead to the same result. Thus even though we have been able to establish that the first-order system can be forced to go from any initial state to any final state in a finite time, we still do not have any way of selecting the "best" control policy to use. In other words, the fact that the system is controllable does not give us any information about the design of a control system.

Another factor of importance is that the definition of controllability applies equally well to stable and unstable systems. The foregoing analysis in no way depends on the sign of the coefficient (a); therefore even first-order plants that are unstable are completely controllable. Although this result is interesting, we would always design the controller to make the system become stable.

Example 8.2-2 Controllability of two isothermal CSTRs in series

If we now consider the problem of two isothermal CSTRs of unequal volume in series (see Figure 8.2-1), the analysis becomes more complex. The system

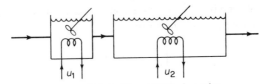

Figure 8.2-1. Two CSTRs in series.

equations are

$$V_1 \frac{dA_1}{dt} = q(A_f - A_1) - k_1 V_1 A_1 \qquad\qquad 8.2\text{-}8$$

$$V_2 \frac{dA_2}{dt} = q(A_1 - A_2) - k_2 V_2 A \qquad\qquad 8.2\text{-}9$$

Again, taking the reactor temperatures as the control variables, letting

$$x_1 = A_1 - A_{1s}, \quad x_2 = A_2 - A_{2s}, \quad u_1 = k_1 - k_{1s}, \quad u_2 = k_2 - k_{2s}$$
$$8.2\text{-}10$$

$$a_{11} = -\left(\frac{q}{V_1} + k_{1s}\right), \quad a_{21} = \frac{q}{V_2}, \quad a_{22} = -\left(\frac{q}{V_2} + k_{2s}\right),$$
$$b_{11} = -A_{1s}, \quad b_{22} = -A_{2s}$$

and linearizing the equations, we obtain

$$\frac{dx_1}{dt} = a_{11} x_1 + b_{11} u_1 \qquad\qquad 8.2\text{-}11$$

$$\frac{dx_2}{dt} = a_{21} x_1 + a_{22} x_2 + b_{22} u_2 \qquad\qquad 8.2\text{-}12$$

There are several possible problems of interest. For a case where we are allowed to adjust the temperature, or the rate constant, in each vessel independently, we intuitively expect that the system will be completely controllable. However, if we are permitted to adjust the temperature only in the second tank, $u_1 = 0$, the linearized system equations show that the control variable, u_2, has no effect on the composition in the first tank. We could hardly expect that this system will be completely controllable. For cases where $u_1 = u_2$ and we are required to change the temperature in both tanks by the same amount, or where $u_2 = 0$ and we can only vary the temperature in the first vessel, it becomes difficult to guess whether or not we will be able to force the system to go from arbitrary initial values, x_{10} and x_{20}, arbitrary final values, $x_1(\theta)$ and $x_2(\theta)$, in a finite time θ. Therefore, we need to develop a mathematical criterion that satisfies our definition of controllability. After this has been accomplished, we will return to this example.

State versus Output Controllability

The previous example illustrates another question that needs to be investigated. For the case where u_1 was equal to zero, we said that we expected that the system would not be completely controllable, because the only control variable, u_2, had no effect on the state variable x_1. However, it still might be possible to force the state variable x_2 between arbitrary levels in a finite time. Since this is the actual output of the system, perhaps controllability of the output, rather than controllability of all of the state variables, should receive our primary consideration. Another example of this type occurs in distillation problems, where, as a rule, we are primarily concerned with the control of the composition on the top and bottom plates and are quite willing to let the compositions on the other plates fluctuate with time. For most plants, the output can be related to the state variables by an expression like

$$\mathbf{z} = \mathbf{Tx} \qquad\qquad 8.2\text{-}13$$

and generally the output is of a lower dimensionality then the state vector. Thus we expect that output controllability will be much easier to achieve than state controllability. The same definition of controllability can be applied, except that we limit our attention to the output variables.

Criterion for Controllability

In order to develop a general method for testing for the controllability of a plant, we consider the equation

$$\dot{\mathbf{x}} = \mathbf{Ax} + \mathbf{Bu} \qquad\qquad 8.2\text{-}14$$

We have shown that the solution of this set of equations (see Eq. 4.7-18) can be written as

$$\mathbf{x}(\theta) = [\exp{(\mathbf{A}\theta)}]\mathbf{x}_0 + [\exp{(-\mathbf{A}\theta)}]\int_0^\theta [\exp{(-\mathbf{A}t)}]\mathbf{Bu}(t)\,dt \qquad 8.2\text{-}15$$

Analogous to the case of first-order systems, we say that the system is completely controllable; that is, we can achieve arbitrary values of the final state vector starting from an arbitrary initial state vector in a finite time, providing that we can find a control vector which will make the vector term

$$\int_0^\theta [\exp{(-\mathbf{A}t)}]\mathbf{Bu}(t)\,dt \qquad\qquad 8.2\text{-}16$$

take on arbitray values. However, if we pursue this approach, it will be necessary to deal with the matrix exponential. This step can be avoided by making a canonical transformation of the original system equations, so that they become uncoupled and we can apply our results for first-order systems directly.

Following the procedure we described previously (see Section 3.3), the canonical equations become

$$\dot{\mathbf{y}} = \mathbf{\Lambda y} + \mathbf{P}^{-1}\mathbf{Bu} \qquad\qquad 8.2\text{-}17$$

or letting

$$\mathbf{P}^{-1}\mathbf{B} = \mathbf{R} \qquad\qquad 8.2\text{-}18$$

we obtain

$$
\begin{pmatrix} \dot{y}_1 \\ \dot{y}_2 \\ \vdots \\ \dot{y}_n \end{pmatrix} = \begin{pmatrix} \lambda_1 & 0 & \cdots & 0 \\ 0 & \lambda_2 & \cdots & 0 \\ \vdots & \vdots & & \vdots \\ 0 & 0 & \cdots & \lambda_n \end{pmatrix} \begin{pmatrix} y_1 \\ y_2 \\ \vdots \\ y_n \end{pmatrix} + \begin{pmatrix} r_{11} & r_{12} & \cdots & r_{1m} \\ r_{21} & r_{22} & \cdots & r_{2m} \\ \vdots & \vdots & & \vdots \\ r_{n1} & r_{n2} & \cdots & r_{nm} \end{pmatrix} \begin{pmatrix} u_1 \\ u_2 \\ \vdots \\ u_m \end{pmatrix} \qquad 8.2\text{-}19
$$

where we are assuming that there are fewer control variables than state variables, $m < n$. Whenever the canonical transformation is possible (i.e., the eigenvalues are distinct), there is a well-defined relationship between the original
state variables x_i and the canonical variables y_j. Thus in order for the state of
the original system to be completely controllable, the canonical variables must
also be completely controllable. Since these equations are completely uncoupled,
we see immediately that if any one of the rows of the **R** matrix is equal to zero,
the control variables will have no effect on the corresponding state variable
and the plant will not be completely controllable. These state variables are
called noncontrollable.

The preceding condition above is necessary but not sufficient, for the fact
that all the state variables may be influenced by one or more of the control
variables still does not ensure us that each of the state variables can be transferred between arbitrary levels in a finite time. It is possible to demonstrate
that the system is completely controllable, however, and, in fact, this can be
accomplished even if there is only a single control variable. For example, if
we consider the equations

$$
\begin{pmatrix} \dot{y} \\ \dot{y}_2 \\ \vdots \\ \dot{y}_n \end{pmatrix} = \begin{pmatrix} \lambda_1 & 0 & \cdots & 0 \\ 0 & \lambda_2 & \cdots & 0 \\ \vdots & \vdots & & \vdots \\ 0 & 0 & & \lambda_n \end{pmatrix} \begin{pmatrix} y_1 \\ y_2 \\ \vdots \\ y_n \end{pmatrix} + \begin{pmatrix} r_1 \\ r_2 \\ \vdots \\ r_n \end{pmatrix} u \qquad 8.2\text{-}20
$$

then for the ith equation in the set we have

$$
\dot{y}_i = \lambda_i y_i + r_i u \qquad 8.2\text{-}21
$$

and this has the solution

$$
y_i(\theta) = e^{\lambda_i \theta} y_{i0} + e^{\lambda_i \theta} \int_0^\theta e^{-\lambda_i t} r_i u \, dt \qquad 8.2\text{-}22
$$

In order for the system to be completely controllable, we need to find a $u(t)$
that satisfies n equations of this type for arbitrary values of $y_i(\theta)$, $y_i(0)$, and θ.
This is equivalent to finding a $u(t)$ that makes the n integrals

$$
\int_0^\theta e^{-\lambda_i t} u(t) \, dt, \qquad i = 1, 2, \cdots, n \qquad 8.2\text{-}23
$$

take on n arbitrarily specified values. By choosing $u(t)$ to be a function that is
piecewise constant and changes value $(n - 1)$ times in the interval $0 \leq t \leq \theta$
(see Figure 8.2-2), we can obtain n arbitrary values for the terms in Eq. 8.2-23.

In fact, an infinite number of other choices for the control function lead to the same result. Hence the system with only a single control variable is completely controllable; similarly, the system with several control variables is also completely controllable, providing that none of the rows of the $R = P^{-1}B$ matrix is zero and that the eigenvalues are distinct.

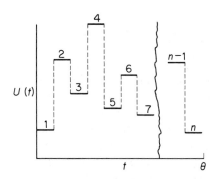

Figure 8.2-2. Control settings.

Example 8.2–3 Controllability of two isothermal CSTRs in series (continued)

The linearized system equations for two isothermal CSTRs in series were found to be

$$\dot{x}_1 = a_{11}x_1 + b_{11}u_1 \qquad\qquad 8.2\text{--}11$$

$$\dot{x}_2 = a_{21}x_1 + a_{22}x_2 + b_{22}u_2 \qquad\qquad 8.2\text{--}12$$

Hence the eigenvalues are

$$\lambda = \tfrac{1}{2}[(a_{11} + a_{22}) \pm \sqrt{(a_{11} + a_{22})^2 - 4a_{11}a_{22}}]$$

or

$$\lambda_1 = a_{11} \qquad \lambda_2 = a_{22} \qquad\qquad 8.2\text{--}24$$

From Appendix A, we find that the relationships between the elements of the eigenvectors are

$$c_{11} = \frac{a_{11} - a_{22}}{a_{21}}c_{21} \qquad c_{12} = 0 \cdot c_{22}$$

Letting $c_{21} = c_{22} = 1$, the matrix of eigenvectors (the transformation matrix) becomes

$$P = \begin{pmatrix} \dfrac{a_{11} - a_{22}}{a_{21}} & 0 \\ 1 & 1 \end{pmatrix} \qquad\qquad 8.2\text{--}25$$

so that its inverse is

$$P^{-1} = \frac{a_{21}}{a_{11} - a_{22}}\begin{pmatrix} 1 & 0 \\ -1 & \dfrac{a_{11} - a_{22}}{a_{21}} \end{pmatrix} \qquad\qquad 8.2\text{--}26$$

Thus the canonical equations are

$$
\begin{pmatrix} \dot{y}_1 \\ \dot{y}_2 \end{pmatrix} = \begin{pmatrix} a_{11} & 0 \\ 0 & a_{22} \end{pmatrix} \begin{pmatrix} y_1 \\ y_2 \end{pmatrix} + \frac{a_{21}}{a_{11} - a_{22}} \begin{pmatrix} b_{11} & 0 \\ -b_{11} & b_{22}\left(\dfrac{a_{11} - a_{22}}{a_{21}}\right) \end{pmatrix} \begin{pmatrix} u_1 \\ u_2 \end{pmatrix} \qquad 8.2\text{-}27
$$

where

$$
R = P^{-1}B = \frac{a_{21}}{a_{11} - a_{22}} \begin{pmatrix} b_{11} & 0 \\ -b_{11} & b_{22}\left(\dfrac{a_{11} - a_{22}}{a_{21}}\right) \end{pmatrix} \qquad 8.2\text{-}28
$$

Now we see that if both control variables are available, none of the rows of the R matrix is equal to zero and the system is completely controllable. Similarly, if $u_1 = u_2$, or the temperature in the second tank cannot be controlled —that is, $u_2 = 0$, or its equivalent, that $b_{22} = 0$—the plant is still completely controllable. However, if the temperature in the second vessel is the only control variable and $u_1 = 0$, or its equivalent, $b_{11} = 0$, the first row of the R matrix contains all zeros and the canonical state variable y_1 is noncontrollable. The relationship between the original system variables and the canonical variables is given by the transformation

$$
\mathbf{x} = \mathbf{P}\mathbf{y} \qquad 8.2\text{-}29
$$

so that

$$
x_1 = \frac{a_{11} - a_{22}}{a_{21}} y_1 \qquad 8.2\text{-}30
$$

$$
x_2 = y_1 + y_2
$$

Since y_1 is noncontrollable, we find that x_1 also is noncontrollable, which is the result we expected to obtain. The output vector is given by the expression

$$
\mathbf{z} = \mathbf{T}\mathbf{x} \qquad 8.2\text{-}13
$$

which can be written either as

$$
\begin{pmatrix} 0 \\ z_2 \end{pmatrix} = \begin{pmatrix} 0 & 0 \\ 0 & 1 \end{pmatrix} \begin{pmatrix} x_1 \\ x_2 \end{pmatrix} \qquad 8.2\text{-}31
$$

or

$$
z = x_2 = (0 \quad 1) \begin{pmatrix} x_1 \\ x_2 \end{pmatrix} \qquad 8.2\text{-}32
$$

Whenever x_2 is controllable, the system output is controllable.

A serious problem is encountered if the volume of the two reactors is equal and $k_{1s} = k_{2s}$ in Eqs. 8.2–10, because, for this case, we find that $a_{11} = a_{22}$ and the eigenvalues are equal rather than distinct. Then the transformation matrix given by Eq. 8.2–25 is singular and the inverse of this matrix, Eq. 8.2–26, does not exist. Problems of this type must be handled by special methods. Our primary purpose here is to present the concept of controllability rather than to describe procedures that can be applied to all possible cases; consequently, we shall not pursue the matter.

Observability

Observability is the dual concept of controllability. It is concerned with the question of whether or not it is possible to obtain all the information about the state of the system by measuring its output, which, as we mentioned earlier, is generally given by the expression

$$z = Tx \qquad\qquad 8.2\text{-}13$$

and is of lower dimensionality than the state vector. For example, from the practical point of view of selecting and designing instruments, it is important to know whether or not the complete dynamic behavior of a distillation column can be inferred from the measured values of the compositions and flows leaving the top and bottom plates. Thus we define observability as

> A system is said to be completely observable if measurements of the output z over some finite time interval contain sufficient information to enable us to identify the state x completely.

Whenever all the elements of the state vector are not observable, we say that they are "hidden" from observation.

In order to develop a quantitative criteria for testing for observability, we first recognize that a complete dynamic description of a plant requires that we include all the natural modes of the system. The simplest way of writing these is in terms of the canonical variables

$$\dot{y} = \Lambda y \qquad\qquad 8.2\text{-}33$$

The original state variables are related to the canonical variables by the transformation

$$x = Py \qquad\qquad 8.2\text{-}34$$

where P is the matrix of right-hand eigenvectors, so that the relationship between the outputs and the canonical variables becomes

$$z = Tx = TPy = Sy \qquad\qquad 8.2\text{-}35$$

where

$$S = TP \qquad\qquad 8.2\text{-}36$$

Expanding this matrix equation, we obtain

$$
\begin{aligned}
z_1 &= s_{11}y_1 + s_{12}y_2 + \cdots + s_{1n}y_n \\
z_2 &= s_{21}y_1 + s_{22}y_2 + \cdots + s_{2n}y_n \\
&\;\;\vdots \\
z_l &= s_{l1}y_1 + s_{l2}y_2 + \cdots + s_{ln}y_n
\end{aligned}
\qquad\qquad 8.2\text{-}37
$$

For the system to be completely observable, the effect of each of the canonical state variables must show up in the outputs. Hence a necessary and sufficient condition for complete observability is that none of the columns of the matrix $S = TP$ is equal to zero. If one or more of the columns are equal to zero, the corresponding canonical state variables are hidden from observation and the system possesses dynamic modes that are not apparent in the measured values of the output.

The relationship between the problem of controllability and observability should now be clear. In the first case, we are interested in seeing if one or more of the state variables are beyond the influence of the inputs; in the second case, we want to find out if one or more of the state variables have no effect on the system outputs. A demonstration that a system is completely controllable or completely observable still does not give us any indication as to how to go about either controlling it or observing it. Plants that are not controllable can often be made controllable by adding more control variables, and plants that are not observable can often be made observable by making more measurements.

Sometimes it is of interest to subdivide the natural modes of a plant into four groups, depending on their controllability and observability characteristics. The first group contains the modes that are both controllable and observable, the second contains those that are controllable but not observable, the third contains those that are observable but not controllable, and the fourth contains those that are neither controllable nor observable (see Figure 8.2-3). It is ap-

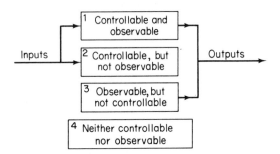

Figure 8.2-3. Classification of system modes.

parent from the figure that the only connection between the input and output occurs in the first group; therefore these are the only modes that can be represented by transfer functions. This fact implies that if we measure the transfer functions for some particular plant only experimentally, we might obtain a misleading picture of the system dynamics and wild fluctuations might be occurring inside of the system which would not be apparent in the measured outputs.

Example 8.2-3 Observability of two isothermal CSTRs in series

As an illustration of the method for determining observability, we return to Example 8.2-2 and consider two isothermal CSTRs in series. The linearized system equations are

$$\dot{x}_1 = a_{11}x_1 \qquad\qquad 8.2\text{--}11$$

$$\dot{x}_2 = a_{12}x_1 + a_{22}x_2 \qquad\qquad 8.2\text{--}12$$

where we have set the control variables, u_1 and u_2, equal to zero. The transformation matrix for the canonical transformation was shown to be

$$\mathbf{P} = \begin{pmatrix} \dfrac{a_{11} - a_{22}}{a_{21}} & 0 \\ 1 & 1 \end{pmatrix} \qquad\qquad 8.2\text{--}25$$

and the system output matrix $\mathbf{T}$ was the row vector

$$\mathbf{T} = (0 \quad 1) \qquad\qquad 8.2\text{--}32$$

Hence the matrix $\mathbf{S} = \mathbf{TP}$ becomes

$$S = (0 \quad 1) \begin{pmatrix} \dfrac{a_{11} - a_{22}}{a_{21}} & 0 \\ 1 & 1 \end{pmatrix} = (1 \quad 1) \qquad\qquad 8.2\text{--}38$$

Neither column of this matrix is equal to zero, and therefore the system is completely observable.

SECTION 8.3   MULTIVARIABLE FEEDFORWARD CONTROL SYSTEMS

As we noted in Chapter 7, feedback control systems can never provide perfect control of a plant because we must wait for the disturbance to cause a change in the output before the measured error signal begins to activate the controller. We showed in Chapter 7 that perfect control could be obtained for processes with a single input and a single output whenever we could measure all the disturbances and had a complete understanding of the process dynamics. The technique we used to achieve this perfect regulation was called feedforward control. It is a simple task to extend the idea of feedforward control to multivariable plants. Linear lumped parameter systems are considered first, for they are the easiest to handle, but then the application to nonlinear plants and to distributed parameter processes is discussed.

Linear Lumped Parameter Systems

We consider a plant described by the set of equations

$$\frac{d\mathbf{x}}{dt} = \mathbf{Ax} + \mathbf{Bu} + \mathbf{Cv} \qquad\qquad 8.3\text{--}1$$

where $\mathbf{x}$ represents the state vector, $\mathbf{u}$ is the control vector, $\mathbf{v}$ is a disturbance vector, and $\mathbf{A}$, $\mathbf{B}$ and $\mathbf{C}$ are constant matrices. If the plant is initially at steady state conditions so that $\mathbf{x} = \mathbf{u} = \mathbf{v} = 0$ at $t < 0$, and for a case where there are as many, or more, control variables as state variables, we can always prevent disturbances from having any effect on the state variables simply by making

$$\mathbf{Bu} + \mathbf{Cv} = 0 \qquad\qquad 8.3\text{--}2$$

Whenever there are exactly as many control variables as state variables and **B** is not a singular matrix, we can obtain explicit expressions for the control system

$$\mathbf{u} = -\mathbf{B}^{-1}\mathbf{Cv} \qquad 8.3\text{-}3$$

and it is clear that a multivariable proportional control system will provide perfect control. In situations where there are more control variables than state variables—that is, $\mathbf{x}$ is an $n \times 1$ column vector, $\mathbf{B}$ is a $n \times r$ matrix, $\mathbf{u}$ is an $r \times 1$ column vector, and $r > n$—all we have to do is to set $r - n$ of the control variables equal to zero, say the last $r - n$, and also set the last $r - n$ columns of **B** equal to zero. This reduced **B** matrix, which we call $\mathbf{B}_r$, now has n rows and n columns, and providing that it is not singular, we can use an expression equivalent to Eq. 8.3-3 to find the n control variables we need to manipulate. Of course, if the reduced **B** matrix is singular, we try setting some of the other original control variables equal to zero and hope that we finally can obtain a $\mathbf{B}_r$ matrix that is nonsingular. For cases where there are fewer control variables than state variables, or where **B** or $\mathbf{B}_r$ is singular, normally it will not be possible to maintain all the outputs constant for arbitrary sets of disturbances. However, it might be possible to keep some of the outputs at their steady state values by following the procedure described below. The most desirable state variables to hold constant must be determined from physical considerations.

If we take the Laplace transform of Eq. 8.3-1 and manipulate the results somewhat, we obtain

$$\tilde{\mathbf{x}} = (s\mathbf{I} - \mathbf{A})^{-1}\mathbf{B}\tilde{\mathbf{u}} + (s\mathbf{I} - \mathbf{A})^{-1}\mathbf{C}\tilde{\mathbf{v}} \qquad 8.3\text{-}4$$

or

$$\mathbf{x} = \mathbf{H}(s)\tilde{\mathbf{u}} + \mathbf{G}(s)\tilde{\mathbf{v}} \qquad 8.3\text{-}5$$

where $\mathbf{H}(s)$ will be a $n \times r$ matrix since $(s\mathbf{I} - \mathbf{A})^{-1}$ is $n \times n$ and **B** is $n \times r$. Using this formulation and considering cases where $r = n$ or $r > n$, we arrive at exactly the same conclusions as above. The case that is still of interest is when $r < n$. Now suppose that we decide to hold r of the state variables constant, say the first r. By setting the last r rows of $\mathbf{H}(s)$ equal to zero so that we obtain a reduced $\mathbf{H}(s)$ matrix, which we call $\mathbf{H}_r(s)$, we can find the control variables that satisfy the equation

$$\tilde{\mathbf{u}} = -\mathbf{H}_R^{-1}(s)\mathbf{G}(s)\tilde{\mathbf{v}} \qquad 8.3\text{-}6$$

Of course, this procedure will fail if $\mathbf{H}_R(s)$ is singular; in that case, we must choose either a different set or a smaller number of state variables to maintain constant. In the first case, we use all the r control variables, whereas, in the second case, we are holding some of the control variables constant. It should be noted that the feedforward control system described by Eq. 8.3-6 contains dynamic elements, for $\mathbf{H}_R(s)$ and $\mathbf{G}(s)$ contain the Laplace parameter. Some simple examples illustrating the method follow.

Example 8.3-1 Feedforward control of a stirred-tank reactor: linearized equations

The equations often used to describe the dynamic behavior of a CSTR are

$$\frac{dA}{dt} = \frac{q}{V}(A_f - A) - kA \tag{8.3-7}$$

$$\frac{dT}{dt} = \frac{q}{V}(T_f - T) + \frac{UA_H Kq_H}{VC_p \rho(1 + Kq_H)}(T_H - T) + \frac{(-\Delta H)}{C_p \rho}kA \tag{8.3-8}$$

Using the definitions given in Example 3.3-1, these equations can be approximated by the set of linear equations

$$\frac{dx_1}{d\tau} = a_{11}x_1 + a_{12}x_2 + b_{11}u_1 + c_{11}x_{1f} \tag{8.3-9}$$

$$\frac{dx_2}{d\tau} = a_{21}x_1 + a_{22}x_2 + b_{21}u_1 + b_{22}u_2 + c_{22}x_{2f} + c_{23}x_{2H} \tag{8.3-10}$$

where x_1 represents the reactant composition, x_2 represents the reactor temperature, u_1 is the feed rate, u_2 is the flow rate of heating fluid, x_{1f} is the feed composition, x_{2f} is the feed temperature, and x_{2H} is the inlet temperature of heating fluid. The equations can be written in vector matrix form as

$$\begin{pmatrix} \dfrac{dx_1}{d\tau} \\ \dfrac{dx_2}{d\tau} \end{pmatrix} = \begin{pmatrix} a_{11} & a_{12} \\ a_{21} & a_{22} \end{pmatrix}\begin{pmatrix} x_1 \\ x_2 \end{pmatrix} + \begin{pmatrix} b_{11} & 0 \\ b_{21} & b_{22} \end{pmatrix}\begin{pmatrix} u_1 \\ u_2 \end{pmatrix} + \begin{pmatrix} c_{11} & 0 & 0 \\ 0 & c_{22} & c_{23} \end{pmatrix}\begin{pmatrix} x_{1f} \\ x_{2f} \\ x_{2H} \end{pmatrix} \tag{8.3-11}$$

or

$$\frac{d\mathbf{x}}{d\tau} = \mathbf{Ax} + \mathbf{Bu} + \mathbf{Cv} \tag{8.3-12}$$

If the system is at steady state conditions initially, $\mathbf{x} = \mathbf{u} = \mathbf{v} = 0$, and at time zero the disturbances, x_{1f}, x_{2f}, and x_{2H}, start to fluctuate in an arbitrary manner, we can prevent these disturbances from having any effect on the state variables by measuring them and designing a feedforward control system, where

$$\mathbf{u} = -\mathbf{B}^{-1}\mathbf{Cv} \tag{8.3-13}$$

or

$$\begin{pmatrix} u_1 \\ u_2 \end{pmatrix} = -\frac{1}{b_{11}b_{22}}\begin{pmatrix} b_{22} & 0 \\ -b_{21} & b_{11} \end{pmatrix}\begin{pmatrix} c_{11} & 0 & 0 \\ 0 & c_{22} & c_{23} \end{pmatrix}\begin{pmatrix} x_{1f} \\ x_{2f} \\ x_{2H} \end{pmatrix}$$

$$= \begin{pmatrix} -\dfrac{c_{11}}{b_{11}}x_{1f} + 0 & + 0 \\ \dfrac{b_{21}c_{11}}{b_{11}b_{22}}x_{1f} - \dfrac{c_{22}}{b_{22}}x_{2f} - \dfrac{c_{23}}{b_{22}}x_{2H} \end{pmatrix} \tag{8.3-14}$$

which is a multivariable proportional controller. This is the procedure described by Eq. 8.3–3 and, of course, exactly the result we obtain if we set $dx_1/d\tau = dx_2/d\tau = x_1 = x_2 = 0$ in Eqs. 8.3–9 and 8.3–10 and solve for u_1 and u_2 in terms of the disturbances.

Now, if we consider a problem where u_1 is considered a constant, or where we set $b_{11} = b_{21} = 0$, the system equations become

$$\begin{pmatrix} \dfrac{dx_1}{d\tau} \\ \dfrac{dx_2}{d\tau} \end{pmatrix} = \begin{pmatrix} a_{11} & a_{12} \\ a_{21} & a_{22} \end{pmatrix}\begin{pmatrix} x_1 \\ x_2 \end{pmatrix} + \begin{pmatrix} 0 & 0 \\ 0 & b_{22} \end{pmatrix}\begin{pmatrix} u_1 \\ u_2 \end{pmatrix} + \begin{pmatrix} c_{11} & 0 & 0 \\ 0 & c_{22} & c_{23} \end{pmatrix}\begin{pmatrix} x_{1f} \\ x_{2f} \\ x_{2H} \end{pmatrix} \qquad 8.3\text{–}15$$

For this case, **B** is a singular matrix and therefore does not possess an inverse. Hence we take the Laplace transform of the equations

$$\begin{pmatrix} s\tilde{x}_1 \\ s\tilde{x}_2 \end{pmatrix} = \begin{pmatrix} a_{11} & a_{12} \\ a_{21} & a_{22} \end{pmatrix}\begin{pmatrix} \tilde{x}_1 \\ \tilde{x}_2 \end{pmatrix} + \begin{pmatrix} 0 & 0 \\ 0 & b_{22} \end{pmatrix}\begin{pmatrix} \tilde{u}_1 \\ \tilde{u}_2 \end{pmatrix} + \begin{pmatrix} c_{11} & 0 & 0 \\ 0 & c_{22} & c_{23} \end{pmatrix}\begin{pmatrix} \tilde{x}_{1f} \\ \tilde{x}_{2f} \\ \tilde{x}_{2H} \end{pmatrix} \qquad 8.3\text{–}16$$

or since we have only a single control variable, we could replace the **B** matrix by a vector **b** and treat the control variable u_2 as a scalar

$$\begin{pmatrix} s\tilde{x}_1 \\ s\tilde{x}_2 \end{pmatrix} = \begin{pmatrix} a_{11} & a_{12} \\ a_{21} & a_{22} \end{pmatrix}\begin{pmatrix} \tilde{x}_1 \\ \tilde{x}_2 \end{pmatrix} + \begin{pmatrix} 0 \\ b_{22} \end{pmatrix}\tilde{u}_2 + \begin{pmatrix} c_{11} & 0 & 0 \\ 0 & c_{22} & c_{23} \end{pmatrix}\begin{pmatrix} \tilde{x}_{1f} \\ \tilde{x}_{2f} \\ \tilde{x}_{2H} \end{pmatrix} \qquad 8.3\text{–}17$$

Solving the equation for $\tilde{\mathbf{x}}$, we obtain

$$\tilde{\mathbf{x}} = (s\mathbf{I} - \mathbf{A})^{-1}\mathbf{b}\tilde{u}_2 + (s\mathbf{I} - \mathbf{A})^{-1}\mathbf{C}\tilde{\mathbf{v}}$$

or

$$\begin{pmatrix} \tilde{x}_1 \\ \tilde{x}_2 \end{pmatrix} = \frac{1}{s^2 - (a_{11} + a_{22})s + a_{11}a_{22} - a_{12}a_{21}}\left[\begin{pmatrix} a_{12}b_{22} \\ (s - a_{11})b_{22} \end{pmatrix}\tilde{u}_2 \right.$$

$$\left. + \begin{pmatrix} c_{11}(s - a_{22}) & c_{12}a_{12} & c_{23}a_{12} \\ c_{11}a_{21} & c_{22}(s - a_{11}) & c_{23}(s - a_{11}) \end{pmatrix}\begin{pmatrix} \tilde{x}_{1f} \\ \tilde{x}_{2f} \\ \tilde{x}_{2H} \end{pmatrix} \right] \qquad 8.3\text{–}18$$

There is only one control variable; thus we can only hope to maintain one of the state variables constant. If we decide that we want to maintain the reactor composition at its original value, $\tilde{x}_1 = 0$, we drop the second equation and solve for the control

$$\tilde{u}_2 = -\frac{1}{a_{12}b_{22}}[c_{11}(s - a_{22})\tilde{x}_{1f} + c_{22}a_{12}\tilde{x}_{2f} + c_{23}a_{12}\tilde{x}_{2H}] \qquad 8.3\text{–}19$$

This expression describes a multivariable, proportional plus derivative controller.

For this case, we see that the matrix $H_R^{-1}(s)$ in Eq. 8.3–6 reduces to the scalar

$$H_R^{-1}(s) = \frac{s^2 - (a_{11} + a_{22})s + a_{11}a_{22} - a_{12}a_{21}}{a_{12}b_{22}} \qquad 8.3\text{–}20$$

It is important to notice that our formal procedure for developing the feedforward control is equivalent to setting the state variables we desire to maintain constant equal to zero and then solving for the control variables that will make this possible. Hence if we let $\tilde{x}_1 = 0$ in Eq. 8.3–16, we obtain

$$0 = a_{12}\tilde{x}_2 + c_{11}\tilde{x}_{1f} \qquad 8.3\text{–}21$$

$$s\tilde{x}_2 = a_{22}\tilde{x}_2 + b_{22}\tilde{u}_2 + c_{22}\tilde{x}_{2f} + c_{23}\tilde{x}_{2H} \qquad 8.3\text{–}22$$

Solving Eq. 8.3–22 for $\tilde{u}_2$,

$$\tilde{u}_2 = \frac{1}{b_{22}}[(s - a_{22})\tilde{x}_2 - c_{22}\tilde{x}_{2f} - c_{23}\tilde{x}_{2H}] \qquad 8.3\text{–}23$$

and using Eq. 8.3–21 to eliminate $\tilde{x}_2$, we arrive at a result that is identical to Eq. 8.3–19. This approach makes it unnecessary to evaluate the matrix $(s\mathbf{I} - \mathbf{A})^{-1}$, which is often a difficult job, and also makes it clear that the uncontrolled state variable $\tilde{x}_2$ will fluctuate proportional to the feed composition disturbances $\tilde{x}_{1f}$ (see Eq. 8.3–21).

Example 8.3–2 Feedforward control of a catalytic cracking unit: linearized equations

As a more complicated example of the synthesis of a feedforward control system, we consider the dynamic model of a catalytic cracking unit described in Example 3.1–4. We assume that we want to maintain the gas composition leaving the reactor y_1, the gas composition leaving the regenerator y_2, and the reactor temperature T_1 constant by manipulating the oil feed rate V_0, the air feed rate V_a, and the catalyst circulation rate W, in order to compensate for disturbances in feed composition y_0, feed temperature T_0, and air temperature T_a. Linearizing the system equations, Eqs. 3.1–38 through 3.1–43, and taking Laplace transforms, we obtain

$$s_1\tilde{y}_1 = a_{11}\tilde{y}_1 + a_{13}\tilde{T}_1 + b_{11}\tilde{V}_0 + c_{11}\tilde{y}_0$$

$$s\tilde{x}_1 = a_{21}\tilde{y}_1 + a_{22}\tilde{x}_1 + a_{23}\tilde{T}_1 + a_{25}\tilde{x}_2 + b_{22}\tilde{W}$$

$$s\tilde{T}_1 = a_{31}\tilde{y}_1 + a_{33}\tilde{T}_1 + a_{36}\tilde{T}_2 + b_{31}\tilde{V}_0 + b_{32}\tilde{W} + c_{32}\tilde{T}_0$$

$$s\tilde{y}_2 = a_{44}\tilde{y}_2 + a_{46}\tilde{T}_2 + b_{43}\tilde{V}_a \qquad 8.3\text{–}24$$

$$s\tilde{x}_2 = a_{52}\tilde{x}_1 + a_{54}\tilde{y}_2 + a_{55}\tilde{x}_2 + a_{56}\tilde{T}_2 + b_{52}\tilde{W}$$

$$s\tilde{T}_2 = a_{63}\tilde{T}_1 + a_{64}\tilde{y}_2 + a_{66}\tilde{T}_2 + b_{62}\tilde{W} + b_{63}\tilde{V}_a + c_{63}\tilde{T}_a$$

where $\tilde{y}_1, \tilde{x}_2, \ldots, \tilde{V}_0 \tilde{W}_1, \ldots$, and $\tilde{y}_0, \tilde{T}_0, \ldots$ represent the transforms of the deviations of the state, control, and disturbance variables, and

$$a_{11} = -\left(\frac{RV_0 T_{1s}}{P_1 H_1} + A_1 e^{-E_1/RT_{1s}}\right)$$

$$a_{13} = \left[\frac{E_1}{RT_{1s}^2}(1 - y_{1s})A_1 e^{-E_1/RT_{1s}} - \frac{V_{0s} R y_{1s}}{P_1 H_1}\right]$$

$$a_{21} = -\frac{0.1 P_1 H_1 A_1 e^{-E_1/RT_{1s}}}{M_1 RT_{1s}} \qquad a_{22} = -\frac{W_s}{M_1}$$

$$a_{23} = \left[\left(\frac{E_1}{RT_{1s}} - 1\right)\frac{0.1 P_1 H_1(1 - y_{1s})A_1 e^{-E_1/RT_{1s}}}{M_1 RT_{1s}^2}\right]$$

$$a_{25} = \frac{W_s}{M_1} \qquad a_{31} = \frac{(\Delta H_1) P_1 H_1 A_1 e^{-E_1/RT_{1s}}}{P_1 H_1 C_{p1} + C_{ps} M_1 RT_{1s}}$$

$$a_{33} = \left[\frac{RC_{p1} V_{0s}(T_{0s} - 2T_{1s}) + RC_{ps} W_s(T_{2s} - 2T_{1s})}{P_1 H_1 C_{p1} + RC_{ps} M_1 T_{1s}}\right.$$

$$\left. - \frac{\{(\Delta H_1) P_1 H_1(1 - Y_{1s})A_1 E_1 e^{-E_1/RT_{1s}}\}/RT_{1s}^2}{P_1 H_1 C_{p1} + RC_{ps} M_1 T_{1s}}\right]$$

$$a_{36} = \frac{RT_{1s} C_{ps} W_s}{P_1 H_1 C_{p1} + RT_{1s} C_{ps} M_1} \qquad a_{44} = -\frac{RT_{2s} V_{as}}{P_2 H_2} - A_2 m e^{-E_2/RT_{2s}}$$

$$a_{46} = \frac{RV_{As}(y_a - y_{2s})}{P_2 H_2} - \frac{mE_2 y_{2s} A_2 e^{-E_2/RT_{2s}}}{RT_{2s}^2}$$

$$a_{52} = \frac{W_s}{M_2} \qquad a_{54} = -\frac{P_2 H_2 A_2 e^{-E_2/RT_{2s}}}{M_2 RT_{2s}} \qquad a_{55} = -\frac{W_s}{M_2}$$

$$a_{56} = \frac{P_2 H_2 y_{2s} A_2 e^{-E_2/RT_{2s}}}{M_2 RT_{2s}^2}\left(\frac{E_2}{RT_{2s}} - 1\right)$$

$$a_{63} = \frac{C_{ps} W_s T_{2s} R}{P_2 H_2 C_{p2} + M_2 C_{ps} RT_{2s}} \qquad a_{64} = -\frac{(\Delta H) P_2 H_2 A_2 e^{-E_2/RT_{2s}}}{P_2 H_2 C_{p2} + M_2 C_{ps} RT_{2s}}$$

$$a_{66} = \frac{RC_{p2} V_{as}(T_{as} - 2T_{2s}) + RC_{ps} W_s(T_1 - 2T_{2s}) + Q}{P_2 H_2 C_{p2} + M_2 C_{ps} RT_{2s}}$$

$$- \frac{(\Delta H_2) P_2 H_2 y_{2s} E_2 A_2 e^{-E_2/RT_{2s}}/RT_{2s}^2}{P_2 H_2 C_{p2} + M_2 C_{ps} RT_{2s}}$$

$$b_{11} = -\frac{RT_{1s} y_{1s}}{P_1 H_1} \qquad b_{22} = \frac{x_{2s} - x_{1s}}{M_1} \qquad b_{31} = \frac{C_{p1} RT_{1s}(T_{0s} - T_{1s})}{P_1 H_1 C_{p1} + C_{ps} M_1 RT_{1s}}$$

$$b_{32} = \frac{C_{ps} RT_{1s}(T_{2s} - T_{1s})}{P_1 H_1 C_{p1} + M_1 C_{ps} RT_{1s}} \qquad b_{43} = \frac{V_{as} RT_{2s}}{P_2 H_2} \qquad b_{52} = \frac{x_{1s} - x_{2s}}{M_2}$$

$$b_{62} = \frac{C_{ps} R(T_{1s} T_{2s} - T_{2s}^2)}{P_2 H_2 C_{p2} + M_2 C_{ps} RT_{2s}} \hspace{3cm} 8.3\text{-}25$$

$$b_{63} = \frac{C_{p2} RT_{2s}(T_{as} - T_{2s})}{P_2 H_2 C_{p2} + M_2 C_{ps} RT_{2s}} \qquad c_{11} = \frac{V_{0s} RT_{1s}}{P_1 H_1}$$

$$c_{32} = \frac{C_{p1} V_{0s} RT_{1s}}{P_1 H_1 C_{p1} + M_1 C_{ps} RT_{1s}} \qquad c_{63} = \frac{C_{p2} V_{as} RT_{2s}}{P_2 H_2 C_{p2} + M_2 C_{ps} RT_{2s}}$$

$$Q = 0$$

Since we want

$$\tilde{y}_1 = \tilde{y}_2 = \tilde{T}_1 = 0$$

Eqs. 8.3–24 reduce to

$$0 = b_{11}\tilde{V}_0 + c_{11}\tilde{y}_0$$

$$s\tilde{x}_1 = a_{22}\tilde{x}_1 + a_{25}\tilde{x}_2 + b_{22}\tilde{W}$$

$$0 = a_{36}\tilde{T}_2 + b_{31}\tilde{V}_0 + b_{32}\tilde{W} + c_{32}\tilde{T}_0$$

$$0 = a_{46}\tilde{T}_2 + b_{43}\tilde{V}_a$$

$$s\tilde{x}_2 = a_{52}\tilde{x}_1 + a_{55}\tilde{x}_2 + a_{56}\tilde{T}_2 + b_{52}\tilde{W}$$

$$s\tilde{T}_2 = a_{66}\tilde{T}_2 + b_{62}\tilde{W} + b_{63}\tilde{V}_a + c_{63}\tilde{T}_a$$

8.3–26

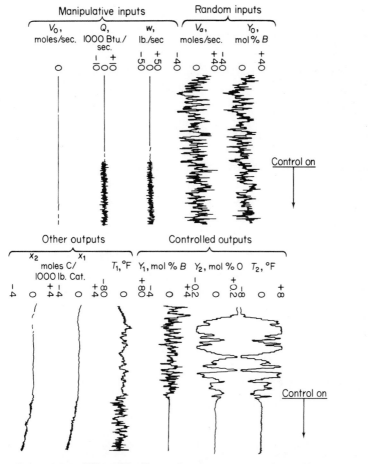

Figure 8-3-1. Effect of feedforward control on a catalytic cracking unit. [Reproduced from W. L. Luyben and D. E. Lamb, *Chem. Eng. Prog. Symposium SER*, **59**, No. 46, 165 (1963), by permission.]

Now, we can eliminate the state variables $\tilde{x}_1$, $\tilde{x}_2$, and $\tilde{T}_2$ from this set of equations and then solve the resulting set of three equations for the three control variables $\tilde{V}_0$, $\tilde{W}$, and $\tilde{V}_a$ in terms of the disturbances $\tilde{y}_0$, $\tilde{T}_0$, and $\tilde{T}_a$. In this way we find that

$$\tilde{V}_0 = -\frac{c_{11}}{b_{11}}\tilde{y}_0 \qquad 8.3\text{--}27$$

$$\tilde{V}_a = -\frac{b_{31}b_{62}c_{11}}{b_{11}(k_1 + k_2 s)}\tilde{y}_0 - \frac{c_{63}b_{32}}{(k_1 + k_2 s)}\tilde{T}_a + \frac{b_{62}c_{32}}{(k_1 + k_2 s)}\tilde{T}_0 \qquad 8.3\text{--}28$$

$$\tilde{W} = \left\{ \left(\frac{b_{31}c_{11}}{b_{11}}\right) \frac{[a_{46}b_{63} + b_{43}(s - a_{66})]}{[a_{46}(k_1 + k_2 s)]}\right\}\tilde{y}_0 - \frac{a_{36}b_{43}c_{63}}{a_{46}b_{32}(k_1 + k_2 s)}\tilde{T}_a$$

$$- \left\{\frac{c_{32}[a_{46}b_{63} + b_{43}(s - a_{66})]}{[a_{46}(k_1 + k_2 s)]}\right\}\tilde{T}_0 \qquad 8.3\text{--}28$$

where

$$k_1 + k_2 s = \frac{\{a_{36}b_{43}b_{62} + b_{32}[a_{46}b_{63} + b_{43}(s - a_{66})]\}}{a_{46}}$$

Thus, we obtain a multivariable controller that is realizable.

A similar control problem was studied by Luyben and Lamb.[1] They chose V_0, Q, and W as the control variables, V_0 and y_0 as the disturbances, and y_1, y_2, and T_2 as the outputs they were attempting to maintain constant. For this problem, the feedforward controller contains only first-order leads except for V_0, which must be set equal to zero. They simulated the complete set of linearized system equations and the controller on an analog computer, and for random inputs obtained the results shown in Figure 8–3.1. It is apparent from the graph that the desired outputs become constant soon after the controller is switched on and that the other outputs fluctuate.

Example 8.3–3 Feedforward control of a distillation column

The equations commonly used to represent the dynamic behavior of a distillation column are essentially the same as those for an absorber or stripping unit. In the rectifying section above the feed plate, a total material balance gives

$$\frac{dh_n}{dt} + \frac{dH_n}{dt} = L_{n+1} - L_n + V_{n-1} - V_n \qquad 8.3\text{--}30$$

or for a case where the vapor holdup is negligible and $V_{n-1} = V_n$,

$$\frac{dH_n}{dt} = L_{n+1} - L_n \qquad 8.3\text{--}31$$

[1] W. L. Luyben and D. E., Lamb, *Chem. Eng. Prog. Symposium Ser.*, **59**, No. 46, 165 (1963).

A component balance gives the expression

$$\frac{dH_n x_n}{dt} = L_{n+1} x_{n+1} - L_n x_n + V_{n-1} y_{n-1} - V_n y_n \qquad 8.3\text{-}32$$

Assuming that the Murphree tray efficiency varies from tray to tray but is independent of time, we can write

$$E_n = \frac{y_n - y_{n-1}}{y_n^* - y_{n-1}} \qquad 8.3\text{-}33$$

Also, for a simple case where the relative volatility is constant, the equilibrium relationship becomes

$$y_n^* = \frac{a x_n}{1 + (\alpha - 1) x_n} \qquad 8.3\text{-}34$$

Similar results are obtained for the stripping section below the feed tray. A total material balance leads to

$$\frac{dH_m}{dt} = L_{m+1} - L_m \qquad 8.3\text{-}35$$

A component balance gives the result

$$\frac{dH_m x_m}{dt} = L_{m+1} x_{m+1} - L_m x_m + V_{m-1} y_{m-1} - V_m y_m \qquad 8.3\text{-}36$$

The Murphree efficiency is

$$E_m = \frac{y_m - y_{m-1}}{y_m^* - y_{m-1}} \qquad 8.3\text{-}37$$

and the equilibrium relationship becomes

$$y_m^* = \frac{\alpha x_m}{1 + (\alpha - 1) x_m} \qquad 8.3\text{-}38$$

For the feed tray, the total material balance is

$$\frac{dH_F}{dt} = L_{F+1} - L_F + V_{F-1} - V_F + F \qquad 8.3\text{-}39$$

where H_F is the holdup on the feed tray, L_{F+1} is the liquid flow entering the feed tray (which is the same as the liquid rate leaving the rectifying section), L_F is the liquid flow rate leaving the feed tray (which is the same as the liquid flow rate entering the stripping section), F is the feed rate to the column, and V_{F-1} and V_F are the vapor flows entering and leaving the feed tray (which are the same as the vapor flow rate $V_m = V_{m-1}$ leaving the stripping section and the vapor flow rate $V_n = V_{n-1}$ in the enriching section). If q_F is the amount of heat required to vaporize one mole of feed divided by the molar latent heat of vaporization and if the vapor holdup is negligible, the relationship between the vapor flows in the rectifying and stripping sections becomes

$$V_F = V_{F-1} + (1 - q_F) F \qquad 8.3\text{-}40$$

A component balance for the feed tray is

$$\frac{dH_F x_F}{dt} = L_{F+1} x_{F+1} + L_F x_F + V_{F-1} y_{F-1} + V_F y_F + F z_F \qquad 8.3\text{-}41$$

where z_F represents the feed composition. The efficiency and equilibrium relationships on the feed tray are similar to those presented before.

Providing that we assume that the condenser is a perfectly mixed system, from a total material balance, we find that

$$\frac{dH_c}{dt} = V_m - L_c \qquad 8.3\text{-}42$$

where the liquid flow rate leaving the condenser, L_c, is the sum of the distillate flow D and the reflux flow L_{N+1}. Since the composition leaving the condenser is equal to the distillate composition and the composition of the reflux stream, and since these must be equal to the composition entering the total condenser

$$x_c = x_D = x_{N+1} = y_N \qquad 8.3\text{-}43$$

from a component balance for the condenser we obtain

$$\frac{dH_c x_{N+1}}{dt} = V_n y_n - L_c x_{N+1} \qquad 8.3\text{-}44$$

A total material balance for the reboiler, again assuming perfect mixing, gives the result

$$\frac{dH_R}{dt} = L_1 - (V_0 + B) \qquad 8.3\text{-}45$$

where $V_0 = V_m = V_{m-1}$ is the vapor flow entering the stripping section and B is the bottoms flow rate. From a component balance, we find that

$$\frac{dH_R x_0}{dt} = L_1 x_1 - B x_0 - V_0 y_0 \qquad 8.3\text{-}46$$

where x_0 is the composition of the exit stream at the bottom. We assume that the vapor and liquid leaving the reboiler are in equilibrium, so that the Murphree tray efficiency is 1.0. Also, the equilibrium relationship becomes

$$y_0 = \frac{\alpha x_0}{1 + (\alpha - 1) x_0} \qquad 8.3\text{-}47$$

The preceding set of equations, together with some functional relationship between tray holdup and liquid flow rate, provides a dynamic model for the column. At steady state conditions, all the time derivative terms become equal to zero, and we can solve the equations by using the McCable-Thiele method. Once the steady state design conditions have been established, and if we are interested only in small deviations from steady state conditions, we can linearize all the equations about steady state in the normal way, take the Laplace transforms of the linearized equations, and after a considerable amount of manipulation, finally determine a set of transfer functions relating the system input and outputs.

In order to simplify the discussion of the design of a feedforward control system for the column, where we attempt to adjust the reflux rate and boil-up rates in such a way as to maintain the overhead and bottom compositions constant despite disturbances in the feed rate and feed composition, we will consider a case where the Murphree tray efficiency is unity and where we write the linearized form of the expression relating tray holdup to liquid flow rate as

$$H_i = a_i L_i \qquad\qquad 8.3\text{-}48$$

Then the component balance for the condenser can be put into the form

$$s\tilde{x}_D = b_1 \tilde{x}_N - b_2 \tilde{x}_D \qquad\qquad 8.3\text{-}49$$

where $\tilde{x}_D$ and $\tilde{x}_N$ represent the Laplace transforms of the deviations, b_1 and b_2 are constants that depend on the original steady state design, and b_1 includes the slope of the equilibrium relationship in the vicinity of the top tray, so that the y_N term in Eq. 8.3-44 can be replaced by x_N. If our feedforward control system maintains the distillate composition constant, $\tilde{x}_D = 0$ and the preceding equation reduces to

$$\tilde{x}_N = 0 \qquad\qquad 8.3\text{-}50$$

The total material balance for the top tray leads to the expression

$$s\tilde{L}_N = b_3(\tilde{R} - \tilde{L}_N)$$

where $\tilde{R}$ is the deviation of the reflux rate, L_{N+1}, from its steady state value. Similarly, the component balance for the top tray becomes

$$s\tilde{x}_N = b_4 \tilde{R} - b_5 \tilde{V}_N + b_6 \tilde{x}_D - b_7 \tilde{x}_N + b_8 \tilde{x}_{N-1} \qquad\qquad 8.3\text{-}51$$

where, again, we have used the appropriate linearization of the equilibrium relationship to eliminate the vapor compositions. Since $\tilde{x}_D = \tilde{x}_N = 0$, we find that

$$\tilde{L}_N = \frac{b_3 R}{s + b_3} \qquad\qquad 8.3\text{-}52$$

$$\tilde{x}_{N-1} = \frac{1}{b_8}(-b_4 \tilde{R} + b_5 \tilde{V}_N) \qquad\qquad 8.3\text{-}53$$

For the next-to-the-top tray, the system equations become

$$s\tilde{L}_{N-1} = b_9 \tilde{L}_N + b_{10} \tilde{L}_{N-1} \qquad\qquad 8.3\text{-}54$$

$$s\tilde{x}_{N-1} = b_{11} \tilde{L}_N - b_{12} \tilde{V}_{N-1} + b_{13} \tilde{x}_N - b_{14} \tilde{x}_{N-1} + b_{15} \tilde{x}_{N-2} \qquad\qquad 8.3\text{-}55$$

or after substituting the previous results and noting that $\tilde{V}_N = \tilde{V}_{N-1} = \tilde{V}_n$,

$$\tilde{L}_{N-1} = \frac{b_9 b_3 \tilde{R}}{(s + b_{10})(s + b_3)} \qquad\qquad 8.3\text{-}56$$

$$\tilde{x}_{N-2} = \frac{1}{b_{15}}\left\{ -\left[\frac{b_{11} b_3}{s + b_3} + \frac{b_4(s + b_{14})}{b_8}\right]\tilde{R} + \left[\frac{b_{12} + b_5(s + b_{14})}{b_8}\right]\tilde{V}_n \right\}$$

$$8.3\text{-}57$$

Now we can continue this procedure tray by tray down the column and eventually obtain expressions for $\tilde{L}_{F+1}$ and $\tilde{x}_{F+1}$. If we then substitute $j\omega$ for the Laplace parameter s and rationalize the complex numbers, the expressions will be

$$\tilde{L}_{F+1} = (\alpha_1 + j\alpha_2)\tilde{R} \qquad\qquad 8.3\text{--}58$$

$$\tilde{x}_{F+1} = (\alpha_3 + j\alpha_4)\tilde{R} + (\alpha_5 + j\alpha_6)\tilde{V}_n \qquad\qquad 8.3\text{--}59$$

$$\tilde{x}_F = (\alpha_7 + j\alpha_8)\tilde{R} + (\alpha_9 + j\alpha_{10})\tilde{V}_n \qquad\qquad 8.3\text{--}60$$

where $\alpha_1, \alpha_2, \cdots, \alpha_{10}$ depend on frequency as well as the constants $b_1, b_2, \cdots$.

In a similar manner, we can start at the bottom of the column, set $\tilde{x}_w = 0$, and move up the column tray by tray. This procedure leads to the expressions

$$\tilde{L}_F = (\alpha_{11} + j\alpha_{12})\tilde{L}_{F-1} \qquad\qquad 8.3\text{--}61$$

$$\tilde{x}_F = (\alpha_{13} + j\alpha_{14})\tilde{V}_m + (\alpha_{15} + j\alpha_{16})\tilde{L}_F \qquad\qquad 8.3\text{--}62$$

$$\tilde{x}_{F-1} = (\alpha_{17} + j\alpha_{18})\tilde{V}_m + (\alpha_{19} + j\alpha_{20})\tilde{L}_{F-1} \qquad\qquad 8.3\text{--}63$$

Also, the equations describing the dynamics of the feed tray can be written

$$\tilde{L}_F = (\alpha_{21} + j\alpha_{22})\tilde{L}_{F+1} + (\alpha_{23} + j\alpha_{24})\tilde{V}_m + (\alpha_{25} + j\alpha_{26})\tilde{V}_n + (\alpha_{27} + j\alpha_{28})\tilde{F}$$
$$8.3\text{--}64$$

$$\tilde{V}_n = \tilde{V}_m + \alpha_{29}\tilde{F} \qquad\qquad 8.3\text{--}65$$

$$\tilde{x}_F = (\alpha_{30} + j\alpha_{31})\tilde{L}_{F+1} + (\alpha_{32} + j\alpha_{33})\tilde{V}_m + (\alpha_{34} + j\alpha_{35})\tilde{V}_n + (\alpha_{36} + j\alpha_{37})\tilde{F}$$
$$+ (\alpha_{38} + j\alpha_{39})\tilde{x}_{F+1} + (\alpha_{40} + j\alpha_{41})\tilde{x}_{F-1} + (\alpha_{42} + j\alpha_{43})\tilde{z}_F \qquad 8.3\text{--}66$$

We can use the these nine equations to solve for the nine unknowns, $\tilde{L}_{F+1}$, $\tilde{R}$, $\tilde{x}_{F+1}$, $\tilde{V}_n$, $\tilde{x}_F$, $\tilde{L}_F$, $\tilde{L}_{F-1}$, $\tilde{V}_m$, and $\tilde{x}_{F-1}$, in terms of the measured inputs $\tilde{F}$ and $\tilde{z}_f$. Of course, all we are actually interested in is the solutions for the reflux rate $\tilde{R}$ and the boil-up rate $\tilde{V}_m$, for these are our control variables. Thus we will obtain expressions like

$$\tilde{R} = (\beta_1 + j\beta_2)\tilde{F} + (\beta_3 + j\beta_4)\tilde{z}_f \qquad\qquad 8.3\text{--}67$$

$$\tilde{V}_m = (\beta_5 + j\beta_6)\tilde{F} + (\beta_7 + j\beta_8)\tilde{z}_f \qquad\qquad 8.3\text{--}68$$

for our feedforward control system.

Luyben and Gerster[2] applied this approach in a numerical study of the separation of a binary mixture, relative volatility $\alpha = 1.4$, in a 40-tray column. It was assumed that there was equimolar overflow within the column, saturated liquid feed and reflux streams, 100 percent tray efficiencies, perfect mixing of liquid on each tray, a reflux ratio of $L/D = 4.0$, a steady state feed rate of 100 moles/min, a steady distillate rate of 50 moles/min, a reboiler holdup of 100 moles, a condenser holdup of 100 moles, a holdup of 30 moles on the stripping trays, and a holdup of 25 moles on the rectifying trays. The flow rate would approximately correspond to that employed in a 15-foot-diameter

2 W. L. Luyben, and J. A. Gerster, *Ind. Eng. Chem. Proc. Design Develop.*, **3**, 374 (1964)

column operating at 80 percent of the flooding velocity for a case where the liquid density was 50 lb/cu ft, the vapor density was 1.1 lb/cu ft, and the holdup on each tray was 3 in. of clear liquid. A digital computer was used to evaluate the effect of frequency on the constants β_1 through β_8 and the amplitude ratios in the feedforward transfer functions given by Eqs. 8.3–67 and 8.3–68. These results were used to prepare a set of Bode plots (see Figure 8.3–2). Since these transfer functions are so complicated, the gain curves were approximated by simple first-order lags, a very poor approximation for situations where there are large resonant peaks, and by combinations of first- and second-order leads and lags. The approximate representations are given in Table 8.3–1. The effectiveness of the two kinds of approximate, dynamic feedforward control systems, as well as a steady state feedforward controller, was evaluated by simulating the equations describing the column-condenser-reboiler system and the various feedforward controllers on an analog computer and then introducing step and pulse changes in F and z_F. The computer results for the distillate and bottoms compositions, x_D and x_w, are shown in Figure 8.3–3, and the control variable changes for the step input case are given in Figure 8.3–4. The graphs indicate that the approximations of the feedforward control transfer functions

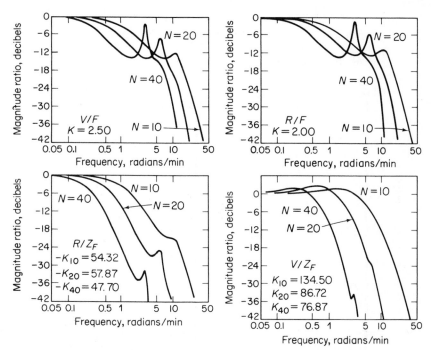

Figure 8.3-2. Bode plots of feedforward control transfer functions. [Reproduced from W. L. Luyben and J. A. Gerster, *Ind. Eng. Chem. Proc. Design Develop.*, **3**, 374 (1964), by permission of the ACS.]

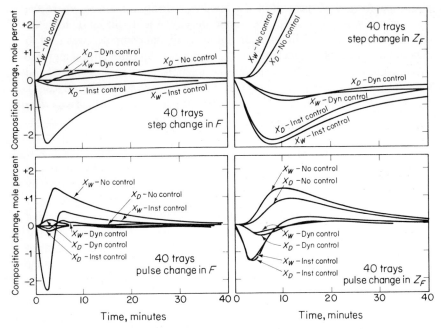

Figure 8.3-3. Column response. [Reproduced from W. L. Luyben and J. A. Gerster, *Ind. Eng. Chem. Proc. Design Develop.*, **3**, 374 (1964), by permission of the ACS.]

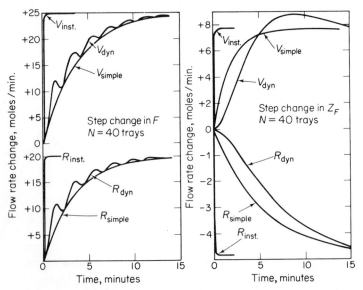

Figure 8.3-4. Control variable changes for step input. [Reproduced from W. L. Luyben and J. A. Gerster, *Ind. Eng. Chem. Proc. Design Develop.*, **3**, 374 (1964), by permission of the ACS.]

are adequate to provide good control. Additional details concerning the effects of changing some of the system parameters can be found in the original paper. Also, they report some results of an experimental study of the feedforward control of a pilot-plant size column.

Of course, if a large analog or digital computer is available, the analytical procedure for determining the feedforward control transfer functions can be avoided. Instead we can program the system equations, solve them numerically

TABLE 8.3-1

APPROXIMATE FEEDFORWARD CONTROL TRANSFER FUNCTION*

$$\frac{\tilde{R}}{\tilde{F}} = \frac{2.0(0.5s + 1)}{(3.333s + 1)[0.3333^2 s^2 + 0.2(0.3333)s + 1]}$$

$$\frac{\tilde{R}}{\tilde{z}_f} = \frac{-47.7}{3.448^2 s^2 + 0.2(3.448)s + 1}$$

$$\frac{\tilde{V}}{\tilde{F}} = \frac{2.50(0.5s + 1)}{(3.704s + 1)[0.3125^2 s^2 + 0.2(0.3125)s + 1]}$$

$$\frac{\tilde{V}}{\tilde{z}_f} = \frac{76.87(5.0s + 1)}{(2.857s + 1)[2.439^2 s^2 + 1.6(2.439)s + 1]}$$

*Reproduced from W. L. Luyben and J. A. Gerster, *Ind. Eng. Chem. Proc. Design Develop.*, 3, 374 (1964), by permissin of the ACS.

TABLE 8.3-2

PARAMETERS FOR CSTR*

$q_s = 2000$ lb/hr, $A_{fs} = 0.5$ lb/lb, $T_{fs} = 530°$R, $C_p = 0.75$ Btu/(lb)(°R)
$V = 2400$ lb, $(-\Delta H) = 600$ Btu/lb, $U = 150$ Btu/(hr)(ft²)(°R), $T_{cs} = 586.7°$R
$E = 30,000$ Btu/lb more, $k_0 = 7.08 \times 10^{10}$ hr^{-1}, $A_H = 100$ ft², $A_s = 0.246$ lb/lb
$T_s = 600°$R, $k_s = 0.86$ hr^{-1}

*Reproduced from W. L. Luyben AICHE Journal, 14, 37 (1968), by permission

for sinusoidal inputs in the disturbance and control variables, and prepare a set of Bode plots relating each of the inputs to the two outputs of interest. By approximating these graphs by simpler transfer functions, using the same procedure as Luyben and Gerster, we can find the feedforward control transfer functions by solving the equations below for $\tilde{R}$ and $\tilde{V}_m$ in terms of $\tilde{F}$ and $\tilde{z}_F$

$$\tilde{x}_D = 0 = H_{11}\tilde{R} + H_{12}\tilde{V}_m + G_{11}\tilde{F} + G_{12}\tilde{z}_F \qquad 8.3\text{-}69$$

$$\tilde{x}_w = 0 = H_{21}\tilde{R} + H_{22}\tilde{V}_m + G_{21}\tilde{F} + G_{22}\tilde{z}_F \qquad 8.3\text{-}70$$

This approach can be applied to any kind of a chemical process.

Feedforward Control of Nonlinear Systems

The simplest procedure for developing a feedforward control system for a nonlinear plant is simply to linearize the dynamic equations and apply the analysis described above. This method will not lead to a perfect control system, for the linearized dynamic equations are only approximations. Hence the outputs can be expected to fluctuate somewhat (e.g., see Figure 8.3–3), but the method is useful because in many cases it will provide compensation for most of the effect of the disturbances on the plant outputs. However, in certain critical situations this might not prove adequate, and the addition of a feedback controller to the feedforward controller based on the linearized system equations still might not give a satisfactory performance. Uder these special circumstances, or if otherwise desirable, we might want to design a nonlinear feedforward controller. The basic idea for this type of system is exactly the same as before, although the algebraic difficulties encountered in the determination of the feedforward control policy become much more profound.

As a simple illustration, we consider the nonlinear CSTR equations

$$\frac{dA}{dt} = \frac{q}{V}(A_f - A) - kA \qquad\qquad 8.3\text{–}7$$

$$\frac{dT}{dt} = \frac{q}{V}(T_f - T) + \frac{UA_H Kq_H}{VC_p\rho(1 + Kq_H)}(T_H - T) + \frac{(-\Delta H)}{C_p\rho}kA \qquad 8.3\text{–}8$$

If we want to design a feedforward control system to maintain the reactor composition and temperature at their steady state design values by manipulating the feed rate and heating flow rate to compensate for measurable fluctuations in the feed composition, feed temperature, and the inlet temperature of heating fluid, we set

$$A = A_s \quad \text{and} \quad T = T_s \qquad\qquad 8.3\text{–}71$$

in the system equations. This implies that

$$\frac{dA}{dt} = \frac{dT}{dt} = 0 \qquad\qquad 8.3\text{–}72$$

so that if we solve the equations for q and q_H, we find that the feedforward control policy is

$$q = \frac{k_s V A_s}{A_f - A_s} = \frac{c_1}{A_f - c_2} \qquad\qquad 8.3\text{–}73$$

$$q_H = \frac{[VC_p\rho/UA_H K(T_H - T_s)][(T_s - T_f)/(A_f - A_s) - (-\Delta H)/C_p\rho]k_s A_s}{1 - [VC_p\rho/UA_H(T_H - T_s)][(T_s - T_f)/(A_f - A_s) - (-\Delta H/C_p\rho]k_s A_s)}$$

$$= \frac{c_3 - c_4 T_f - c_5 A_f}{1 + c_6 T_f + c_7 T_H + c_8 A_f + c_9 T_H A_f} \qquad\qquad 8.3\text{–}74$$

where A_f, T_f, and T_H are measured functions of time and c_1 through c_9 are known constants. We could also call this a heat and a material balance controller. No dynamic elements are required, which is similar to the corresponding case for the linearized equations (see Example 8.3–1). The nonlinearities in the control policy are simple enough that the control system would be inexpensive to build.

The results are more complicated if we are allowed to manipulate only a single control variable, say q_H. For this case, we can only maintain one of the outputs constant; and if we require that $A = A_s$, so that $dA/dt = 0$, we can use Eq. 8.3–7 to find a relationship between the fluctuating feed composition and reactor temperature

$$T = \frac{-E/R}{\ln\left[(q_s/k_0 V)(A_f - A_s/A_s)\right]} \qquad 8.3\text{–}75$$

Now we can use this result and its derivative to eliminate T from Eq. 8.3–8 and thereby obtain a relationship between q_H and the disturbances A_f, T_f, and T_H. Clearly, this result will be much more complicated than the multivariable controller and will require a knowledge of the derivative of A_f, as well as the

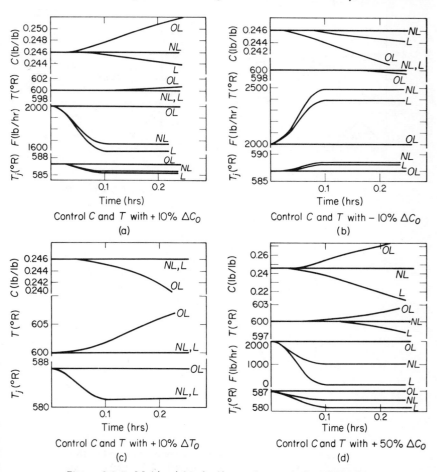

Figure 8.3-5. Multivariable feedforward control of a CSTR. [Reproduced from W. L. Luyben, *AIChE Journal*, **14**, 37 (1968), by permission.]

values of A_f, T_f, and T_H. This is analogous to our result for the linearized
system (see Eq. 8.3–19). If we attempt to apply this technique to the catalytic
cracking unit or distillation column problems, generally we expect to obtain
feedforword control policies that are too complicated to be of value.

Example 8.3–4 Nonlinear feedforward control of a CSTR

Luyben[3] studied the feedforward control of a CSTR for a case where there
was a cooling coil in the reactor. He considered the system parameters given in
Table 8.3–2 and assumed that the feed rate and the coolant inlet temperature

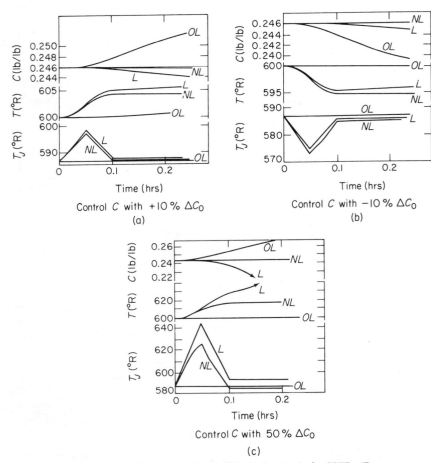

Figure 8.3-6. Single variable feedforward control of a CSTR. [Re-
produced from W. L. Luyben, *AIChE Journal*, **14**, 37 (1968), by permission.]

[3] W. L. Luyben, *AIChE Journal*, **14**, 37 (1968)

were the control variables. For this case, the energy equation becomes

$$\frac{dT}{dt} = \frac{q}{V}(T_f - T) - \frac{U_a}{VC_p\rho}(T - T_c) + \frac{(-\Delta H)}{C_p\rho}kA \qquad 8.3\text{-}76$$

and the feedforward controller is Eq. 8.3–73 and

$$T_c = T_s - \frac{k_s VA_s}{U_a}\left[(-\Delta H) - \frac{C_p\rho(T_s - T_f)}{A_f - A_s}\right] \qquad 8.3\text{-}77$$

His results of a numerical simulation of the uncontrolled plant (OL), the plant with a nonlinear (NL), and a linear (L), multivariable feedforward control system for an S-shaped disturbance lasting 0.1 hour are shown in Figure 8.3–5. It is apparent from the graphs that the nonlinear controller always gives perfect control and that the discrepancy between the linearized and nonlinear systems increases as the magnitude of the disturbance increases. Luyben also considered the use of just the inlet coolant temperature to control the reactor composition. The control policy for this case becomes

$$T_c = \left[1 + \frac{q_s C_p\rho}{U_a}\right]\left\{\frac{-E/R}{\ln\left[q_s(A_f - A_s)/k_0 VA_s\right]}\right\} - \left(\frac{q_s C_p\rho T_f}{U_a}\right)$$

$$- \frac{(-\Delta H)q_s(A_f - A_s)}{U_a} + \left[\frac{q_s VC_p\rho R}{U_a E(A_f - A_s)}\right]$$

$$\times \left[\frac{-E/R}{\ln\left[q_s(A_f - A_s)/k_0 VA_s\right]}\right]^2 \frac{dA_f}{dt} \qquad 8.3\text{-}78$$

The results are shown in Figure 8.3–6. Again the nonlinear controller is superior. It is interesting to note that for the large disturbance, the linear controller arrives at a new steady state with T_c changed in the wrong direction. The discontinuity in the slope of T_c is caused by the S-shaped input disturbance.

Distributed Parameter Systems

Since the linearized dynamic equations describing a distributed parameter process can often be solved for the matrix of transfer functions relating the system inputs and outputs—that is,

$$\tilde{\mathbf{x}} = \mathbf{M}\tilde{\mathbf{u}} + \mathbf{N}\tilde{\mathbf{v}} \qquad 8.3\text{-}79$$

the design of a feedforward control system for a distributed parameter plant can be obtained using the same procedure as described earlier. Normally it will not be possible to extend the analytical method to the design of nonlinear feedforward controllers, for when the time derivatives are set equal to zero, we are still left with one or more ordinary, nonlinear differential equations that would have to be solved numerically for every possible value, or combination of values, of the disturbances. Also, difficulties are encountered because distributed parameter systems often exhibit a dead-time response; hence the feedforward controller will not be realizable. However, there is a great incentive to synthesize feedforward control systems for distributed parameter plants

because feedback controllers, which do not become activated until the disturbance has been in the plant for some time, give an unsatisfactory response. A
considerable amount of research needs to be done on this topic.

Limitations of Feedforward Control

Even though the use of feedforward control systems seems to make the
design of feedback controllers obsolete at first glance, it must be remembered
that perfect control will be obtained only if

1. There is a control variable for each output we desire to maintain constant.
2. All disturbances are measurable.
3. All the system dynamics are known exactly.
4. The control policy is realizable—that is, the controller cannot contain
   an s^2 or $e^{t_a s}$ term.

Perhaps a more dramatic way of remembering these limitations is to recognize that conceptually it is possible to attach enough measuring devices and
feedforward controllers to enable a blind man to drive from one city to another,
but it would not be wise for you to sell him an insurance policy. Thus a feedback control system should always be used to supplement the feedforward
controller. However, the great advantages of using feedforward control to compensate for most of the effects of disturbances have been overlooked for a great
number of years, and many people have tended to rely only on feedback control. This is like driving a car and focusing on the white line only 5 yards in front
of the car instead of looking ahead to anticipate disturbances—that is, defensive
driving.

SECTION 8.4   CONTROL OF A COMPLETE CHEMICAL PLANT

Our study of control-system synthesis techniques has been limited to the
control of a single process unit. Of course, we know that in any practical situation a large number of these units are connected together. Also, in most cases,
unreacted material and unused energy are recycled back to a point near where
they enter the plant, so that the equipment arrangement appears to be very
complex; that is, the outputs from some units become the inputs to others,
located both upstream and downstream from the particular unit under consideration. We could attempt to overcome this difficulty by controlling every unit
in the total plant configuration and then hoping that despite the fact we can
never achieve perfect control, we will still be able to achieve a satisfactory
performance. However, intuitively we expect that there should be a better
approach. Unfortunately this area of control theory has received little attention
in the literature. Some of the material that appears to be applicable to the
design of control systems for large plants is briefly reviewed below.

The Effect of Inventory and Storage on Control System Design

We still have not considered dynamic interactions between the units in a plant or the techniques that can be used to isolate various pieces of equipment. In addition, we have overlooked certain features of the classical design problem. For example, most plants contain storage facilities at one or more points in the major process stream, as well as for the raw materials and products, so that if some piece of equipment fails, there will still be a certain amount of material available that can be processed in the operable section of the plant. With this procedure, the breakdown of a relatively minor piece of equipment, such as a pump, will not necessitate the shutdown of a whole petroleum refinery. Of course, the storage tanks introduce additional capacity of material, and possibly energy, into the plant configuration, and this will affect the dynamic response of the total system. The size, number, and location of the storage facilities should be selected to achieve the "best" operability of the plant, which means that we must be able to achieve a satisfactory performance from both a steady state and a dynamic point of view.

Another illustration of a situation where dynamic considerations are used to help fix the size of a piece of equipment is a flash drum. Generally it is assumed that the feed entering the unit instantaneously separates into a vapor-liquid mixture at equilibrium. If this assumption is valid, then, conceptually at least, the separation can be accomplished in an infinitesimally small volume with an infinitesimally small capital cost. A similar result is obtained for all equilibrium stages, and, according to a conventional steady state analysis, only rate-limited processes need to be of a finite size. However, in actual practice, the diameter of the flash drum is usually specified so that the vapor velocity will be less than the terminal velocity of a liquid droplet with a 50-micron-particle diameter. Experience has shown that this "rule-of-thumb" leads to an acceptable entrainment loss. The height of the drum is picked to allow 3 to 4 ft for demister equipment and a sufficient surge capacity so that it takes 3 to 5 min for the liquid to flow between the level taps of a level controller. In other words, a surge capacity is included in the design specification.

The deliberate hold-up of material within a plant needs to be considered, along with the design of control systems for the plant, in order to achieve a satisfactory dynamic performance. Until recently[1] there was little attempt to treat the relationship between these factors in a quantitative way. Buckley noted that the early control strategy of using individual control loops to force variables to their steady state operating levels as fast as possible usually failed because of the interactions between adjacent loops. These interactions often caused plants to oscillate or to become unstable, and even when the controllers were detuned enough to prevent the oscillations, it was still a difficult job to operate the plant.

[1] P. S. Buckley, *Techniques of Process Control*, Ch. 13, Wiley, N.Y., 1964.

In order to avoid this difficulty with interaction, many companies adopted the idea of attempting to maintain every value of flow rate, temperature, pressure, etc., constant, so that the product quality would always remain constant. Buckley calls this approach *set-point environmental control*, and it should be recognized that most plants are still designed on this basis. Of course, it is not actually possible to manipulate the flow rate to or from a unit in order to compensate for some disturbance and at the same time maintain the flow rate constant. Also, with perfect set-point environmental control, the holdup in each unit and storage vessel would always remain constant; whereas, in practice, it will make a difference whether we use the inlet or outlet flow rate as a control variable for a unit.

Buckley has proposed a new approach to control system design, which he calls *dynamic process control*, and which does not depend on the inconsistency just described. By considering the purpose of control systems from a plant manager's standpoint, it becomes clear that he must make it possible to produce material with the desired product quality specifications (composition, viscosity, color) at the desired production rate. The product quality requirement means that some of the controllers must act as regulators to prevent disturbances from affecting the quality. However, the production rate requirement can be satisfied by a completely different strategy. We first recognize that we only need to make the time-average production rate equal to the desired value, providing that we can draw some product from storage facilities on a temporary basis. This step can be accomplished if we add another requirement that we always maintain the inventories in the storage facilities between some specified maximum and minimum limits. Also, as a third requirement, we must be willing to change the flow rates to or from any piece of equipment sufficiently slowly that we do not cause a drastic upset in some other unit. A control system satisfying these three requirements for production rate is called a *material balance control*. This is an appropriate term because no effort is directed toward maintaining the levels in certain units constant (set-point control), but, instead, we attempt to make the inflows equal to the outflows in order to eliminate the accumulation and achieve a steady state material balance.

Virtually all our previous discussion of control system synthesis was concerned with product quality controllers. Our results indicated that these regulators acted as high-pass filters for disturbances; that is, the gain of the transfer function was very small at both low and high frequencies (see Figures 7.5–27 and 8.4–1). As we shall show below, the level control systems commonly used for material balance control are low-pass filters—the gain is large at low frequencies and very small at high frequencies (see Figure 8.4–1). Buckley realized that this difference in behavior could be used to develop a new control strategy for chemical plants. Thus by selecting the break frequencies of the material balance controllers to be an order of magnitude lower than the resonant frequency of the product quality controllers, the two kinds of control systems

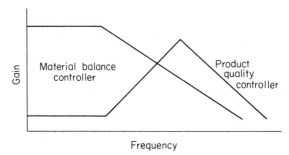

Frequency

Figure 8.4-1. Gain plot for controllers.

will not interact. In other words, by interspersing storage units and level control systems in the appropriate way between units where product quality controllers are located, we can effectively eliminate all the dynamic interaction between units and provide essentially complete compensation for disturbances.

Buckley also pointed out that the overall operability of a plant is strongly dependent on the way in which the intermediate material balance controllers function. At present, plants are designed so that the material balance control is in the direction of flow for most of the units. However, his analysis indicated that by reversing this normal practice and controlling the intermediate storages in the direction opposite to the flow, a significant improvement in performance can be obtained. The following examples illustrate the design of material balance controllers for single units and the effect of arranging intermediate material balance controllers in the same and the opposite direction to the flow.

Example 8.4–1 Material balance control for single units

Develop all the possible closed-loop transfer functions for the two level control systems presented in Figure 8.4–2. Assume proportional control in both cases. Then show that these level controllers satisfy the requirements discussed above for a material balance controller.

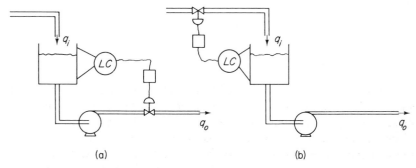

(a)   (b)

Figure 8.4-2. Level control systems: (a) control in direction of flow; (b) control in direction opposite to flow.

Solution

A material balance on the tank gives the expression

$$\frac{d\rho V}{dt} = \rho q_i - \rho q_0 \qquad \text{8.4-1}$$

where V = volume of liquid in the tank, ρ = density, and q = flow rate. At steady state conditions, $q_i = q_0$; thus Eq. 8.4-1 can also be used to represent the deviations of the flows and volume from their steady state values. Using this interpretation and noting that the fluid density usually is a constant, we obtain

$$\frac{dV}{dt} = q_i - q_0 \qquad \text{8.4-2}$$

For a level control system, we change one of the flow rates, either the inlet or outlet, proportional to the deviation of the level (or volume) in the tank from its desired value. Thus for case a, control in the direction of flow, we can write the Laplace transform of the system equation as

$$s\tilde{V} = \tilde{q}_i - \tilde{q}_0 \qquad \text{8.4-3}$$

and the transform of the control law as

$$\tilde{q}_0 = k(\tilde{V} - \tilde{V}_{\text{sp}}) \qquad \text{8.4-4}$$

where

$$k = \frac{(q_0)_{\max}}{V_T} \qquad \text{8.4-5}$$

and $(q_0)_{\max}$ = the maximum possible outflow, V_T = the difference between the maximum and minimum storages, and $\tilde{V}_{\text{sp}}$ = the level (or volume) set point. After some simple algebraic manipulation of Eqs. 8.4-3 and 8.4-4, we obtain the three closed-loop transfer functions.

Case a

$$\frac{\tilde{V}}{\tilde{q}_i} = \frac{1}{s + k} \qquad \text{8.4-6}$$

$$\frac{\tilde{V}}{\tilde{V}_{\text{sp}}} = \frac{k}{s + k} \qquad \text{8.4-7}$$

$$\frac{\tilde{q}_0}{\tilde{q}_i} = \frac{k}{s + k} \qquad \text{8.4-8}$$

Thus we see that all these closed-loop tranfer functions are first-order lags and that level control systems will act as low-pass filters. Furthermore, if we make $V_T/(q_0)_{\max}$ sufficiently large, so that k is small, we can satisfy all three of our requirements for a material balance controller: the inflow will be equal to the outflow on a long-term basis, the inventory will be maintained between its maximum and minimum limits, and the manipulated flow will change only gradually. The first and last of these conditions are true control functions, whereas the inventory specification is actually a constraint.

It should also be noted that the transfer functions for the closed-loop system are very different from the previous results we obtained for regulators. In fact, the design procedures for regulating controllers are not applicable, and the closed-loop system can never be unstable.

The system equation for case b, control in the direction opposite to flow, is identical to Eq. 8.4-3 but now the control law becomes

$$\tilde{q}_i = -k(\tilde{V} - \tilde{V}_{sp}) \qquad\qquad 8.4-9$$

Then the system transfer functions are

$$\frac{\tilde{V}}{\tilde{q}_0} = \frac{-1}{s+k} \qquad\qquad 8.4-10$$

$$\frac{\tilde{V}}{\tilde{V}_{sp}} = \frac{k}{s+k} \qquad\qquad 8.4-11$$

$$\frac{\tilde{q}_i}{\tilde{q}_0} = \frac{k}{s+k} \qquad\qquad 8.4-12$$

which are similar to those obtained previously.

Example 8.4-2 Overall material balance control systems

Consider the two overall material balance control systems shown in Figure 8.4-3 and determine whether it is better to control the intermediate storages in the direction of flow (case a) or the direction opposite to the flow (case b).

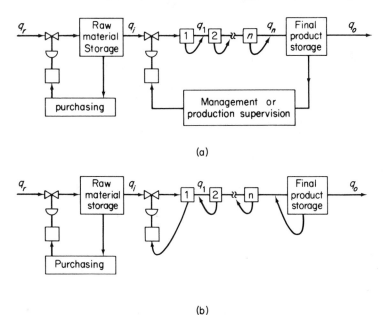

(a)

(b)

Figure 8.4-3. Overall control systems with (a) intermediate material balance control in the direction of flow; (b) intermediate material balance in the direction opposite to flow.

Solution

First we must recognize that both the raw material storage and the product must be controlled in a direction opposite to the flow. Thus for case a, where the intermediate storages are controlled in the direction of flow, we must have a feedback loop from the product storage back to the first stage of the plant. Assuming that all the intermediate storage units are the same size and letting q_R = raw material flow rate, q_n = flow from the nth stage, q_i = flow to the first stage, q_0 = product flow, V_n = volume contained in the nth stage, V_R = volume in raw material storage, V_p = volume in product storage, $\tau_n = V_n/q_n$, $\tau_R = V_R/q_R$, and $k_p = (q_0)_{\max}/V_p$, we can use our previous results to write the following set of equations for case a:

Case a

Raw material storage, from Eq. 8.4–12

$$\tilde{q}_R = \frac{1}{\tau_R s + 1} \tilde{q}_i \qquad\qquad 8.4\text{--}13$$

n intermediate stages in series with control in the direction of flow, from Eq. 8.4–8

$$\tilde{q}_n = \left(\frac{1}{\tau_n s + 1}\right)^n \tilde{q}_i \qquad\qquad 8.4\text{--}14$$

Accumulation in product storage

$$s\tilde{V}_n = \tilde{q}_n - \tilde{q}_0 \qquad\qquad 8.4\text{--}15$$

Control law for product storage

$$\tilde{q}_i = k_p \tilde{V}_n \qquad\qquad 8.4\text{--}16$$

Combining these equations, we obtain the overall transfer function

$$\frac{\tilde{q}_R}{\tilde{q}_0} = \frac{(\tau_R s + 1)(\tau_n s + 1)^n}{[(s/k_p)(\tau_n s + 1)^n + 1]} \qquad\qquad 8.4\text{--}17$$

The denominator of this transfer function is of high order; therefore we will have to be careful of our choice of k_p in order to avoid stability problems. Since we know that $k_p = (q_0)_{\max}/V_p$, and since $(q_0)_{\max}$ is fixed by the maximum production capacity of the plant, the limit of stable operation for the plant depends on the size of the product storage tank. Using the stability techniques discussed earlier, Buckley showed that at the stability limit we obtain the following values relating the number of stages to the ratio of the product to the intermediate storage holdup:

n	1	2	3	6
V_p/V_n	4	7	12	27

Even though this analysis is greatly oversimplified, it does indicate that we must have a large product storage facility if the intermediate storage is controlled in the direction of flow.

Now if we consider control of the intermediate storages in the direction opposite to the flow, case b, the equation describing each unit is similar to Eq. 8.4-12. Thus we obtain the set of expressions

Case b
Raw material storage

$$\tilde{q}_R = \frac{1}{\tau_R s + 1} \tilde{q}_i \qquad\qquad 8.4\text{-}18$$

n intermediate stages

$$\tilde{q}_i = \left(\frac{1}{\tau_n s + 1}\right)^n \tilde{q}_n \qquad\qquad 8.4\text{-}19$$

Product storage

$$\tilde{q}_n = \frac{1}{(s/k_p) + 1} \tilde{q}_0 \qquad\qquad 8.4\text{-}20$$

After eliminating the intermediate variables, we find that the overall transfer function is

$$\frac{\tilde{q}_R}{\tilde{q}_0} = \frac{1}{(\tau_R s + 1)(\tau_n s + 1)^n [(s/k_p) + 1]} \qquad\qquad 8.4\text{-}21$$

For this configuration, we can choose any value we want for k_p or V_p without ever encountering a stability limit. It is clear, then, that this is the most desirable control scheme.

Decomposition Techniques for Large Systems

Even though an appropriately designed set of alternating product quality and material balance controllers can be used to eliminate certain kinds of disturbances entering a plant, it is doubtful that they will completely damp out all fluctuations. In particular, there is no reason to expect that energy will not accumulate at different points in the system. Hence it is still quite likely that there will be some dynamic interactions between different units and between control systems.

Unfortunately, at present, there is no analytical technique that can be used to account for all these potential interactions. In fact, the problem of calculating the steady state behavior of a complicated plant is currently the topic of several research studies,[2] and we know from Chapters 2 and 3 that this is a much easier problem than determining the dynamic behavior. Some insight into the nature of the difficulties encountered in the steady state analysis can be gained by considering the schematic diagram for a plant that produces benzene hexachloride[3] (see Figure 8.4-4).

[2] See D. F. Rudd, and C. C. Watson, *Strategy of Process Engineering*, Ch. 12, Wiley, N.Y., 1968 and Ch. 8 of the reference given in footnote 6 in Section 7.4.

[3] F. G. Vilbrandt, and C. E. Dryden, *Chemical Engineering Plant Design* (4th ed.), p. 47, McGraw-Hill, N.Y., 1959.

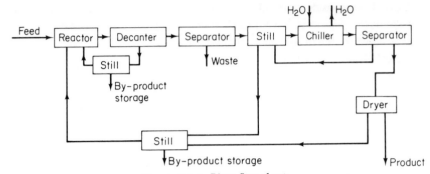

Figure 8.4-4. Plant flow sheet.

Now if we assume that we have one or more equations describing each unit, so that once the inputs to a unit have been specified we can compute the output from that unit, we would hope that specifying all of the inputs to the plant would allow us to compute all the outputs and all the intermediate values. However, the presence of the recycle streams means that the caclulation procedure is far from straightforward. In other words, even if we know the value of the feed stream to the reactor, we still cannot calculate the reactor output until we also specify the outputs from the two stills that are inputs to the reactor. Of course, the inputs to these stills depend on the reactor output, as modified by a number of intervening units. For a case where all the input-output relationships for each unit were linear, or certain kinds of nonlinear equations, we could algebraically manipulate the equations to develop specific input-output relationships of interest. Again our results from Chapters 2 and 3 indicate that the system equations will generally be nonlinear and often will be ordinary or partial differential equations; thus the algebraic elimination procedures will probably fail. Consequently, it will be necessary to use trial-and-error numerical methods to find solutions to the problem.

The most efficient way of ordering the calculations for the trial-and-error procedure is to assume values for the inputs to the decanter and chiller. With this information, it is possible to use the equations for each piece of equipment to compute the values of every other stream in a straightforward manner. Then if the computed values of the decanter and chiller inputs do not match the assumed values, we must make new assumptions and repeat the procedure. Starting at most other places in the flow sheet will not be as efficient, for it will be necessary to assume the values of more than two streams.

An appropriate ordering of the calculations can be determined by inspection for this simple plant. More complicated processes require a more sophisticated approach, but recent research results indicate that graph theory and other techniques make it possible to find solutions in a systematic way. However, if we attempt to extend the analysis to the dynamic operation of the plant, we encounter much greater difficulties. For this case, the inputs to each unit normally are functions of time rather than particular values of variables. Therefore, for the present, we must treat all the system equations simultaneously, although we are hopeful that research studies will lead to improved techniques.

Simulation

One procedure for studying the dynamics and control of a large plant is simply to linearize all the system equations and to consider them all together as a single system of very high order. In many situations it will be possible to neglect oscillations passing around recycle loops, because the damping provided by each element in the loop will be sufficient to reduce the amplitude of the fluctuations to zero. When these simplifications are not valid, it is necessary to solve the complete set of nonlinear equations, or some approximation of this set of equations, numerically. We want to be able to determine how the addition of various kinds of control systems will affect the dynamic performance of the plant; therefore an analog computer is often used for the study. Some examples of this kind of simulation study are described below.

Example 8-4.3   Simulation of a nitric acid plant

DuPont of Canada Ltd., as well as a number of other companies, manufactures nitric acid by the oxidation of ammonia. The reaction is carried out in the vapor phase under pressure. A schematic diagram of the plant is shown in Figure 8.4-5. From this diagram we see that ammonia is stored in two horizontal tanks. The storage tanks contain steam coils, which are used to provide sufficient heat to increase the vapor pressure above the liquid ammonia so that the liquid flows to the vaporizer unit. The liquid flow rate is regulated by a cascade controller, where the set point of a flow controller is changed according to pressure variations within the vaporizer. Steam flowing through coils within the vaporizer is used to vaporize the liquid ammonia.

The pressure of ammonia vapor entering the mixing section of the reactor is controlled by a pressure-reducing valve in the line from the vaporizer. Also, the flow rate of ammonia is adjusted to maintain the reactor temperature approximately constant. Compressed air, which has been preheated in the shell side of a heat exchanger, is added to the ammonia vapor in the mixing section of the reactor. When the reaction mixture encounters the catalyst in the reactor, an exothermic reaction takes place. The product stream, containing oxides of nitrogen, water vapor, and nitrogen, is cooled in the tube side of the exchanger used to preheat the air. Additional cooling takes place in a cooler condenser. An acid-separation unit is used to remove condensate, and then the oxides of nitrogen are sent to an absorber, where they combine with the water to form nitric acid. The acid product is pumped from the bottom of the tower to storage tanks.

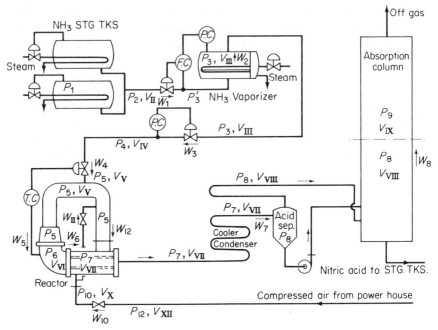

Figure 8.4-5. Flow sheet for nitric acid production. [Reproduced from R. N. Boyd and V. J. Bakanowski, *Engineering Journal*, **43**, No. 12, 2 (1960).]

DuPont's attempt to operate the plant, with the control systems described above, was not always successful. The ammonia storage tanks were located about 800 ft from the vaporizer unit, and pressure surges, similar to water hammer, were observed in this long feed line whenever switching from one feed tank to the other took place. These pressure fluctuations caused pressure changes in the reactor, which, in turn, caused the reactor yield to be lower than the design value. Consequently, Boyd and Bakanowski[4] of DuPont undertook a simulation study of the plant and its control systems.

First they derived a set of 14 first-order, nonlinear differential eqations and 28 algebraic relationships, which they thought would adequately describe the dynamic performance of the plant. The equations were programmed for an analog computer and required 108 operational amplifiers. Next, the response of the model to arbitrary disturbances was studied on the computer. For example, they assumed that when switching from one feed tank to the other occurred, a disturbance of 10 psi/sec entered the plant. The response of the liquid flow of ammonia to the vaporizer to this disturbance is shown in Figure 8.4-6. It is apparent from this graph that if the pressure drop across the liquid

[4] R. N. Bode and V. J. Bakanowski, *Engineering Journal*, **43**, No. 12, 2 (1960)

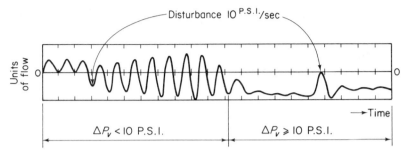

Figure 8.4-6. Liquid flow rate response. [Reproduced from R. N. Boyd and V. J. Bakanowski, *Engineering Journal*, **43**, No. 12, 2 (1960).]

ammonia flow control valve is less than 10 psi, the corresponding pressure oscillations in the long ammonia feed line are not damped. Alternately, whenever the pressure drop across this valve exceeds 10 psi, the fluctuations are quickly damped.

Figure 8.4–7 shows the effect of this disturbance on the reactor temperature. The response to the disturbance entering at *A* corresponds to the original control system. Adjustment of the controller gains gives the results for the disturbance entering at *B* and *C*. After studying the system dynamics in this way and verifying the model with experimental data from the plant, it was decided to install additional surge capacity in the ammonia piping system. This modification gave the results shown in Figure 8.4–8, where the response to disturbances entering at times *D*, *E*, *F*, and *G* are shown for various controller gain settings.

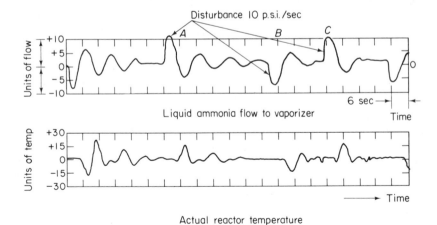

Figure 8.4-7. Reactor temperature response. [Reproduced from R. N. Boyd and V. J. Bakanowski, *Engineering Journal*, **43**, No. 12, 2 (1960).]

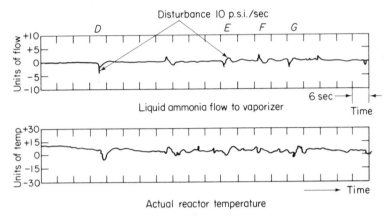

Figure 8.4-8. Reactor temperature response for modified plant. [Reproduced from R. N. Royd and V. J. Bakanowski, *Engineering Journal*, **43**, No. 12, 2 (1960).]

Example 8-4.4 *Simulation of solvent dewaxing plant*

Bettes and Wright,[5] at Standard Oil Company of Indiana, found that the solvent-recovery section of a solvent-dewaxing plant was unstable and therefore used a simulation method to develop a new control system. The original plant and its controllers are shown in Figure 8.4–9. The product stream is used to preheat the wax-solvent mixture in a heat exchanger. This hot feed is evaporated at atmospheric pressure to remove a portion of the solvent. A weir is placed in the evaporator, so that one end of the shell acts like a surge tank. The vapor mixture leaving the evaporator is heated with steam and is flashed at a pressure of about 35 psig. Additional steam heating and a third flash, at atmospheric pressure, are used to remove most of the remaining solvent from the wax product. The vapor from the pressure flash is used as the heat source for the primary evaporation. Also, in order to minimize utility requirements, the plant was designed to have a tight heat balance.

The original plant control systems attempted to maintain the total feed rate constant by using a flow controller and to maintain the pressure in the pressure-flash tower constant by manipulating the flow of vapor from this unit. With this arrangement, the effective condensing surface for pressure-flash vapors varied as the flow controller changed the flow rate. Level controllers were installed as shown in Figure 8.4–9.

It was known that there would be frequent variations in the solvent concentration of the feed mixture, due to an intermittent filter-washing process upstream from the plant. At low-flow rates and high-feed concentrations of solvent, it was observed, even during start-up of the plant, that the overall

[5] R. S. Bettes and L. T. Wright, *Oil and Gas Journal*, **58**, No. 17, 202 (1960).

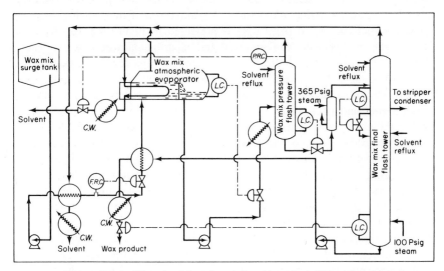

Figure 8.4-9. Flow sheet for solvent-dewaxing plant. [Reproduced from R. S. Bettes and L. T. Wright, *Oil and Gas Journal*, **58**, No. 17, 202 (1960).]

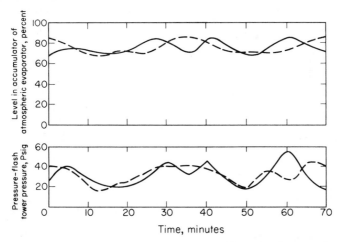

Figure 8.4-10. System response. [Reproduced from R. S. Bettes and L. T. Wright, *Oil and Gas Journal*, **58**, No. 17, 202 (1960).]

stability of the system was poor. Experimental data indicated that there were large, undamped oscillations in pressure, temperature, and level at various places in the plant. A typical operating record is shown in Figure 8.4-10.

Rather than attempt an experimental study to determine which one (or more) of the thermal loops in the plant was causing the instability and how to modify the control systems to obtain an acceptable dynamic performance, Bettes and Wright used a simulation method. A comparison of their analog computer model output and actual plant data is presented in Figure 8.4–10. Although the agreement between the theoretical and experimental results is not exact, a number of dynamic tests indicated that a similar behavior could be observed for both systems. Then the analog computer model was used to demonstrate that the troublesome loop was from the atmospheric evaporator, to the pressure flash tower, and back again. The dynamic components in this loop that can affect the plant stability are composition lags in the evaporator (including the receiving section beyond the weir), transportation lags in the piping to the pressure flash tower, thermal lags in the exchangers, and instrumentation lags. It is a simple matter to use the analog simulation to study the plant response to changes in each one of these elements. Also, the start-up of the plant, as well as the response to various upsets, can be determined quickly for a wide variety of conditions

The results of the analog simulation work indicated that the best dynamic performance would be obtained by reassigning the control functions of three controllers, changing the operating range of the level controller on the atmospheric evaporator, and eliminating the surge effect of the weir in the atmospheric evaporator by cutting a large hole in it. The revised flow sheet showing the controllers is given in Figure 8.4–11. With these modifications, operation of the plant has been smooth, and even large fluctuations in the solvent feed composition are quickly damped.

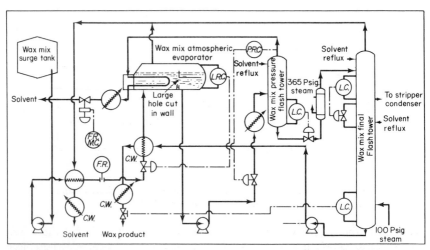

Figure 8.4-11. Revised flow sheet. [Reproduced from R. S. Bettes and L. T. Wright, *Oil and Gas Journal*, **58**, No. 17, 202 (1960).]

SECTION 8.5 SUMMARY

In this chapter we have considered the design of multivariable controllers. Most of the discussion was limited to linear systems. Primarily this emphasis was due to the fact that many of the techniques have been developed so recently that they have not been tested to any great extent in industrial plants; consequently, it is not certain whether the nonlinear charactersitics of the process will be the limiting feature or whether some other factor, such as the actual implementation of the controller, poses the major difficulty. This situation should be clarified in the next few years as more results become available.

Our discussion of multivariable controllers commenced with cascade control loops, because they were referred to in the previous chapter. In cascade control we use two measured signals (one near the source of the primary disturbances and one at the plant output), two controllers (a very fast one to compensate for the disturbances and another to adjust the set point of the fast controller to correct for output deviations), but only a single process variable that we manipulate. An obvious generalization of the cascade controller is to include a second manipulative input, so that we have two separate, single-loop control systems. However, these control loops interact, which makes it difficult to extend the classical servomechanism control theory to a multivariable process. For linear lumped parameter processes it is possible to devise a scheme whereby the single-loop controllers appear not to interact, and for this case we can use the criteria given in Chapter 7 to find the controller settings of each loop individually. An alternate procedure is to design a noninteracting control system more directly by considering the dynamic modes of the system. However, we cannot decide whether one of these control systems is better than the other until we introduce a quantitative definition of what "best" means for cases where interaction is present. This question provides the basis for the next chapter.

In order to be certain that it is indeed possible to use a control system of some kind to make a process have desirable dynamic characteristics, we introduced criteria for the controllability and observability of a plant. We said that the plant was completely controllable if we could adjust the control variables in such a way that we could force the plant to go from any initial state to any final state in a finite time. The dual concept of complete observability means that measurements of the system outputs, which normally are fewer in number than the state variables, over some finite-time inverval contain sufficient information to allow us to identify the state completely. Both concepts are most useful when we are dealing with very large and complicated plants, where our intuition is not always reliable.

Feedback control systems can never provide perfect control, for we must wait until we observe an error at the plant output before we adjust the setting of the control variable. Conversely, feedforward controllers allow perfect regulation, at least on a conceptual basis. However, in order for the analysis to be valid, there must be a control variable for each output we desire to maintain constant, all disturbances must be measurable, the plant dynamics must be known exactly, and the final control policy must be realizable. For the case of linear lumped parameter processes, the feedforward control systems are often simple to implement. It is possible to extend the approach to nonlinear and distributed parameter processes, but then implementation becomes a difficult job. Thus good engineering practice dictates that a feedforward control system should always be augmented by a feedback controller.

Even though it is difficult to extend the ideas of servomechanism theory to a unit having multiple inputs and outputs, this problem is much simpler than the control of a complete chemical plant where there are a large number of interconnected units. For problems of this magnitude, a simulation study is often warranted, although in some cases it is possible to use analytical methods. Additional research effort is needed in this area.

EXERCISES

1.(B*) Design a feedforward control system for the problem of a pair of parallel reactions occurring in a nonisothermal CSTR that was described in Exercise 35, p. 220, Volume 1. The variables that can be manipulated are the feed rate and the flow rate of the heat-transfer fluid; those we would most like to maintain constant are the composition of component B and temperature; and those we expect to fluctuate and can measure are feed temperature and the inlet temperature of the heat-transfer fluid. Discuss any potential limitations of your design.

2.(B*) Design a feedforward control system for the gas absorption unit containing two plates described in Exercises 9, p. 106, and 20, p. 215, Volume 1. Select the liquid and vapor flow rates as the control variables that would prevent the exit vapor and liquid compositions from changing irrespective of fluctuations in either the entering vapor or liquid composition.

3.(C*) Design a feedforward control system for the reactor-separator plant described in Exercise 7, p. 105, Volume 1. Choose the feed rate and the recycle flow rate as control variables and attempt to maintain the reactant and product concentrations in the rich phase leaving the separator constant. Consider the possibility of feed composition disturbances and rate constant variations caused by changes in the ambient temperature.

4.(B*) Apply the criteria for observability and controllability to the CSTR reactor problem presented in Exercise 1 above.

5.(B*) Determine the observability and controllability of the gas absorber referred to in Exercise 2 above.

6.(C*) Discuss the observability and controllability of each unit, and the overall plant, for the reactor and separator system referred to in Exercise 3 above.

7.(B) Use the results of Example 7.5-10 to desgin a cascade control system for the complex reaction mixture discussed in detail in Chapter 7.

8.(B*) Consider the plate, gas absorption column discussed in Exercise 2 and attempt to design a cascade control system for the unit. Explain the reasons for your selection of the primary and secondary control loops. Also discuss the extension of your analysis to columns containing more than two plates.

9.(B) Determine the stability of the two single-loop control systems described Example 8.4-1.

10.(B*) Design a noninteracting control system for the CSTR reactor problem presented in Exercise 1 above. Discuss your results, emphasizing the potential advantages or disadvantages of a noninteracting system.

11.(B*) Design a noninteracting control system for the gas absorption unit mentioned in Exercise 2 above. Discuss the significance of your design.

12.(C*) Apply the technique of modal analysis to any one of the units described in Exercises 1 through 3.

13.(13) Consider any one of the units described in the first three exercises above and use the procedure recommended by Bristol to choose pairs of inputs and outputs for single-loop controllers so that the amount of interaction would be minimized.

14.(B) Write a critical review of the NH_3 plant simulation study by G. M. Armstrong and L. R. Olsen, *Chem. Eng.* **69**, 135, September 3, 1962.

Optimal
Control 9
Theory

In our previous discussion of the design of optimum steady state control systems (Section 2.3), we measured the disturbances entering a plant and then computed the values of the control variables that yielded the maximun profit from the process (the state variables were eliminated from the profit expression by using the material and energy balances describing the system). Since we are adjusting the control settings depending on observed changes in the disturbances, we recognize this to be a feedforward control policy. For cases where it was not possible to measure all the input disturbances, we modified our approach and measured some of the process outputs. By combining these measured values with the system equations, we could compute the unknown inputs and the control settings corresponding to the most profitable operation. Obviously this is a feedback strategy. Thus we see that whenever the system dynamics are negligible, it does not make any difference whether a feedback or a feedforward policy is implemented, and in both cases we try to maximize the profit.

From our study of dynamic control systems in Chapters 7 and 8, we know that there is considerable difference between feedforward and feedback control schemes for dynamic processes. Perfect regulation can never be achieved with a feedback controller, for we must wait until an error is observed at the output from the plant before even attempting to provide compensation. Theoretically it is possible to obtain perfect regulation with a feedforward controller, although the limitations of the theory almost always prevent this from being true in practice. However, it is important to recognize that both kinds of controllers are regulators, which attempt to maintain the plant outputs as close as possible to the optimum steady state design conditions. Basically this attempt to regulate the system is a different approach than we used for optimum steady state control, where we deliberately changed the operating level in order always to maximize the process profitability.

It would be highly desirable to formulate dynamic control problems that were analogous to the optimum steady state control philosophy. Unfortunately these problems seem to be beyond our capability at this time. Normally we cannot measure all the disturbances entering the plant because of a lack of adequate instruments, measurement costs, and the fact that we cannot even identify all the disturbances for many systems. In some cases it might be possible to specify a statistical description of the disturbances, but methods for predicting the response of nonlinear processes to stochastic inputs are in their infancy. Without information concerning the nature of the disturbances, we are forced to make our observations of fluctuations either at points within the system or at the output. However, the nonlinear nature of the differential equations used to describe the system usually prevents us from computing the unknown input signals. Similarly, we cannot determine the most profitable performance of the plant corresponding to those inputs. Thus, we are forced to base our designs of control systems on a somewhat less significant goal.

For most plants, the disturbances fluctuate around some constant average value. This is the value used in the optimum steady state design procedure. We might anticipate that any advantage we would obtain when the disturbance was higher (or lower) than this mean operating level would be compensated for by poorer performance when the fluctuation was in the opposite direction. Thus, we could say that on a time-average basis we would like to operate as close as possible to the optimum steady state design condition. This is the argument that provided the basis for our design of regulating control systems. Previously we judged the performance of these controllers on the basis of empirical criteria, such as decay ratio and gain margin, but now we want to consider the design of optimal regulating controllers. In other words, we want to design a multivariable feedback controller to force the plant to return to the optimum steady state conditions in the best possible way.

We start our discussion with the different definitions that can be used for the "best" operating policy. Then a procedure for determining the optimal control law is developed, and numerous applications of the method are described. The theory has not been tested extensively, but the limitations are noted whenever possible.

SECTION 9.1 PERFORMANCE CRITERIA

Obviously the first difficulty encountered in the design of an optimal control system is to define the word optimal. In other words, we must have an explicit definition of what we mean by the "best" control. This is in contrast to our previous efforts, where we hoped to obtain a satisfactory response in terms of such quantities as decay ratio, settling time, offset, and phase and gain margins.

A number of criteria have been proposed as measures of the performance of control systems for single-input, single-output plants. For example,

Integral of the absolute error, $\text{IAE} = \int |e| \, dt$

Integral of time and absolute error, $\text{ITAE} = \int t|e| \, dt$

Integral of squared error, $\text{ISE} = \int e^2 \, dt$

Integral of time and squared error, $\text{ITSE} = \int te^2 \, dt$

are among the many that have been studied. Intuitively we would expect that a well-designed control system would keep the difference between the actual and desired outputs—the error—small and therefore lead to smaller values of the criteria above than a poorly designed controller. An integral criterion is used in order to penalize the controller the whole time an error is present; an e^2 term adds higher penalties whenever the error is large; and multiplying an error signal by t introduces large penalties if the error tends to persist.

Gibson and co-workers[1] made an extensive investigation of electrical engineering control problems for numerous functions similar to those above. Unfortunately their computations indicate that no single performance index could be used to specify the quality of a controller. This conclusion is based on observed contradictions in their results; that is, when a poor performance was indicated by the criterion, the response characteristics measured in terms of bandwidth, decay ratio, gain and phase margin were known to be excellent. Although this result is discouraging, it should not keep us from exploring the possibility of optimal control. This is particularly true for multivariable controllers, where our "rules of thumb" concerning the desirability of the response characteristics of single-input, single-output plants are no longer applicable. In other words, we do not know that a certain amount of interaction between the control variables is necessarily a disadvantage.

We might expect that it would be much easier to select a performance criterion for chemical process problems, for we almost always want to maximize the profit obtainable from the plant (or minimize the operating cost). However, it should be remembered that we must also consider the cost of the

[1] J. E., Gibson et al., *Trans AIEE*, **80**, Part II, 65 (1961).

control system in the final economic analysis. In other words, if it requires a large on-line computer to implement the control policy that maximizes profit, whereas we could obtain 95 percent of this increase by using a suboptimal policy that is simple to implement with a conventional device, the latter case might have a higher economic return. Therefore in our investigation of optimal control theory, we will consider a number of performance criteria, including an integral square error criterion

$$P = \frac{1}{2} \int_0^\theta (c_{11} x_1^2 + c_{12} x_2^2 + \ldots + c_{21} u_1^2 + c_{22} u_2^2 + \ldots) \, dt$$

$$= \frac{1}{2} \int_0^\theta x^T C_1 x + u^T C_2 u) \, dt \qquad\qquad 9.1\text{-}1$$

which includes deviations of both the state and control variables from their optimum steady state values over the time interval of interest, θ; a minimum time criterion

$$P = \int_0^\theta dt \qquad\qquad 9.1\text{-}2$$

which is just the time it takes the plant to return to the desired steady state output after a disturbance enters the system; and a simple profit model, which for a stirred-tank reactor problem might have the form

$$P = \int_0^\theta [c_1 q(A_f - A) - c_2 q A_f - c_3 q_H] \, dt \qquad\qquad 9.1\text{-}3$$

where the first term represents the value of the product, the second is the cost of raw materials, and the third is the cost of the utilities.

SECTION 9.2   A PROBLEM IN VARIATIONAL CALCULUS

Once we have specified a performance criterion, it is a simple matter to define an optimal control problem. For example, if we consider a lumped parameter system described by the set of equations

$$\frac{dx_i}{dt} = f_i(x_1, x_2, \ldots, x_n; u_1, u_2, \ldots u_m), \qquad i = 1, 2, \ldots, n \qquad 9.2\text{-}1$$

we want to determine how we should manipulate the control variables, $u_1(t)$, $u_2(t), \ldots, u_m(t)$, as a function of time in order to force the plant to go from some initial state, $x_1(0) = x_{10}, x_2(0) = x_{20}, \ldots, x_n(0) = x_{n0}$, back to the optimum steady state design conditions, $x_1(\theta) = x_2(\theta) = \ldots = x_n(\theta) = 0$, in such a way that we minimize the performance index, for example, in a minimum amount of time

$$P = \int_0^\theta dt \qquad\qquad 9.2\text{-}2$$

It should not be surprising that we will obtain a different control policy for different performance criteria, and therefore we will have to solve several cases and compare the results.

The problem stated above is very different in a mathematical sense from our previous problem of optimum steady state control. For the steady state control system, we looked for the single values of the control variables that minimized the instantaneous operating cost. This was a problem that could be solved in many instances by the application of the classical methods of calculus. However, now we are attempting to find functions—that is, how to change the control variables as a function of time—in order to minimize a performance index. In this case the performance index is called a functional (a function of a function) and the problem is one that can be solved using the calculus of variations.

As an illustration of the difference between these two kinds of problem, let us consider the design of a control system for a simple first-order process

$$\frac{dx}{dt} = Ax + Bu \qquad\qquad 9.2\text{-}3$$

We assume that the plant has been designed to operate at some optimun steady state, and at time zero we measure the output $x(0) = x_0$, and find that it is not at the desired value. Next, we ask how we should adjust the control variable u as a function of time to return to the optimum steady state, $x(\theta) = 0$. in such a way as to minimize the performance index

$$P = \frac{1}{2} \int_0^\theta (x^2 + c^2 u^2)\, dt \qquad\qquad 9.2\text{-}4$$

where c^2 represents the relative cost of deviations of the control variable to the state variable.

A Problem in Calculus

One possible approach would be to admit that we have no idea of how to solve this problem, but we will assume that a proportional controller will be adequate for the job

$$u = Mx \qquad\qquad 9.2\text{-}5$$

and simply try to find the controller gain that minimizes the performance index. With this assumption, the state equation becomes

$$\frac{dx}{dt} = (A + BM)x \qquad\qquad 9.2\text{-}6$$

which can be solved to give

$$x = x_0 \exp (A + BM)t \qquad\qquad 9.2\text{-}7$$

It is clear that the system will never approach the desired final condition,

$x(\theta) = 0$, as time goes on unless

$$A + BM < 0 \qquad\qquad 9.2\text{-}8$$

This condition can always be satisfied by using a negative feedback control system

$$M < 0 \qquad\qquad 9.2\text{-}9$$

and making the gain large enough.

Now our performance index can be written as

$$P = \frac{1}{2} \int_0^\theta (1 + c^2 M^2) x^2 \, dt$$

$$= \frac{x_0^2}{2} (1 + c^2 M^2) \int_0^\theta \exp\left[2(A + BM)t\right] dt$$

$$= \frac{x_0^2(1 + c^2 M^2)}{4(A + BM)} \{\exp\left[2(A + BM)\theta\right] - 1\} \qquad 9.2\text{-}10$$

According to Eq. 9.2-7, we much choose θ to be greater than two or three time constants of the controlled system

$$\theta \gg 3(A + BM)^{-1} \qquad\qquad 9.2\text{-}11$$

in order to return to the desired final steady state within that time. Alternately, we could replace the value of θ appearing as the upper limit of the integral by $\theta = \infty$. For either case, the first term in brackets in Eq. 9.2-10 will be negligible.

Thus the performance index becomes

$$P = \frac{-x_0^2(1 + c^2 M^2)}{4(A + BM)} \qquad\qquad 9.2\text{-}12$$

To find the value of the controller gain M that minimizes this performance function, all we have to do is set the derivative of P with respect to M equal to zero and solve for the optimum value of M

$$\frac{dP}{dM} = 0 = -\frac{x_0^2}{4} \left[\frac{(A + BM)2c^2 M - (1 + c^2 M^2)B}{(A + BM)^2}\right] \qquad 9.2\text{-}13$$

This result shows that the optimum gain must be independent of the magnitude of the initial error. The optimum gain is

$$M = -\frac{A}{B} \left[1 + \sqrt{1 + \frac{B^2}{c^2 A^2}}\right] \qquad\qquad 9.2\text{-}14$$

where the positive root is taken in order to ensure that Eq. 9.2-9 will be satisfied and the system will be stable.

Hence from Eq. 9.2-5 the optimum control policy will be

$$u = -\frac{A}{B} \left[1 + \sqrt{1 + \frac{B^2}{c^2 A^2}}\right] x \qquad\qquad 9.2\text{-}15$$

or after substituting Eq. 9.2–9,

$$u = -\frac{A}{B}\left[1 + \sqrt{1 + \frac{B^2}{c^2 A^2}}\right] x_0 \exp\left(-A\sqrt{1 + \frac{B^2}{c^2 A^2}}\right)t \qquad 9.2\text{–}16$$

Even though this last expression tells us how to change the control variable as a function time, the result given by Eq. 9.2–15 actually is more useful because it can be implemented as a feedback controller rather than as a feedforward system. It will become more apparent that we will often try to put the solution of optimal control problems in a form where they give us a feedback control policy, instead of a feedforward scheme asked for by the problem formulation. The advantages of a feedback system should be clear from the previous two chapters.

A Problem in Variational Calculus

The preceding example was reduced to a problem in calculus because we specified that a proportional feedback controller would be used and then proceeded to find the controller gain that minimized the performance index. However, it would be of much greater interest to find some way of determining the relationship between the control variable and time, or the state variable, without imposing any restrictions on this control policy. For this very simple problem, we might attempt to find such a relationship by using the classical results of the calculus of variations. With this approach, we must have our performance index in the form

$$P = \int_0^\theta F(x, \dot{x}, t)\, dt \qquad 9.2\text{–}17$$

so that we can apply what is known as the Euler-Lagrange equation

$$\frac{\partial F}{\partial x} = \frac{d}{dt}\left(\frac{\partial F}{\partial \dot{x}}\right) \qquad 9.2\text{–}18$$

where again

$$\dot{x} = \frac{dx}{dt} \qquad 9.2\text{–}19$$

to determine the optimal policy. We will discuss this equation in more detail in the next section.

The state equation and performance index for our example are

$$\dot{x} = Ax + Bu \qquad 9.2\text{–}3$$

$$P = \frac{1}{2}\int_0^\theta (x^2 + c^2 u^2)\, dt \qquad 9.2\text{–}4$$

so that we can obtain the desired form by solving Eq. 9.2–3 for u and using this result to eliminate the control variable from Eq. 9.2–4.

$$P = \frac{1}{2} \int_0^\theta \left[x^2 + \frac{c^2}{B^2} (\dot{x} - Ax)^2 \right] dt \qquad 9.2\text{--}20$$

Now, the Euler-Lagrange equation states that the relationship between x and $\dot{x}$ must be

$$x - \left(\frac{Ac^2}{B^2} \right)(\dot{x} - Ax) = \frac{d}{dt} \left[\frac{c^2}{B^2} (\dot{x} - Ax) \right]$$

$$= \frac{c^2}{B^2} (\ddot{x} - A\dot{x}) \qquad 9.2\text{--}21$$

or

$$\ddot{x} - \frac{B^2}{c^2} \left(1 + \frac{A^2 c^2}{B^2} \right) x = 0 \qquad 9.2\text{--}22$$

The general solution of this second-order equation is

$$x = c_1 \exp \left[- \left(\frac{B}{c} \sqrt{1 + \frac{A^2 c^2}{B^2}} \right) t \right] + c_2 \exp \left(\frac{B}{c} \sqrt{1 + \frac{A^2 c^2}{B^2}} \right) t \right] \qquad 9.2\text{--}23$$

In order for the plant to return to the desired steady state as time progresses, we see that we must choose $\theta = \infty$ and we must have that $c_2 = 0$. Similarly, from the initial condition, we find that $c_1 = x_0$. Thus

$$x = x_0 \exp \left[- \left(\frac{B}{c} \sqrt{1 + \frac{A^2 c^2}{B^2}} \right) t \right] \qquad 9.2\text{--}24$$

This result gives the optimal path that the system must follow, but we are more interested in determining the optimal control policy. We can accomplish this by using Eq. 9.2–3.

$$u = \frac{\dot{x} - Ax}{B} = - \left(\frac{1}{c} \sqrt{1 + \frac{A^2 c^2}{B^2}} + \frac{A}{B} \right) x_0 \exp \left[- \left(\frac{B}{c} \sqrt{1 + \frac{A^2 c^2}{B^2}} \right) t \right]$$

$$= - \frac{A}{B} \left(1 + \sqrt{1 + \frac{B^2}{c^2 A^2}} \right) x_0 \exp \left[- \left(A \sqrt{1 + \frac{B^2}{c^2 A^2}} \right) t \right] \qquad 9.2\text{--}25$$

Alternately, if we substitute Eq. 9.2–24, we obtain

$$u = - \frac{A}{B} \left(1 + \sqrt{1 + \frac{B^2}{c^2 A^2}} \right) x \qquad 9.2\text{--}26$$

which shows that the optimal control policy is to use a proportional feedback controller, where the optimum gain of the controller is identical to that obtained previously, Eq. 9.2–15. Hence the variational approach gives us an expression for the form of the optimal control law, as well as the optimum settings for the controller gain.

It is interesting to note that as the relative cost of the control action c approaches zero, the optimum controller gain approaches infinity. This is similar to the result we obtained in our earlier study of proportional control for first-order systems; that is, an infinite gain gives a very fast response and eliminates offset. Thus the optimal control problem reveals that controller gains are restricted not only by stability limitations but also because of cost considerations.[1]

Equivalence of Lagrange and Mayer Problems

The way we have formulated the variational problem for a lumped parameter system

$$\frac{dx_i}{dt} = f_i(x_1, \ldots, x_n; u_1, \ldots u_m) \qquad 9.2\text{-}27$$

is to look for the control variables, as a function of time, which minimize a performance index that is given as an integral; for example,

$$P = \int_0^\theta dt \qquad 9.2\text{-}28$$

This is referred to as the Lagrange problem in variational calculus. We could just as well have asked to determine the $u_j(t)$ that minimized the functional

$$P = \theta \qquad 9.2\text{-}29$$

Whenever the performance index is written in terms of the terminal conditions of the process, the formulation is called a Mayer problem in variational calculus. However, it should be recognized that it is a simple matter to convert one type of problem into the other by introducing a new state variable. For example, instead of minimizing the integral

$$P = \frac{1}{2}\int_0^\theta (c_{11}x^2 + \ldots + c_{1n}x_n^2 + c_{21}u_1^2 + \ldots + c_{2m}u_m^2)\, dt \qquad 9.2\text{-}30$$

we could let

$$\dot{x}_{n+1} = c_{11}x_1^2 + \ldots + c_{1n}x_n^2 + c_{21}u_1^2 + \ldots + c_{2m}u_m^2 \qquad 9.2\text{-}31$$

with

$$x_{n+1}(0) = 0 \qquad 9.2\text{-}32$$

and then minimize

$$P = x_{n+1}(\theta) = \int_0^\theta \frac{dx_{n+1}}{dt}\, dt \qquad 9.2\text{-}33$$

Since it is always possible to make this conversion, we will not bother to distinguish between the problems and will always use the integral formulation.

---

[1] Although this argument is intuitively satisfying, it relies on the form chosen for the performance index. Other performance measures do not always give the same result. This point will be discussed in additional detail later in the chapter.

SECTION 9.3 PONTRYAGIN'S MINIMUM PRINCIPLE

Although the Euler-Lagrange equation was found very useful for the example in the previous section, this is not a general result. For most chemical process problems, it is essentially impossible to put the state equations and performance index into the proper form. Also, it is very difficult to use the classical, variational techniques to handle problems where there are hard constraints on the control or state variables. This is an unfortunate circumstance because we know that most control variables for chemical plants correspond to flow rates, and these certainly have hard limits; that is, the valve settings can vary only between the fully open and completely closed positions. Thus we need to find some new approach for developing the necessary conditions describing the optimal control action.

The method we will develop is called Pontryagin's minimum principle. A number of ways can be used to obtain the desired results, and the reader is encouraged to look up some of the other methods[1,2,3] to see if he can find one that is closer to his own intuition and to obtain a firmer foundation of optimization techniques. Our derivation[4] is presented in terms of a system with two state variables and two control variables, such as a nonisothermal stirred-tank reactor, but the approach is not limited to this case.

Derivation of the Minimum Principle

We consider a plant described by two differential equations

$$\frac{dx_1}{dt} = f_1(x_1, x_2, u_1, u_2) \qquad\qquad 9.3\text{-}1$$

$$\frac{dx_2}{dt} = f_2(x_1, x_2, u_1, u_2) \qquad\qquad 9.3\text{-}2$$

which has been designed to operate at some optimum steady state condition, and at time zero we measure some deviation from this steady state

$$x_1(0) = x_{10} \qquad x_2(0) = x_{20} \qquad\qquad 9.3\text{-}3$$

and try to find out how we should change the control variables, $u_1(t)$ and $u_2(t)$, as a function of time to force the plant to return to the desired steady state

$$x_1(\theta) = x_2(\theta) = 0 \qquad\qquad 9.3\text{-}4$$

[1] L. T. Fan, *The Continuous Maximum Principle*, Ch. 2, Wiley, N. Y., 1966.

[2] R. Luus, and L. Lapidus, *Optimal Control of Engineering Process*, Blaisdell, Waltham, Mass., 1967.

[3] L. B. Koppel, *Introduction to Control Theory*, Ch. 6, Prentice-Hall, Englewood Cliffs, N. J., 1968.

[4] Taken from J. M. Douglas and M. M. Denn, *Ind. Eng. Chem.*, **57**, No. 11, 18 (1965).

in such a way that we minimize the performance index

$$P = \int_0^\theta F(x_1, x_2, u_1, u_2)\, dt \qquad\qquad 9.3\text{-}5$$

In order to get a start on the problem, let us suppose that we arbitrarily choose two functions for the control variables and call these $\bar{u}_1(t)$ and $\bar{u}_2(t)$ (see Figure 9.3-1). Now, we know the initial conditions for the plant, Eq. 9.3-3, so that we can solve the state equations, Eqs. 9.3-1 and 9.3-2, and find

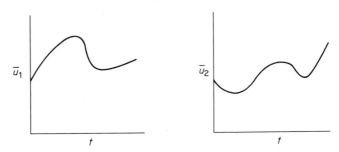

Figure 9.3-1. Original choice for control variables.

$x_1(t)$ and $x_2(t)$. These values correspond to $\bar{u}_1$ and $\bar{u}_2$, and we call them $\bar{x}_1$ and $\bar{x}_2$. Of course, our choice for the control functions is somewhat limited because we eventually want the plant to return to the desired steady state conditions (see Eq. 9.3-4). However, we will assume that this can always be accomplished, either by setting the control variables equal to zero after some time and letting the stable plant return on its own or by using the method described in Section 8.2. Once we have obtained the values of the state variables corresponding to these control functions, we can evaluate the performance index and obtain a single measure of the performance of the system,

Next, we suppose that we make some very small change in our original choices for the control variables (see Figure 9.3-2).

$$u_1(t) = \bar{u}_1(t) + \delta u_1(t) \qquad u(t) = \bar{u}_2(t) + \delta u_2(t) \qquad 9.3\text{-}6$$

These new choices, along with the initial conditions on the state variables, can

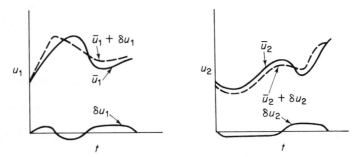

Figure 9.3-2. Change in control variables.

be used to solve the state equations again, so that we obtain some new values for $x_1(t)$ which we call $\bar{x}_1(t) + \delta x_1(t)$, and similarly for x_2. Now we have all the information required to evaluate the performance index a second time, and we can see if our new choices are superior to the original guesses. Obviously what we would like to do is either to keep on changing the control functions until we find the set that gives the minimum value for the performance index or to establish a set of conditions that corresponds to the best choice.

If we compare the results obtained in the first case with the second by subtracting them, we can write

$$\frac{d(\bar{x}_1 + \delta x_1)}{dt} - \frac{d\bar{x}_1}{dt} = \frac{d\,\delta x_1}{dt} = f_1(\bar{x}_1 + \delta x_1, \bar{x}_2 + \delta x_2, \bar{u}_1 + \delta u_1, \bar{u}_2 + \delta u_2)$$

$$-f_1(\bar{x}_1, \bar{x}_2, \bar{u}_1, \bar{u}_2) \qquad 9.3\text{-}7$$

$$\frac{d(\bar{x}_2 + \delta x_2)}{dt} - \frac{d\bar{x}_2}{dt} = \frac{d\,\delta x_2}{dt} = f_2(\bar{x}_1 + \delta x_1, \bar{x}_2 + \delta x_2, \bar{u}_1 + \delta u_1, \bar{u}_2 + \delta u_2)$$

$$-f_2(\bar{x}_1, \bar{x}_2, \bar{u}_1, \bar{u}_2) \qquad 9.3\text{-}8$$

since the functions $\bar{x}_1 + \delta x_1, \bar{x}_2 + \delta x_2, \bar{u}_1 + \delta u_1, \bar{u}_2 + \delta u_2$, and $\bar{x}_1, \bar{x}_2, \bar{u}_1, \bar{u}_2$ must both satisfy the state equations. Thus we can obtain explicit expressions for the changes in the state variables. Similarly, the difference in the performance index is

$$\delta P = P(\bar{u}_1 + \delta u_1, \bar{u}_2 + \delta u_2) - P(\bar{u}_1, \bar{u}_2)$$

$$= \int_\theta^{\theta + \delta\theta} F(\bar{x}_1 + \delta x_1, \bar{x}_2 + \delta x_2, \bar{u}_1 + \delta u_1, \bar{u}_2 + \delta u_2)\,dt$$

$$- \int_0^\theta F(\bar{x}_1, \bar{x}_2, \bar{u}_1, \bar{u}_2)\,dt \qquad 9.3\text{-}9$$

where the possibility of the system requiring a different time to return to the desired steady state has been included in the upper limit of the first integral. Providing that the changes δu_1 and δu_2 are small enough, we can expand the first terms in the functions on the right-hand sides of Eqs. 9.3–7 through 9.3–9 in a Taylor series around the original paths and neglect all second- and higher-order terms. This procedure leads to the results

$$\frac{d\delta x_1}{dt} = \frac{\partial f_1}{\partial x_1}\delta x_1 + \frac{\partial f_1}{\partial x_2}\delta x_2 + \frac{\partial f_1}{\partial u_1}\delta u_1 + \frac{\partial f_1}{\partial u_2}\delta u_2 + \dots \qquad 9.3\text{-}10$$

$$\frac{d\delta x_2}{dt} = \frac{\partial f_2}{\partial x_1}\delta x_1 + \frac{\partial f_2}{\partial x_2}\delta x_2 + \frac{\partial f_2}{\partial u_1}\delta u_1 + \frac{\partial f_2}{\partial u_2}\delta u_2 + \dots \qquad 9.3\text{-}11$$

$$\delta P = \int_0^\theta \left(\frac{\partial F}{\partial x_1}\delta x_1 + \frac{\partial F}{\partial x_2}\delta x_2 + \frac{\partial F}{\partial u_1}\delta u_1 + \frac{\partial F}{\partial u_2}\delta u_2 + \dots\right) dt + (F|_{t=\theta})\,\delta\theta$$

$$9.3\text{-}12$$

where the last expression is obtained by breaking the first integral on the right-hand side of Eq. 9.3–9 into an interval from zero to θ and from θ to $\theta + \delta\theta$, combining the first of these with the second integral in Eq. 9.3–9 and carrying out the Taylor series expansion to give the integral term in Eq. 9.3–12, and, finally, approximating the integral from θ to $\theta + \delta\theta$ by the last team in Eq. 9.3–12.

In order to combine the state equations with the performance index, we subtract the appropriate time derivatives from both sides of Eqs. 9.3–10 and 9.3–11, multiply each equation by a Lagrange multiplier (which is similar to the multipliers we use to handle constraints in normal calculus problems except that here they are a function of time), integrate both equations from zero to θ, and then add the equations. After collecting terms, we obtain

$$
\begin{aligned}
\delta P = \int_0^\theta \Bigg[& \left(\frac{\partial F}{\partial x_1} + \lambda_1 \frac{\partial f_1}{\partial x_1} + \lambda_2 \frac{\partial f_2}{\partial x_1} \right) \delta x_1 - \lambda_1 \frac{d\,\delta x_1}{dt} \\
& + \left(\frac{\partial F}{\partial x_2} + \lambda_1 \frac{\partial f_1}{\partial x_2} + \lambda_2 \frac{\partial f_2}{\partial x_2} \right) \delta x_2 - \lambda_2 \frac{d\,\delta x_2}{dt} \\
& + \left(\frac{\partial F}{\partial u_1} + \lambda_1 \frac{\partial f_1}{\partial u_1} + \lambda_2 \frac{\partial f_2}{\partial u_1} \right) \delta u_1 \\
& + \left(\frac{\partial F}{\partial u_2} + \lambda_1 \frac{\partial f_1}{\partial u_2} + \lambda_2 \frac{\partial f_2}{\partial u_2} \right) \delta u_2 \Bigg] dt + F|_{t=\theta}\,\delta\theta \qquad \text{9.3–13}
\end{aligned}
$$

Unfortunately this expression contains both the variations δx_1 and δx_2, as well as their time derivatives. However, we can eliminate the derivatives if we integrate by parts

$$
\int_0^\theta \lambda_1 \left(\frac{d\,\delta x_1}{dt} \right) dt = \lambda_1(\theta)\,\delta x_1(\theta) - \lambda_1(0)\,\delta x_1(0) - \int_0^\theta \delta x_1 \left(\frac{d\,\lambda_1}{dt} \right) dt \quad \text{9.3–14}
$$

and we obtain a similar result for $\int_0^\theta \lambda_2 (d\,\delta x_2/dt)\,dt$. Substituting these expressions, we find that

$$
\begin{aligned}
\delta P = \int_0^\theta \Bigg[& \left(\frac{d\lambda_1}{dt} + \frac{\partial F}{\partial x_1} + \lambda_1 \frac{\partial f_1}{\partial x_1} + \lambda_2 \frac{\partial f_2}{\partial x_1} \right) \delta x_1 \\
& + \left(\frac{d\lambda_2}{dt} + \frac{\partial F}{\partial x_2} + \lambda_1 \frac{\partial f_1}{\partial x_2} + \lambda_2 \frac{\partial f_2}{\partial x_2} \right) \delta x_2 \\
& + \left(\frac{\partial F}{\partial u_1} + \lambda_1 \frac{\partial f_1}{\partial u_1} + \lambda_2 \frac{\partial f_2}{\partial u_1} \right) \delta u_1 \\
& + \left(\frac{\partial F}{\partial u_2} + \lambda_1 \frac{\partial f_1}{\partial u_2} + \lambda_2 \frac{\partial f_2}{\partial u_2} \right) \delta u_2 \Bigg] dt \\
& + F(\theta)\,\delta\theta - \lambda_1(\theta)\,\delta x_1(\theta) - \lambda_2(\theta)\,\delta x_2(\theta) + \lambda_1(0)\,\delta x_1(0) \\
& + \lambda_2(0)\,\delta x_2(0) \qquad\qquad\qquad\qquad\qquad\qquad\qquad\qquad\qquad\quad \text{9.3–15}
\end{aligned}
$$

The way in which we have formulated the problem allows us to impose arbitrary variations in δu_1 and δu_2, and these changes cause the variations in δx_1 and δx_2. On an intuitive basis, we would expect that it would be much easier to determine the effect that changes in the control variables have on the change in the performance index if the variations in the state variables did not appear in the result. However, the Lagrange multiplier, or adjoint, variables are completely arbitrary functions, which we introduced; and since we can select these in any way we please, we might as well make the choice that leads to the greatest simplification of the equation. Therefore we let

$$\frac{d\lambda_1}{dt} = -\left(\frac{\partial F}{\partial x_1} + \lambda_1 \frac{\partial f_1}{\partial x_1} + \lambda_2 \frac{\partial f_2}{\partial x_1}\right) \qquad 9.3\text{-}16$$

$$\frac{d\lambda_2}{dt} = -\left(\frac{\partial F}{\partial x_2} + \lambda_1 \frac{\partial f_1}{\partial x_2} + \lambda_2 \frac{\partial f_2}{\partial x_2}\right) \qquad 9.3\text{-}17$$

With this choice, the terms multiplying δx_1 and δx_2 disappear from the integrand, although we now have to solve a pair of coupled differential equations in order to find the functions $\lambda_1(t)$ and $\lambda_2(t)$. Also, we are going to have to find a pair of boundary conditions for this pair of differential equations in order to complete the definitions of the adjoint variables.

The expression for the change in the performance index, Eq. 9.3-15, also has terms containing $\delta x_1(0)$ and $\delta x_2(0)$. We have considered a problem where the initial values of x_1 and x_2 are measured, so that any kind of control policy we select cannot change these initial values. This means that both $\delta x_1(0)$ and $\delta x_2(0)$ must be equal to zero, and these terms can be dropped from the equation.

With these simplifications, Eq. 9.3-15 becomes

$$\delta P = \int_0^\theta \left[\left(\frac{\partial F}{\partial u_1} + \lambda_1 \frac{\partial f_1}{\partial u_1} + \lambda_2 \frac{\partial f_2}{\partial u_1}\right) \delta u_1 + \left(\frac{\partial F}{\partial u_2} + \lambda_1 \frac{\partial f_1}{\partial u_2} + \lambda_2 \frac{\partial f_2}{\partial u_2}\right) \delta u_2\right] dt$$

$$+ F(\theta)\,\delta\theta - \lambda_1(\theta)\,\delta x_1(\theta) - \lambda_2(\theta)\,\delta x_2(\theta) \qquad 9.3\text{-}18$$

If it would be possible to find some way of eliminating the final three terms in this expression, we would have an equation relating the change in the system performance to the arbitrary variations we can impose on the control variables. It is tempting to state that the final values of x_1 and x_2 are fixed (Eq. 9.3-4) and therefore that $\delta x_1(\theta)$ and $\delta x_2(\theta)$ must be equal to zero. However, in order always to be able to return to the desired steady state operating condition, we might have to wait different lengths of time,[5] and therefore we can expect that $\delta x_1(\theta)$ and $\delta x_2(\theta)$ are related to $\delta\theta$. As a first approximation, we can use Eq. 9.3-1 to write

$$x_1(\theta + \delta\theta) = x_1(\theta) + (f_1|_{t=\theta})\,\delta\theta$$

$$= [\bar{x}_1(\theta) + \delta x_1(\theta)] + f_1(\theta)\,\delta\theta \qquad 9.3\text{-}19$$

Since the final value of x_1 is fixed, we must have that the new value is equal to the original value we obtained, or

$$x_1(\theta + \delta\theta) = \bar{x}_1(\theta) \qquad 9.3\text{-}20$$

---

[5] Some additional modifications are necessary if the final time is equal to infinity.

This condition requires that

$$\delta x_1(\theta) = -f_1(\theta)\, \delta\theta \qquad\qquad 9.3\text{-}21$$

A similar result is obtained for $\delta x_2(\theta)$. Hence we find that

$$\delta P = \int_0^\theta \left[\left(\frac{\partial F}{\partial u_1} + \lambda_1 \frac{\partial f_1}{\partial u_1} + \lambda_2 \frac{\partial f_2}{\partial u_1} \right) \delta u_1 + \left(\frac{\partial F}{\partial u_2} + \lambda_1 \frac{\partial f_1}{\partial u_2} + \lambda_2 \frac{\partial f_2}{\partial u_2} \right) \delta u_2 \right] dt$$

$$+ (F + \lambda_1 f_1 + \lambda_2 f_2)|_{t=\theta}\, \delta\theta \qquad\qquad 9.3\text{-}22$$

Our whole purpose in this development is to try to find a set of necessary conditions corresponding to an optimal control policy. We know, by definition, that once we have established the policy that minimizes the performance index, any change we impose will make δP increase. Providing that the plant is completely controllable, we can make our changes in the control variables over slightly different time intervals, as well as changing their magnitude, before forcing the process to return to the desired final condition, and in this way we can treat $\delta\theta$ as an independent variable. In other words, we could impose the changes so that $\delta\theta$ was either a positive or a negative quantity. According to Eq. 9.3-22, if we change the sign of $\delta\theta$, we expect the sign of δP to change. We cannot allow this to happen for changes around the optimal value, however, for we know that δP can only increase for variations in that neighborhood. Therefore the only way we can resolve this conflict is to make certain that

$$F + \lambda_1 f_1 + \lambda_2 f_2 = 0 \qquad \text{at } t = \theta \qquad\qquad 9.3\text{-}23$$

This is the stopping condition for our optimal process.

Unconstrained Control Variables

We have finally reduced our expression for the change in the performance index to the form

$$\delta P = \int_0^\theta \left[\left(\frac{\partial F}{\partial u_1} + \lambda_1 \frac{\partial f_1}{\partial u_1} + \lambda_2 \frac{\partial f_2}{\partial u_1} \right) \delta u_1 + \left(\frac{\partial F}{\partial u_2} + \lambda_1 \frac{\partial f_1}{\partial u_2} + \lambda_2 \frac{\partial f_2}{\partial u_2} \right) \delta u_2 \right] dt$$

$$9.3\text{-}24$$

which contains only variations in the control variables. As long as we are free to change these variables in any manner we please (i.e., the control variables are not constrained), we might as well make the choice that gives us the greatest insight into the behavior of the plant with the least amount of effort. A particularly apt choice is

$$\delta u_1 = -\epsilon \left(\frac{\partial F}{\partial u_1} + \lambda_1 \frac{\partial f_1}{\partial u_1} + \lambda_2 \frac{\partial f_2}{\partial u_1} \right) \qquad\qquad 9.3\text{-}25$$

$$\delta u_2 = -\epsilon \left(\frac{\partial F}{\partial u_2} + \lambda_1 \frac{\partial f_1}{\partial u_2} + \lambda_2 \frac{\partial f_2}{\partial u_2} \right) \qquad\qquad 9.3\text{-}26$$

where ϵ is a small positive constant. Substituting these expressions in to Eq.

9.3–24, we obtain

$$\delta P = -\epsilon \int_0^\theta \left[\left(\frac{\partial F}{\partial u_1} + \lambda_1 \frac{\partial f_1}{\partial u_1} + \lambda_2 \frac{\partial f_2}{\partial u_1} \right)^2 + \left(\frac{\partial F}{\partial u_2} + \lambda_1 \frac{\partial f_1}{\partial u_2} + \lambda_2 \frac{\partial f_2}{\partial u_2} \right)^2 \right] dt$$

9.3–27

The integrand of this result is always positive, which means that the integral must always be positive and δP must always be less than or equal to zero. However, if we apply these changes in the neighborhood of the optimal solution —that is, the solution which gives a minimum value for P—we know that any change must make δP increase. Again, we can only resolve this discrepancy by requiring that

$$\frac{\partial F}{\partial u_1} + \lambda_1 \frac{\partial f_1}{\partial u_1} + \lambda_2 \frac{\partial f_2}{\partial u_1} = 0 \qquad\qquad 9.3\text{–}28$$

$$\frac{\partial F}{\partial u_2} + \lambda_1 \frac{\partial f_1}{\partial u_2} + \lambda_2 \frac{\partial f_1}{\partial u_2} = 0 \qquad\qquad 9.3\text{–}29$$

These are the necessary conditions for the optimal control policy. It should should be noted that they are algebraic relationships, which must be satisfied at every instant of time.

In order to review the method, we have two state equations

$$\frac{dx_1}{dt} = f_1(x_1, x_2, u_1, u_2) \qquad\qquad 9.3\text{–}1$$

$$\frac{dx_2}{dt} = f_2(x_1, x_2, u_1, u_2) \qquad\qquad 9.3\text{–}2$$

and two equations for the adjoint variables

$$\frac{d\lambda_1}{dt} = -\left(\frac{\partial F}{\partial x_1} + \lambda_1 \frac{\partial f_1}{\partial x_1} + \lambda_2 \frac{\partial f_2}{\partial x_1} \right) \qquad\qquad 9.3\text{–}16$$

$$\frac{d\lambda_2}{dt} = -\left(\frac{\partial F}{\partial x_2} + \lambda_1 \frac{\partial f_1}{\partial x_2} + \lambda_2 \frac{\partial f_2}{\partial x_2} \right) \qquad\qquad 9.3\text{–}17$$

which have to be solved simultaneously with the necessary conditions for the optimal control action

$$\frac{\partial F}{\partial u_1} + \lambda_1 \frac{\partial f_1}{\partial u_1} + \lambda_2 \frac{\partial f_2}{\partial u_1} = 0 \qquad\qquad 9.3\text{–}28$$

$$\frac{\partial F}{\partial u_2} + \lambda_1 \frac{\partial f_1}{\partial u_2} + \lambda_2 \frac{\partial f_2}{\partial u_2} = 0 \qquad\qquad 9.3\text{–}29$$

The boundary conditions for the problem are

$$x_1(0) = x_{10} \qquad x_2(0) = x_{20} \qquad\qquad 9.3\text{–}3$$

and

$$x_1(\theta) = 0 \qquad x_2(\theta) = 0 \qquad\qquad 9.3\text{–}4$$

which provide four boundary conditions for the state and adjoint equations.

Also, the "stopping" condition

$$F + \lambda_1 f_1 + \lambda_2 f_2 = 0 \qquad\qquad 9.3\text{-}23$$

must be satisfied at $t = \theta$.[6]

These relationships are valid only for the particular problem under consideration, although it is a simple job to modify them to cover more general situations. Our main interest is the optimal design of regulating controllers; therefore we will not bother to discuss these modifications except to state that if the final values of the state variables are not specified, we let

$$\lambda_1(\theta) = 0 \qquad \lambda_2(\theta) = 0 \qquad\qquad 9.3\text{-}30$$

to remove the terms containing $\delta x_1(\theta)$ and $\delta x_2(\theta)$ from Eq. 9.3–18. For either kind of problem, we have a set of mixed boundary conditions; that is, some are known at $t = 0$ while others are known at $t = \theta$. Thus we expect to encounter difficulties if we attempt to solve the equations numerically. More thorough discussions of the appropriate modifications for other kinds of problems, as well as computational techniques, can be found in books by Denn[7] and others.[8]

Definition of the Hamiltonian

The equations describing optimal control problems are relatively complicated; consequently, it is often convenient to introduce a function, called the the Hamiltonian, which is defined by the equation

$$H = F + \lambda_1 f_1 + \lambda_2 f_2 \qquad\qquad 9.3\text{-}31$$

Then the differential equations of interest can be put into the canonical form

$$\dot{x}_1 = \frac{\partial H}{\partial \lambda_1} \qquad \dot{x}_2 = \frac{\partial H}{\partial \lambda_2} \qquad\qquad 9.3\text{-}32$$

$$\dot{\lambda}_1 = -\frac{\partial H}{\partial x_1} \qquad \dot{\lambda}_2 = -\frac{\partial H}{\partial x_2} \qquad\qquad 9.3\text{-}33$$

the necessary conditions become

$$\frac{\partial H}{\partial u_1} = 0 \qquad \frac{\partial H}{\partial u_2} = 0 \qquad\qquad 9.3\text{-}34$$

and the stopping condition is

$$H = 0 \qquad \text{at } t = \theta \qquad\qquad 9.3\text{-}35$$

These relationships can be verified by substituting the defintion of H and comparing the results with the equations listed in the summary above. It is a simple matter to show that the Hamiltonian must be a constant along an opti-

---

[6] Unless the operating time is infinitely large.

[7] M.M., Denn, *Introduction to Optimization by Variational Methods,* McGraw-Hill, N.Y., 1969.

[8] See footnotes 1 through 3, p. 257.

mal path, since

$$\frac{dH}{dt} = \frac{\partial H}{\partial x_1}\dot{x}_1 + \frac{\partial H}{\partial x_2}\dot{x}_2 + \frac{\partial H}{\partial \lambda_1}\dot{\lambda}_1 + \frac{\partial H}{\partial \lambda_2}\dot{\lambda}_2 + \frac{\partial H}{\partial u_1}\dot{u}_1 + \frac{\partial H}{\partial u_2}\dot{u}_2 \qquad 9.3\text{-}36$$

Substituting Eqs. 9.3–31. through 9.3–34, we obtain

$$\frac{dH}{dt} = \frac{\partial H}{\partial x_1}\frac{\partial H}{\partial \lambda_1} + \frac{\partial H}{\partial x_2}\frac{\partial H}{\partial \lambda_2} - \frac{\partial H}{\partial \lambda_1}\frac{\partial H}{\partial x_1} - \frac{\partial H}{\partial \lambda_2}\frac{\partial H}{\partial x_2} = 0 \qquad 9.3\text{-}37$$

which means that H must remain constant. Furthermore, from the stopping condition, Eq. 9.3–35, we know that final value of the Hamiltonian must be equal to zero. Therefore it must always be equal to zero.

$$H = 0 \qquad \text{for all time} \qquad 9.3\text{-}38$$

For systems of arbitrary order, we can write the equations for the optimal control problem as

$$H = F + \sum_{n=1}^{N} \lambda_n f_n = 0 \qquad 9.3\text{-}39$$

$$\dot{x}_i = \frac{\partial H}{\partial \lambda_i} = f_i, \qquad i = 1, 2, \ldots, N \qquad 9.3\text{-}40$$

$$\dot{\lambda}_i = -\frac{\partial H}{\partial x_i} = -\frac{\partial F}{\partial x_i} - \sum_{n=1}^{N} \lambda_n \frac{\partial f_n}{\partial x_i}, \qquad i = 1, 2, \ldots, N \qquad 9.3\text{-}41$$

$$\frac{\partial H}{\partial u_j} = 0 = \frac{\partial F}{\partial u_j} + \sum_{n=1}^{N} \lambda_n \frac{\partial f_n}{\partial u_j}, \qquad j = 1, 2, \ldots, m \qquad 9.3\text{-}42$$

$$x_i(0) = x_{i0} \qquad x_i(\theta) = 0 \quad \text{or} \quad \lambda_i(\theta) = 0, \qquad i = 1, 2, \ldots, N \qquad 9.3\text{-}43$$

Example 9.3–1 Optimal control of a first-order system

In order to demonstrate the application of the minimum principle, we will again consider the problem studied in the previous section.

Solution

The state equation is

$$\dot{x} = Ax + Bu \qquad 9.2\text{-}3$$

and the performance index[9] is

$$P = \frac{1}{2}\int_0^{\theta=\infty} (x^2 + c^2 u^2)\, dt \qquad 9.2\text{-}4$$

Hence we can formulate the Hamiltonian as

$$H = \tfrac{1}{2}(x^2 + c^2 u^2) + \lambda(Ax + Bu) \qquad 9.3\text{-}44$$

The adjoint equation is

$$\dot{\lambda} = -\frac{\partial H}{\partial x} = -x - A\lambda \qquad 9.3\text{-}45$$

[9] From our previous results we anticipate that $\theta = \infty$.

and the necessary condition for the optimal control action is

$$\frac{\partial H}{\partial u} = 0 = c^2 u + B\lambda$$

or

$$u = -\frac{B}{c_2}\lambda \qquad\qquad 9.3\text{–}46$$

If we use this result to eliminate the control variable from the state equation, the state and adjoint equations become

$$\dot{x} = Ax - \frac{B^2}{c^2}\lambda \qquad\qquad 9.3\text{–}47$$

$$\dot{\lambda} = -x - A\lambda \qquad\qquad 9.3\text{–}48$$

This pair of first-order differential equations are linear and have constant coefficients. Consequently, we can use standard techniques to eliminate the adjoint variable and obtain a second-order equation in the state variable. Thus we find that

$$\ddot{x} - \left(A^2 + \frac{B^2}{c^2}\right)x = 0 \qquad\qquad 9.3\text{–}49$$

which is identical to Eq. 9.2–22, derived from the Euler-Lagrange equation. The boundary conditions have not been changed; hence the remainder of the solution will also be identical to the previous result.

$$u = -\frac{A}{B}\left(1 + \sqrt{1 + \frac{B^2}{c^2 A^2}}\right)x \qquad\qquad 9.3\text{–}50$$

Again we see that the variational approach gives us the proper form for the optimal control law, as well as the optimum value of the controller gain. Also, the proportional control action means that we will obtain an asymptotic return to steady state conditions so that the final time will be $\theta = \infty$.

Example 9.3–2 Derivation of the Euler-Lagrange equation

Since we have shown that we obtain exactly the same solution of a simple optimal control problem using the Euler-Lagrange equation and the minimum principle, the two approaches must be equivalent, at least for some problems. Thus we might expect that we could use the minimum principle to derive the Euler-Lagrange equation. In order to demonstrate that this is possible, we will consider the classical problem in variational calculus, where we seek a function $x(t)$ that will minimize the integral

$$P = \int_0^\theta F(x, \dot{x}, t)\, dt \qquad\qquad 9.3\text{–}51$$

Solution

Unfortunately this is not the form that we used to derive the minimum principle; therefore we must look for some transformations in order to reduce it to the proper form. The first thing we must recognize is that for any given initial condition, $x(0)$, the function $x(t)$ can be uniquely determined if its derivative is known. Thus we can let our decision or control variable, u, be equal to $\dot{x}$. Another difficulty is encountered because time appears in an explicit manner in the integrand. However, if we define a new state variable $\dot{x}_2 = 1$, with $x_2(0) = 0$, we remove this difficulty. Now, letting $x_1 = x$, we can write the problem as

$$\dot{x}_1 = u \qquad\qquad\qquad 9.3\text{–}52$$

$$\dot{x}_2 = 1 \qquad x_2(0) = 0 \qquad 9.3\text{–}53$$

$$P = \int_0^\theta F(x_1, x_2, u)\, dt \qquad 9.3\text{–}54$$

which is exactly the form we need.

The Hamiltonian becomes

$$H = F + \lambda_1 u + \lambda_2 \qquad 9.3\text{–}55$$

The adjoint equations are

$$\dot{\lambda}_1 = -\frac{\partial H}{\partial x_1} = -\frac{\partial F}{\partial x_1} \qquad 9.3\text{–}56$$

$$\dot{\lambda}_2 = -\frac{\partial H}{\partial x_2} = -\frac{\partial F}{\partial x_2} \qquad 9.3\text{–}57$$

or in terms of the original variables,

$$\dot{\lambda}_1 = -\frac{\partial F}{\partial x} \qquad \dot{\lambda}_2 = -\frac{\partial F}{\partial t} \qquad 9.3\text{–}58$$

The necessary condition for the optimal control action is

$$\frac{\partial H}{\partial u} = 0 = \frac{\partial F}{\partial u} + \lambda_1 \qquad 9.3\text{–}59$$

or in terms of the original variables,

$$\lambda_1 = -\frac{\partial F}{\partial \dot{x}} \qquad 9.3\text{–}60$$

Integrating Eq. 9.3–56, or the first expression in Eq. 9.3–58, we obtain

$$\lambda_1(t) = \lambda_1(0) - \int_0^t \frac{\partial F}{\partial x}\, dt \qquad 9.3\text{–}61$$

Combining this with Eq. 9.3–60, we find that

$$\frac{\partial F}{\partial \dot{x}} = \int_0^t \frac{\partial F}{\partial x}\, dt + c \qquad 9.3\text{–}62$$

where the initial value of the adjoint variable has been replaced by an arbitrary constant c. This expression is called the equation of du Bois-Reymond. If we take its time derivative, we obtain the Euler-Lagrange equation

$$\frac{d}{dt}\left(\frac{\partial F}{\partial \dot{x}}\right) = \frac{\partial F}{\partial x} \qquad\qquad 9.3\text{-}63$$

Also, it is possible to show that the appropriate boundary conditions are

$$x(0) = \text{given} \quad \text{or} \quad \lambda_1(0) = \left(\frac{\partial F}{\partial \dot{x}}\bigg|_{t=0}\right) = 0$$

$$x(\theta) = \text{given} \quad \text{or} \quad \lambda_1(\theta) = \left(\frac{\partial F}{\partial \dot{x}}\bigg|_{t=\theta}\right) = 0 \qquad 9.3\text{-}64$$

In some cases, $x(t)$ might have a sharp corner, or cusp, so that its derivative will not be continuous. Then the integrand $F(x, \dot{x}, t)$ will not be continuous at this point and the Euler-Lagrange equation will not be applicable. However, the function $\int (\partial F/\partial x)\, dt$ will remain a continuous function because the value of an integral is independent of what happens at a single point. Therefore, from Eq. 9.3-62, we see that $\partial F/\partial \dot{x}$ must also be continuous. This result is called the "Erdmann-Weierstrass corner condition."

Constrained Control Variables

For the one simple optimal control problem we have studied, we found that we should use a proportional feedback controller

$$u = -\frac{A}{B}\left(1 + \sqrt{1 + \frac{B^2}{c^2 A^2}}\right) x \qquad\qquad 9.3\text{-}50$$

If the relative cost of control, c, is fairly small, the controller gain will be very high; and if we encounter a fairly large deviation x, we will require a very large amount of control, u. However, for most chemical processes, the control variables are chosen as flow rates of various streams, and we know that these are limited to values between where the valve is completely opened or closed. In other words, despite the fact that an optimal control law might call for a large control action, we know that the actual control action must lie within a bounded region

$$u_{j*} \leq u_j \leq u_j^* \qquad\qquad 9.3\text{-}65$$

where u_j^* represents the smallest allowable value of u_j and u_j^* is the upper bound. Of course, this means that the necessary conditions for the optimal control law are not necessarily always correct and that we must find some way of modifying our procedure to account for hard constraints on the control variables.

After a careful review of our previous derivation, we find that all the steps are independent of whether or not the control variables are bounded, until we get to the point where we have the expression

$$\delta P = \int_0^\theta \left(\frac{\partial H}{\partial u_1}\delta u_1 + \frac{\partial H}{\partial u_2}\delta u_2\right) dt \qquad\qquad 9.3\text{-}24$$

(where we have introduced the definition of the Hamiltonian, Eq. 9.3–31, into the original Eq. 9.3–24 in order to simplify the notation). At this point we made the choices

$$\delta u_1 = -\epsilon \frac{\partial H}{\partial u_1} \qquad \delta u_2 = -\epsilon \frac{\partial H}{\partial u_2} \qquad\qquad 9.3\text{–}66$$

because they gave us a very convenient expression to work with. This procedure will always be satisfactory, providing that it does not lead to values above the upper bound or below the lower bound. Hence the only modifications we are required to make are for cases where one or both control variables lie on a boundary. For example, if we suppose that we make the changes given by Eq. 9.3–66 but find that eventually u_1 becomes equal to its maximum value, u_1^*, we know that the only change we are allowed to make in u_1 is to decrease its value

$$\delta u_1 < 0 \qquad\qquad 9.3\text{–}67$$

We can still let

$$\delta u_1 = \epsilon \frac{\partial H}{\partial u_1} < 0 \qquad\qquad 9.3\text{–}68$$

but we cannot make any statements as yet about the sign of ϵ because we do not know the sign of $\partial H/\partial u_1$. Taking the simplest case of a single control variable, or letting $\delta u_2 = 0$ in a multivariable problem, we can substitute Eq. 9.3–68 into Eq. 9.3–24, to obtain

$$\delta P = \epsilon \int_0^\theta \left(\frac{\partial H}{\partial u_1}\right)^2 dt \qquad\qquad 9.3\text{–}69$$

If we make this kind of a change in the neighborhood of a solution that minimizes the performance index, we know that δP must always increase. The integrand in Eq. 9.3–69 is always positive, and therefore, for this case, ϵ must be positive. Thus we find from Eq. 9.3–68 that

$$\frac{\partial H}{\partial u_1} < 0 \qquad\qquad 9.3\text{–}70$$

Remembering that the only changes we are making in u_1 is to decrease its value from the upper bound, we see from Figure 9.3–3 that the Hamiltonian

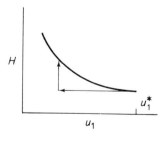

Figure 9.3-3. Bounded control variables.

must be increasing. However, this implies that the Hamiltonian had its minimum value when the control variable was set to its maximum value in the optimal solution. Thus the Hamiltonian is a minimum when the control action is saturated.

For the opposite case, where u_1 is at the lowest bound, u_1^*, we reach the same conclusion. We know that the only change we are allowed to make is to increase u_1

$$\delta u_1 > 0 \qquad\qquad 9.3\text{-}71$$

and we let

$$\delta u_1 = \epsilon \frac{\partial H}{\partial u_1} > 0 \qquad\qquad 9.3\text{-}72$$

where the sign of ϵ cannot be specified as yet. Substituting this variation into Eq. 9.3–24 and letting $\delta u_2 = 0$, we obtain

$$\delta P = \epsilon \int_0^\theta \left(\frac{\delta H}{\delta u_1}\right)^2 dt \qquad\qquad 9.3\text{-}73$$

We know that in the neighborhood of an optimal path, δP must be greater than zero. Therefore we again find that ϵ must be positive and, from Eq. 9.3–72, that

$$\frac{\partial H}{\partial u_1} > 0 \qquad\qquad 9.3\text{-}74$$

By drawing a sketch similar to Figure 9.3–3 and remembering that now we are allowing only positive changes in u_1, we see that the Hamiltonian must increase when we make any change in the control variable from its lower bound, providing that the lower bound is on the optimal path. In other words, the Hamiltonian is also a minimum along the lower boundary. Similar results are obtained for the other control variable u_2.

To summarize our results, then, we determine the optimal path by setting the partial derivatives of the Hamiltonian with respect to the control variables equal to zero in the interior of the allowable control region (this makes the Hamiltonian stationary). By extending the analysis and considering second-derivative terms, it is possible to show that the minimum of P is obtained when the Hamiltonian is minimized. Of course, for some problems we might obtain only a local minimum or a saddlepoint instead of a global minimum.[10] Also, similar to normal calculus problems, there may be several local maxima or minima encountered in a problem. However, if the optimal control action coincides with a constraint, we have been able to demonstrate that the Hamiltonian is a minimum. A similar approach[11] can be used to modify the equations

[10] See Sections 6.8 and 6.9 of reference 7 on p. 264.
[11] See Section 6.16 of reference 7 on p. 264.

when there are constraints on the state variables or mixed constraints involving the state and control variables.

It is interesting to note that the Hamiltonian must always be equal to zero, even if the control variables are maintained constant at one of their extreme values along an optimal path. Instead of $\partial H/\partial u_1 = \partial H/\partial u_2 = 0$ in Eq. 9.3–36, we must have $\dot{u}_1 = \dot{u}_2 = 0$, so that the Hamiltonian is always constant, and, from Eq. 9.3–35 we know that the constant must be equal to zero.

Dynamic Programming

In the preceding discussion we showed that both the classical method of variational calculus (the Euler-Lagrange equation) and Pontryagin's minimum principle can sometimes be used to solve optimal control problems. However, as we will demonstrate below, dynamic programming can also be used to derive the necessary conditions. Again we consider the state equations

$$\frac{dx_1}{dt} = f_1(x_1, x_2, u_1, u_2) \qquad\qquad 9.3\text{–}1$$

$$\frac{dx_2}{dt} = f_2(x_1, x_2, u_1, u_2) \qquad\qquad 9.3\text{–}2$$

and the performance index

$$P = \int_0^\theta F(x_1, x_2, u_1, u_2)\, dt \qquad\qquad 9.3\text{–}5$$

with the boundary conditions given by Eqs. 9.3–4 and 9.3–5. Now, at any time, we can break the process up into increments: from t to $t + \Delta$ and from $t + \Delta$ to θ, where t can cover the range from zero to θ (see Figure 9.3–4).

$$P = \int_t^{t+\Delta} F(x_1, x_2, u_1, u_2)\, dt + \int_{t+\Delta}^\theta F(x_1, x_2, u_1, u_2)\, dt \qquad 9.3\text{–}75$$

According to Bellman's *principle of optimality*,[12] an optimum policy must have

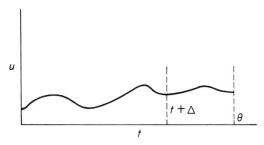

Figure 9.3-4. Time intervals for control variables.

[12] R. E. Bellman, *Dynamic Programming*, P. 83, Princeton Unversity Press, Princeton, N. J., 1957.

the property that no matter what control we impose in the early stages, we can never hope to achieve an optimal policy unless the control action in the later stage is optimal with respect to the state of the process resulting from the early decisions. Another way of phrasing this statement was given by Aris[13,14] as: "If you don't do the best you can with what you happen to have got, you'll never do the best you might have done with what you should have had." In order to put these statements into a mathematical form, we will define a function that represents the optimal value of the performance index over some interval ending at θ

$$g(x_1, x_2, t) = \underset{u_1, u_2}{\text{Min}} \int_t^\theta F(x_1, x_2, u_1, u_2)\, dt \qquad \text{9.3-76}$$

where $g(x_1, x_2, t)$ tells us that the value of this performance index depends on the starting time for the interval, t, and the state variables at that point, x_1 and x_2. Eventually we want to find $g(x_{10}, x_{20}, 0)$.

From Eq. 9.3–75, we can write that the optimal policy starting at some time t is

$$g(x_1, x_2, t) = \underset{u_1, u_2}{\text{Min}} \left[\int_t^{t+\Delta} F(x_1, x_2, u_1, u_2)\, dt \right.$$
$$\left. + \int_{t+\Delta}^\theta F(x_1, x_2, u_1, u_2)\, dt \right] \qquad \text{9.3-77}$$

However, according to the principle of optimality, the second integral in the equation above must correspond to an optimal policy with respect to the initial values at this interval, in order for there to be any chance that the overall policy is optimal. This means that we can write

$$g(x_1, x_2, t) = \underset{u_1, u_2}{\text{Min}} \left[\int_t^{t+\Delta} F(x_1, x_2, u_1, u_2)\, dt \right.$$
$$\left. + g(x_1|_{t+\Delta}, x_2|_{t+\Delta}, t + \Delta) \right] \qquad \text{9.3-78}$$

By picking a point where Δ is very small—that is, t is very close to $t + \Delta$— we can approximate the functions on the right-hand side of this equation by their Taylor series expansions and neglect all second- and higher-order terms

$$\int_t^{t+\Delta} F(x_1, x_2, u_1, u_2)\, dt = F|_t \Delta + \cdots = F(t) \Delta + \cdots \qquad \text{9.3-79}$$

$$g(x_1|_{t+\Delta}, x_2|_{t+\Delta}, t + \Delta)$$
$$= g(x_1|_t, x_2|_t, t) + \frac{\partial g}{\partial x_1}\frac{dx_1}{dt}\Delta + \frac{\partial g}{\partial x_2}\frac{dx_2}{dt}\Delta + \cdots \qquad \text{9.3-80}$$

---

[13] R. Aris et al., *Fundamental Ideas and Applications of Optimization Techniques in Design and Control*, p. 53, AIChE Special Lecture Series, Cleveland, Ohio, May 1961.

[14] Numerous applications are presented in R. Aris, *The Optimal Design of Chemical Reactors*, Academic Press, N. Y., 1961.

or after substituting Eqs. 9.3–1 and 9.3–2 into this last expression

$$g(x_1|_{t+\Delta}, x_2|_{t+\Delta}, t + \Delta) = g(x_1, x_2, t) + \frac{\partial g}{\partial x_1} f_1 \Delta + \frac{\partial g}{\partial x_2} f_2 \Delta + \cdots$$

$$9.3\text{–}81$$

Substituting these results into Eq. 9.3–78 gives

$$g(x_1, x_2, t) = \underset{u_1, u_2}{\text{Min}} \left[F(t) \Delta + g(x_1, x_2, t) + \frac{\partial g}{\partial x_1} f_1 \Delta + \frac{\partial g}{\partial x_2} f_2 \Delta + \cdots \right]$$

$$9.3\text{–}82$$

or since $g(x_1, x_2, t)$ is not explicity dependent on u_1 and u_2

$$0 = \underset{u_1, u_2}{\text{Min}} \left[F(t) \Delta + \frac{\partial g}{\partial x_1} f_1 \Delta + \frac{\partial g}{\partial x_2} f_2 \Delta + \cdots \right] \qquad 9.3\text{–}83$$

Now, dividing by Δ and taking the limit as Δ goes to zero (which will guarantee that all the second- and higher-order terms in the Taylor expansions approach zero, because they are at least of the order Δ^2), we find that

$$\underset{u_1, u_2}{\text{Min}} \left(F + \frac{\partial g}{\partial x_1} f_1 + \frac{\partial g}{\partial x_2} f_2 \right) = 0 \qquad 9.3\text{–}84$$

This is a first-order, quasi-linear, partial differential equation, which actually provides several pieces of information. First, the equation itself must be satisfied

$$F + \frac{\partial g}{\partial x_1} f_1 + \frac{\partial g}{\partial x_2} f_2 = 0 \qquad 9.3\text{–}85$$

Also, the minimum value of the function must satisfy the equations (assuming that we have an unconstrained problem)

$$\frac{\partial F}{\partial u_1} + \frac{\partial g}{\partial x_1} \frac{\partial f_1}{\partial u_1} + \frac{\partial g}{\partial x_2} \frac{\partial f_2}{\partial u_1} = 0 \qquad 9.3\text{–}86$$

$$\frac{\partial F}{\partial u_2} + \frac{\partial g}{\partial x_1} \frac{\partial f_1}{\partial u_2} + \frac{\partial g}{\partial x_2} \frac{\partial f_2}{\partial u_2} = 0 \qquad 9.3\text{–}87$$

where again we have used the fact that $g(x_1, x_2, t)$ does not depend on u_1 or u_2 in an explicit manner.

In order to put the equations in a more recognizable form, we need to develop some additional results. Taking the partial derivatives of Eq. 9.3–85 with respect to x_1 and x_2 gives

$$\frac{\partial F}{\partial x_1} + \frac{\partial g}{\partial x_1} \frac{\partial f_1}{\partial x_1} + \frac{\partial g}{\partial x_2} \frac{\partial f_2}{\partial x_1} + \frac{\partial^2 g}{\partial x_1^2} f_1 + \frac{\partial^2 g}{\partial x_1 \partial x_2} f_2 = 0 \qquad 9.3\text{–}88$$

$$\frac{\partial F}{\partial x_2} + \frac{\partial g}{\partial x_1} \frac{\partial f_1}{\partial x_2} + \frac{\partial g}{\partial x_2} \frac{\partial f_2}{\partial x_2} + \frac{\partial^2 g}{\partial x_1 \partial x_2} f_1 + \frac{\partial^2 g}{\partial x_2^2} f_2 = 0 \qquad 9.3\text{–}89$$

We note that

$$\frac{d}{dt}\left(\frac{\partial g}{\partial x_1}\right) = \frac{\partial^2 g}{\partial x_1^2}\frac{dx_1}{dt} + \frac{\partial^2 g}{\partial x_1 \partial x_2}\frac{dx_2}{dt} = \frac{\partial^2 g}{\partial x_1^2}f_1 + \frac{\partial^2 g}{\partial x_1 \partial x_2}f_2 \qquad 9.3\text{-}90$$

$$\frac{d}{dt}\left(\frac{\partial g}{\partial x_2}\right) = \frac{\partial^2 g}{\partial x_1 \partial x_2}\frac{dx_1}{dt} + \frac{\partial^2 g}{\partial x_2^2}\frac{dx_2}{dt} = \frac{\partial^2 g}{\partial x_1 \partial x_2}f_1 + \frac{\partial^2 g}{\partial x_2^2}f_2 \qquad 9.3\text{-}91$$

where we have used the state equations, Eqs. 9.3–1 and 9.3–2, to eliminate the time derivatives of the state variables. Using these last results to simplify Eqs. 9.3–88 and 9.3–89, we obtain

$$\frac{\partial F}{\partial x_1} + \frac{\partial g}{\partial x_1}\frac{\partial f_1}{\partial x_1} + \frac{\partial g}{\partial x_2}\frac{\partial f_2}{\partial x_1} + \frac{d}{dt}\left(\frac{\partial g}{\partial x_1}\right) = 0 \qquad 9.3\text{-}92$$

$$\frac{\partial F}{\partial x_2} + \frac{\partial g}{\partial x_1}\frac{\partial f_1}{\partial x_2} + \frac{\partial g}{\partial x_2}\frac{\partial f_2}{\partial x_2} + \frac{d}{dt}\left(\frac{\partial g}{\partial x_2}\right) = 0 \qquad 9.3\text{-}93$$

Now if we let

$$\lambda_1 = \frac{\partial g}{\partial x_1} \qquad \lambda_2 = \frac{\partial g}{\partial x_2} \qquad 9.3\text{-}94$$

the equations become

$$\frac{d\lambda_1}{dt} = -\left(\frac{\partial F}{\partial x_1} + \lambda_1\frac{\partial f_1}{\partial x_1} + \lambda_2\frac{\partial f_2}{\partial x_1}\right) \qquad 9.3\text{-}95$$

$$\frac{d\lambda_2}{dt} = -\left(\frac{\partial F}{\partial x_2} + \lambda_1\frac{\partial f_1}{\partial x_2} + \lambda_2\frac{\partial f_2}{\partial x_2}\right) \qquad 9.3\text{-}96$$

which are indentical to our original adjoint equations, Eqs. 9.3–16 and 9.3–17. Similarly, the necessary conditions for the optimal control, Eqs. 9.3–86 and 9.3–87, become

$$\frac{\partial F}{\partial u_1} + \lambda_1\frac{\partial f_1}{\partial u_1} + \lambda_2\frac{\partial f_2}{\partial u_1} = 0 \qquad 9.3\text{-}97$$

$$\frac{\partial F}{\partial u_2} + \lambda_1\frac{\partial f_1}{\partial u_2} + \lambda_2\frac{\partial f_2}{\partial u_2} = 0 \qquad 9.3\text{-}98$$

which are identical to our previous results for an unconstrained problem, Eqs. 9.3–28 and 9.3–29. Also, Eq. 9.3–85 becomes

$$F + \lambda_1 f_1 + \lambda_2 f_2 = 0 \qquad 9.3\text{-}99$$

which is the condition that the Hamiltonian is equal to zero. Since these expressions are valid for any time t, they must be true for all times. Thus the two approaches give identical results for unconstrained control problems. It is interesting to note that this approach tells us that the Lagrange multipliers, or adjoint variables, are just the gradients of the optimum performance criterion (see Eq. 9.3–94).

It is important to recognize that the preceding derivation is not valid unless the partial derivatives $\partial g/\partial x_1$ and $\partial g/\partial x_2$ are continuous; that is, the manipulations described in Eqs. 9.3–90 and 9.3–91 are not valid otherwise. However, there are a number of systems of interest where this restriction is not satisfied. Hence in some cases, there are major differences between a dynamic programming approach and the application of the minimum principle.

SECTION 9.4 LINEAR FEEDBACK CONTROL SYSTEMS

Now that we have developed a general mathematical framework for solving optimal control problems for lumped parameter systems, we want to use this approach to design control systems for chemical processes. We are particularly interested in developing multivariable controllers for nonlinear plants, but we will start by considering the linearized system equations. This procedure has the advantage of simplicity, in addition to making it possible to compare our results with the control systems developed using the servomechanism approach outlined in the two previous chapters. Perhaps the simplest way of presenting this material is to discuss a number of examples in detail. Stirred-tank reactor problems are used for most of the examples, primarily because they are relatively easy to manage and at the same time are highly nonlinear, but some other examples are also discussed.

Example 9.4–1 Linear feedback control of a CSTR

The linearized equations for a stirred-tank reactor may be written in the form

$$\dot{x}_1 = a_{11} x_1 + a_{12} x_2 \qquad\qquad 9.4\text{–}1$$

$$\dot{x}_2 = a_{21} x_1 + a_{22} x_2 + b_{22} u_2 \qquad\qquad 9.4\text{–}2$$

where x_1 and x_2 represent dimensionless deviations of the composition and temperature from their optimum steady state design values, u_2 represents the deviation of the heating flow rate, which is the only control variable for this problem, and the constants are defined in Example 3.3–1 or Section 7.1. We consider the performance index

$$P = \frac{1}{2} \int_0^\theta (c_1 x_1^2 + c_2 x_2^2 + c_3 u_2^2) \, dt \qquad\qquad 9.4\text{–}3$$

and look for the function $u_2(t)$, which forces the reactor to go from some initial measured deviation, $x_1(0) = x_{10}$ and $x_2(0) = x_{20}$, back to the optimum steady state design, $x_1(\theta) = x_2(\theta) = 0$, in such a way that the performance index is minimized. This problem represents an extension of the results presented in Example 9.3–1 to second-order systems.

Solution

It turns out that we can simplify the algebra if we introduce new state variables by making the transformation

$$x_1 = -y_1 \qquad\qquad\qquad 9.4\text{-}4$$

$$x_2 = \frac{a_{11}}{a_{12}} y_1 + \frac{1}{a_{12}} y_2 \qquad\qquad 9.4\text{-}5$$

The new state equations become

$$\dot{y}_1 = y_2 \qquad\qquad\qquad 9.4\text{-}6$$

$$\dot{y}_2 = -(a_{11}a_{22} - a_{12}a_{21})y_1 - (a_{11} + a_{22})y_2 + a_{12}b_{22}u_2 \qquad 9.4\text{-}7$$

the corresponding initial and final conditions are

$$y_1(0) = y_{10} \qquad y_2(0) = y_{20} \qquad\qquad 9.4\text{-}8$$

$$y_1(\theta) = 0 \qquad y_2(\theta) = 0 \qquad\qquad 9.4\text{-}9$$

and the performance index is

$$P = \frac{1}{2} \int_0^{\theta} \left[\left(c_1 + c_2\frac{a_{11}^2}{a_{12}^2} \right) y_1^2 + \frac{c_2}{a_{12}^2} y_2^2 + \frac{2c_2 a_{11}}{a_{12}^2} y_1 y_2 + c_3 u_2^2 \right] dt \quad 9.4\text{-}10$$

The form of the equations given by Eqs. 9.4-6 and 9.4-7 is particularly convenient for studying certain types of higher-order systems. For example, if we have an equation

$$\frac{d^n x}{dt^n} + a_{n-1}\frac{d^{n-1} x}{dt^{n-1}} + \cdots + a_1 \frac{dx}{dt} + a_0 x = bu \qquad 9.4\text{-}11$$

we can let

$$x_1 = x$$

$$x_2 = \frac{dx_1}{dt} = \frac{dx}{dt}$$

$$x_3 = \frac{dx_2}{dt} = \frac{d^2 x}{dt^2}$$

$$\vdots$$

$$x_n = \frac{dx_{n-1}}{dt} = \frac{d^{n-1} x}{dt^{n-1}} \qquad\qquad 9.4\text{-}12$$

so that we can replace the nth-order differential equation by the n first-order equations

$$\frac{dx_1}{dt} = x_2$$

$$\frac{dx_2}{dt} = x_3$$

$$\vdots \qquad\qquad\qquad\qquad\qquad 9.4\text{-}13$$

$$\frac{dx_{n-1}}{dt} = x_n$$

$$\frac{dx_n}{dt} = -a_0 x_1 - a_1 x_2 - \cdots - a_{n-1} x_n + bu$$

and now we can apply our formulation of Pontryagin's minimum principle. Thus for our second-order system, the Hamiltonian becomes

$$H = \frac{1}{2}\left[\left(c_1 + c_2\frac{a_{11}^2}{a_{12}^2}\right)y_1^2 + \left(\frac{c_2}{a_{12}^2}\right)y_2^2 + \left(\frac{2c_2 a_{11}}{a_{12}^2}\right)y_1 y_2 + c_3 u_2^2\right] + \lambda_1 y_2$$

$$- \lambda_2[(a_{11}a_{22} - a_{12}a_{21})y_1 + (a_{11} + a_{22})y_2 - a_{12}b_{22}u_2] \qquad 9.4\text{-}14$$

and the adjoint equations are

$$\dot{\lambda}_1 = -\frac{\partial H}{\partial y_1} = (a_{11}a_{22} - a_{12}a_{21})\lambda_2 - \left(c_1 + c_2\frac{a_{11}^2}{a_{12}^2}\right)y_1 - \left(\frac{c_2 a_{11}}{a_{12}^2}\right)y_2 \quad 9.4\text{-}15$$

$$\dot{\lambda}_2 = -\frac{\partial H}{\partial y_2} = -\lambda_1 + (a_{11} + a_{22})\lambda_2 - \left(\frac{c_2 a_{11}}{a_{12}^2}\right)y_1 - \left(\frac{c_2}{a_{12}^2}\right)y_2 \qquad 9.4\text{-}16$$

The optimal control is obtained when

$$\frac{\partial H}{\partial u_2} = 0 = c_3 u_2 + a_{12}b_{22}\lambda_2 \qquad 9.4\text{-}17$$

or

$$u_2 = -\frac{a_{12}b_{22}}{c_3}\lambda_2 \qquad 9.4\text{-}18$$

Now that we have obtained the equations describing the optimal control problem, we could attempt to find a solution by using Eq. 9.4-18 to eliminate the control variable from the state equations, Eq. 9.4-6 and 9.4-7, and looking for a solution of the four first-order, linear differential equations given by the state and adjoint equations, Eqs. 9.4-6, 9.4-7, 9.4-15, and 9.4-16, which satisfy the boundary conditions given by Eqs. 9.4-8 and 9.4-9. This is the procedure we followed in Example 9.3-1. However, the difficulties associated with solving four equations or, in general, $2n$ equations for an nth-order system can be avoided by simply assuming a solution having the form

$$\lambda_1 = m_{11}y_1 + m_{12}y_2 \qquad 9.4\text{-}19$$

$$\lambda_2 = m_{12}y_1 + m_{22}y_2 \qquad 9.4\text{-}20$$

where m_{11}, m_{12}, and m_{22} are unknown constants. Of course, if our assumed solution cannot be made to satisfy the optimization equations, we will not be able to consider it a valid solution, and it will be necessary either to try a different assumed solution or to look for a direct solution of the equations. we should also mention that this solution will be valid only if the final time in the performance index, Eq. 9.4-10, is taken to be $\theta = \infty$, since we obtain an asymptotic return to the desired steady state operating point.

When we assume that Eqs. 9.4-19 and 9.4-20 are valid at every instant of time, we are also implying that

$$\dot{\lambda}_1 = m_{11}\dot{y}_1 + m_{12}\dot{y}_2 \qquad 9.4\text{-}21$$

$$\dot{\lambda}_2 = m_{12}\dot{y}_1 + m_{22}\dot{y}_2 \qquad 9.4\text{-}22$$

Substituting the state equations, Eqs. 9.4–6 and 9.4–7, we have

$$\dot{\lambda}_1 = m_{11}y_2 - m_{12}[(a_{11}a_{22} - a_{12}a_{21})y_1 + (a_{11} + a_{22})y_2 - a_{12}b_{22}u_2]$$

$$\dot{\lambda}_2 = m_{12}y_2 - m_{22}[(a_{11}a_{22} - a_{12}a_{21})y_1 + (a_{21} + a_{22})y_2 - a_{12}b_{22}u_2]$$

We can eliminate u_2 from these equations and write them in terms of y_1 and y_2 only, by substituting Eq. 9.4–20 into Eq. 9.4–18. In this way we obtain

$$\dot{\lambda}_1 = -m_{12}\left(a_{11}a_{22} - a_{12}a_{21} + \frac{a_{12}^2 b_{22}^2}{c_3}m_{12}\right)y_1$$

$$+ \left[m_{11} - m_{12}\left(a_{11} + a_{22} + \frac{a_{12}^2 b_{22}^2}{c_3}m_{22}\right)\right]y_2 \qquad 9.4\text{–}23$$

$$\dot{\lambda}_2 = -m_{22}\left(a_{11}a_{22} - a_{12}a_{21} + \frac{a_{12}^2 b_{22}^2}{c_3}m_{12}\right)y_1$$

$$+ \left[m_{12} - m_{22}\left(a_{11} + a_{22} + \frac{a_{12}^2 b_{22}^2}{c_3}m_{22}\right)\right]y_2 \qquad 9.4\text{–}24$$

However, we can also use the adjoint equations, Eqs. 9.4–15 and 9.4–16, along with the assumed solutions, Eqs. 9.4–19 and 9.4–20, to develop relationships between $\dot{\lambda}_1, \dot{\lambda}_2$ and the state variables. This approach leads to the expressions

$$\dot{\lambda}_1 = \left[m_{12}(a_{11}a_{22} - a_{12}a_{21}) - \left(c_1 + c_2\frac{a_{11}^2}{a_{12}^2}\right)\right]y_1$$

$$+ \left[m_{22}(a_{11}a_{22} - a_{12}a_{21}) - c_2\frac{a_{11}}{a_{12}^2}\right]y_2 \qquad 9.4\text{–}25$$

$$\dot{\lambda}_2 = \left[-m_{11} + m_{12}(a_{11} + a_{22}) - c_2\frac{a_{11}}{a_{12}^2}\right]y_1$$

$$+ \left[-m_{12} + m_{22}(a_{11} + a_{22}) - \frac{c_2}{a_{12}^2}\right]y_2 \qquad 9.4\text{–}26$$

Our assumed solution can be valid only if both sets of results are identical. For arbitrary values of y_1 and y_2, we can make the equations the same by individually equating the coefficients of the y_1 and y_2 terms. This leads to three independent relationships in the three unknowns m_{11}, m_{12}, and m_{22}.

$$m_{12}^2\frac{a_{12}^2 b_{22}^2}{c_3} + 2m_{12}(a_{11}a_{22} - a_{12}a_{21}) - \left(c_1 + c_2\frac{a_{11}^2}{a_{12}^2}\right) = 0 \qquad 9.4\text{–}27$$

$$m_{12}m_{22}\frac{a_{12}^2 b_{22}^2}{c_3} - m_{11} + m_{12}(a_{11} + a_{22})$$

$$+ m_{22}(a_{11}a_{22} - a_{12}a_{21}) - c_2\frac{a_{11}}{a_{22}^2} = 0 \qquad 9.4\text{–}28$$

$$m_{22}^2\frac{a_{12}^2 b_{22}^2}{c_3} + 2m_{22}(a_{11} + a_{22}) - 2m_{12} - \frac{c_2}{a_{12}^2} = 0 \qquad 9.4\text{–}29$$

Only three independent equations are obtained because the coefficient of the y_2 term in the equation for $\dot{\lambda}_1$ gives the same result as the coefficient of the y_1 term in the equation for $\dot{\lambda}_2$. If we had not taken $m_{12} = m_{21}$ in our assumed solution, Eqs. 9.4–19 and 9.4–20, our analysis would have given us this result. It is a simple matter to solve the foregoing set of quadratic equations. The value of m_{12} can be found by using Eq. 9.4–27; next, m_{22} can be determined from Eq. 9.4–29, and, finally, m_{11} can be established by using Eq. 9.4–28. The positive roots are taken so that we will obtain a negative feedback control system. If we had not made the linear transformation given by Eqs. 9.4–4 and 9.4–5, the set of quadratic equations we would have obtained are coupled, and it would have been a more difficult task to calculate the unknown constants in the assumed solution.

After these constants have been determined, we can find the optimal control policy. From Eqs. 9.4–18 and 9.4–20 we have

$$u_2 = -\frac{a_{12}b_{22}}{c_3}\lambda_2$$

$$= -\frac{a_{12}b_{22}}{c_3}(m_{12}y_1 + m_{22}y_2) \qquad 9.4\text{–}30$$

or after substituting the inverse of the transformation given by Eqs. 9.4–4 and 9.4–5, we obtain

$$u_2 = \frac{a_{12}b_{22}}{c_3}[(m_{12} - a_{11}m_{22})x_1 - (a_{12}m_{22})x_2] \qquad 9.4\text{–}31$$

In other words, we find that we should use a proportional feedback control system, but that we should use a linear combination of the composition and temperature deviations as the input to the controller. This is in contrast to the servomechanism approach, where we were forced to choose the deviations of only one of these variables as the error signal, since the theory only applies to a single-input, single-output system.[1] With this generalized proportional controller, we can show that the system approaches the desired final steady state operating conditions in an exponential manner, so that it will take an infinite length of time to return to steady state.

The optimal control policy also gives us the optimal gains for the controller, providing that the constants c_1, c_2, and c_3 in the performance index have been specified. Unfortunately, a certain amount of subjective judgment is required in choosing these quantities, for a profit expression has not been used as the performance criterion. However, it would be possible to compare the results obtained from the optimal control problem with controllers designed on the basis of servomechanism theory—that is, a decay ratio of $\frac{1}{4}$ a phase margin of 45 deg, and so on. Hopefully, then, some "rules of thumb" could be developed for selecting the relative weighting factors in the performance index. If rules of this type are not available, however, it seems reasonable to choose these

---

[1] Actually, for this simple problem we could use the state equation, Eq. 9.4–2, to eliminate x_1 and replace our multivariable, proportional control law by a proportional plus derivative controller acting only on temperature. However, this procedure will fail when there are several control variables.

weighting factors so that the maximum errors of the state and control variables have the same order of magnitude.

TABLE 9.4-1

SYSTEM PARAMETERS FOR A CSTR WITH A COOLING COIL*

Parameters:

$A_f = 0.0065$, $C_p = 1.0$, $E = 14000$, $(-\Delta H) = 27000$, $K = 0.2$, $k_0 = 7.86 \times 10^{12}$

$q_s = 10$, $q_{cs} = 5$, $\rho = 1.0$, $T_c = 340$, $T_f = 350$, $V = 1000$

Steady state solution:

$A_s = 15.31 \times 10^{-5}$ $T_s = 460.91$

Constants in performance index:

$c_1 = 84.5$, $c_2 = 6.16$, $c_3 = 10^{-3}$

---

*Reproduced from J. M. Douglas, and M. M. Denn, *Ind. Eng. Chem.*, **57**, No. 11, 18 (1965), by permission.

Another difficulty with the preceding result is that it is based on the linearized system equations. For the system parameters given in Table 9.4–1,[2] it is a straightforward task to find the constants in the optimal control law. If we solve the original set of state equations, Eqs. 9.4–1 and 9.4–2, with this control policy, for a variety of initial conditions, we can find the paths that the system must follow as it returns to the optimum steady state design condition in an optimal manner. Instead of plotting the composition and temperature against time, we choose to plot the composition against the corresponding value of temperature at the same time to obtain a phase plane plot (we can mark off time intervals as a parameter on this plot if desirable). This procedure leads to the results shown in Figure 9.4–1, where, according to Eqs. 3.3–7 and 3.3–11, the dimensionless composition and temperature are given by the expressions

$$z_1 = \frac{A_s}{A_{fs}} + x_1 \qquad z_2 = \frac{T_s C_p \rho}{(-\Delta H)A_{fs}} + x_2 \qquad 9.4\text{-}32$$

It is apparent from the graph that some of the paths pass through the region of negative compositions, which we know is impossible from a physical standpoint. This discrepancy is due to the fact that the linearized system equations cannot be used to describe the nonlinear plant for large deviations from the equilibrium point. It would be possible to determine the region where the linearized equations were valid, using perturbation theory, but it would be more desirable to look for a way of treating the original nonlinear equations. This kind of an extension will be discussed in additional detail later in this section.

Pearson[3] has presented a detailed discussion of the method described above to systems of higher order—that is, having the form of Eq. 9.4–11 or 9.4–13. In addition, he pointed out that the quadratic equations containing the un-

---

[2] This example is taken from reference 4 on p. 257.

[3] J. D. Pearson, *J. Electron, Control*, **13**, 453 (1962).

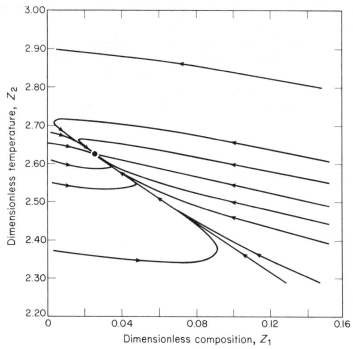

Figure 9.4-1. Phase plane plot for a linear feedback controller. [Reproduced from J. M. Douglas and M. M. Denn, *Ind. Eng. Chem.*, **57**, No. 11, 18 (1965) by permission of the ACS.]

known values of the optimal controller gain, m_{12} and m_{22}, are simple enough that they can be solved on-line with a small computer. Then if one of the system parameters changes with time for some reason, we could feed this information to the computer and automatically calculate new optimal controller gains. This kind of a system would be called an *adaptive controller*, for it adapts to changes in the system parameters or environment.

Example 9.4–2 A Multivariable linear feedback controller

The previous example represents a major extension of simple servomechanism theory because we use the controller to compensate for changes in both composition and temperature. We are providing more information to the controller, and therefore we expect that it should be able to do a better job. This is particularly true when one of the outputs might be very close to its desired value while the other is far away. Our optimal controller would call for a large amount of control in this case, whereas a servomechanism system might call for hardly any control.

Using this same reasoning, we expect that it would be possible to get an even better performance if we changed more than one control variable whenever we measure deviations from the optimum steady state design. A general procedure that can be used for this purpose, for the case of linear systems and an integral square error performance index, was developed by Kalman.[4] The solution will be presented in matrix form in order to simplify the algebra. However, it is quite similar to the previous problem, and a reader who gets confused with the notation should attempt to work out the results for a linearized CSTR problem. For that case, the system equations are

$$\dot{x}_1 = a_{11}x_1 + a_{12}x_2 + b_{11}u_1 \qquad\qquad 9.4\text{-}33$$

$$\dot{x}_2 = a_{21}x_1 + a_{22}x_2 + b_{21}u_1 + b_{22}u_2 \qquad 9.4\text{-}34$$

and a generalized performance index[5] is

$$P = \frac{1}{2} \int_0^{\theta=\infty} (c_{11}x_1^2 + 2c_{12}x_1x_2 + c_{22}x_2^2$$

$$+ r_{11}u_1^2 + 2r_{12}u_1u_2 + r_2^2u_2^2)\,dt \qquad 9.4\text{-}35$$

Solution

In matrix notation, we can write

$$\dot{\mathbf{x}} = \mathbf{A}\mathbf{x} + \mathbf{B}\mathbf{u} \qquad\qquad 9.4\text{-}36$$

and

$$P = \frac{1}{2} \int_0^{\theta=\infty} (\mathbf{x}^T\mathbf{C}\mathbf{x} + \mathbf{u}^T\mathbf{R}\mathbf{u})\,dt \qquad 9.4\text{-}37$$

where the superscript T stands for the transpose of the matrix and $\mathbf{C}$ and $\mathbf{R}$ are symmetric, positive definite matrices.

We can write the Hamiltonian, a scalar, as

$$H = \tfrac{1}{2}(\mathbf{x}^T\mathbf{C}\mathbf{x} + \mathbf{u}^T\mathbf{R}\mathbf{u}) + \boldsymbol{\lambda}^T\mathbf{A}\mathbf{x} + \boldsymbol{\lambda}^T\mathbf{B}\mathbf{u} \qquad 9.4\text{-}38$$

The canonical equations become

$$\dot{\mathbf{x}} = \frac{\partial H}{\partial \boldsymbol{\lambda}} = \mathbf{A}\mathbf{x} + \mathbf{B}\mathbf{u} \qquad\qquad 9.4\text{-}36$$

$$\dot{\boldsymbol{\lambda}} = -\frac{\partial H}{\partial \mathbf{x}} = -\mathbf{C}\mathbf{x} - \mathbf{A}^T\boldsymbol{\lambda} \qquad 9.4\text{-}39$$

where we are differentiating with respect to a vector. Similarly, the optimal control law is

$$\frac{\partial H}{\partial \mathbf{u}} = 0 = \mathbf{R}\mathbf{u} + \mathbf{B}^T\boldsymbol{\lambda}$$

---

[4] See, for example, R. E. Kalman, "The Theory of Optimal Control and the Calculus of Variations," *RIAS Report 61-3*, 1961.

[5] Again, we choose $\theta = \infty$ because we want to reach the origin as time progresses. A discussion of the finite-time problem is presented in Exercise 11, p. 351.

or

$$\mathbf{u} = -\mathbf{R}^{-1}\mathbf{B}^T\boldsymbol{\lambda} \qquad 9.4\text{-}40$$

where the superscript (-1) stands for the inverse of the matrix, which we assume exists. This is not always the case, however; in fact, in our previous example

$$\mathbf{R} = \begin{pmatrix} r_{11} & r_{12} \\ r_{12} & r_{22} \end{pmatrix} = \begin{pmatrix} 0 & 0 \\ 0 & c_3 \end{pmatrix} \qquad 9.4\text{-}41$$

which does not have an inverse. (Of course, for this case, $\mathbf{R}$ is not positive definite either.)

Now, for the case where $\theta = \infty$, we assume that we can find a solution of the preceding equations that has the form

$$\boldsymbol{\lambda} = \mathbf{M}\mathbf{x} \qquad 9.4\text{-}42$$

where $\mathbf{M}$ is a symmetric matrix, but its elements are unknown. This solution implies that

$$\dot{\boldsymbol{\lambda}} = \mathbf{M}\dot{\mathbf{x}} \qquad 9.4\text{-}43$$

or after substituting the state equations

$$\dot{\boldsymbol{\lambda}} = \mathbf{M}(\mathbf{A}\mathbf{x} + \mathbf{B}\mathbf{u}) \qquad 9.4\text{-}44$$

Substituting our assumed solution, Eq. 9.4-42, into the optimal control law, Eq. 9.4-40, and using this result to eliminate $\mathbf{u}$ from the equation above, we obtain

$$\dot{\boldsymbol{\lambda}} = \mathbf{M}\mathbf{A}\mathbf{x} - \mathbf{M}\mathbf{B}\mathbf{R}^{-1}\mathbf{B}^T\mathbf{M}\mathbf{x} \qquad 9.4\text{-}45$$

However, we can also develop an expression for $\dot{\boldsymbol{\lambda}}$ in terms of the state variables by substituting our assumed solution into the adjoint equations, Eq. 9.4-39

$$\dot{\boldsymbol{\lambda}} = -\mathbf{C}\mathbf{x} - \mathbf{A}^T\mathbf{M}\mathbf{x} \qquad 9.4\text{-}46$$

Obviously our assumed solution will be valid only if these two expressions are identical; and for arbitrary values of $\mathbf{x}$, this means that we must have

$$-\mathbf{M}\mathbf{B}\mathbf{R}^{-1}\mathbf{B}^T\mathbf{M} + \mathbf{M}\mathbf{A} + \mathbf{A}^T\mathbf{M} + \mathbf{C} = 0 \qquad 9.4\text{-}47$$

This is a coupled set of quadratic equations, similar to Eqs. 9.4-27 through 9.4-29, which must be solved to find the elements of $\mathbf{M}$. Once these values have been established, we can find the optimal feedback control law from Eqs. 9.4-40 and 9.4-42

$$\mathbf{u} = -\mathbf{R}^{-1}\mathbf{B}^T\boldsymbol{\lambda} = -\mathbf{R}^{-1}\mathbf{B}^T\mathbf{M}\mathbf{x} \qquad 9.4\text{-}48$$

Hence we find that the optimal controller is a multivariable, proportional feedback system, which is simple to implement, using either analog or digital computing components. (Actually, any hardware that can add signals can be used.)

Since the problem just described is so similar to the preceding example, it will suffer from the same potential shortcomings. The only additional difficulty is finding techniques for solving the set of coupled, quadratic equations given by Eq. 9.4–47. One approach is to use a nonlinear root-finding routine on a digital computer. An alternate method is to add a time derivative term to the equation

$$\mathbf{M} - \mathbf{MBR}^{-1}\mathbf{B}^T\mathbf{M} + \mathbf{MA} + \mathbf{A}^T\mathbf{M} + \mathbf{C} = 0 \qquad 9.4\text{–}49$$

and then to solve this set of Riccati equations numerically in order to find the steady state solutions, which are merely the roots. Kalman has suggested that this second approach is generally simpler, for it can be shown that the set of differential equations given by Eq. 9.4–49 is always stable, so that the numerical solutions will always converge to the proper values.

Example 9.4–3 Effect of disturbances on the optimal control policy

In any real plant problem, there will always be disturbances that enter the system and make it move away from the optimum steady state design conditions. The whole purpose of the control system is to compensate for the effect that these disturbances have on the plant. During our study of servomechanism control problems, we were often able to make a quantitative assessment of how well the controller performed its job, and we should be able to do the same thing for optimal control problems. We will limit our discussion to a first-order system, but a solution for an nth-order plant with a single control variable has been published by Denn.[6]

Solution

The system equation is taken as

$$\dot{x} = Ax + Bu + v(t) \qquad 9.4\text{–}50$$

where $v(t)$ represents the disturbance, which may be a deviation in feed composition, or the temperature, of an isothermal CSTR. The performance index is

$$P = \frac{1}{2}\int_0^{\theta=\infty} (x^2 + c^2u^2)\,dt \qquad 9.2\text{–}4$$

Since time appears explicitly in the state equation, we make the change of variables

$$x_1 = x \quad \dot{x}_2 = 1 \quad x_2(0) = 0 \qquad 9.4\text{–}51$$

so that the state equations become

$$\dot{x}_1 = Ax_1 + Bu + v(x_2) \qquad 9.4\text{–}52$$

$$\dot{x}_2 = 1 \qquad x_2(0) = 0 \qquad 9.4\text{–}53$$

and the performance index becomes

$$P = \frac{1}{2}\int_0^{\theta=\infty} (x_1^2 + c^2u^2)\,dt \qquad 9.4\text{–}54$$

[6] See Ch. 4. of reference 7, p. 261.

Hence the Hamiltonian can be written as

$$H = \tfrac{1}{2}(x_1^2 + c^2 u^2) + \lambda_1[Ax_1 + Bu + v(x_2)] + \lambda_2 \qquad 9.4\text{-}55$$

the adjoint equations are

$$\dot{\lambda}_1 = -x_1 - A\lambda_1 \qquad 9.4\text{-}56$$

$$\dot{\lambda}_2 = -\lambda_1 \frac{\partial v}{\partial x_2} = -\lambda_1 \frac{\partial v}{\partial t} = -\lambda_1 \dot{v} \qquad 9.4\text{-}57$$

and the optimal control law is

$$\frac{\partial H}{\partial u} = 0 = c^2 u + B\lambda_1$$

or

$$u = -\frac{B}{c^2}\lambda_1 \qquad 9.4\text{-}58$$

After examining these equations, we find that λ_2 and x_2 are not really needed for the solution; therefore, after substituting Eq. 9.4-58 into Eq. 9.4-52, we can drop the subscripts and write

$$\dot{x} = Ax - \frac{B^2}{c^2}\lambda + v(t) \qquad 9.4\text{-}59$$

$$\dot{\lambda} = -x - A\lambda \qquad 9.4\text{-}60$$

If we look for a solution of this set of linear equations directly, by eliminating the adjoint variable λ, we obtain

$$\ddot{x} - \left(A^2 + \frac{B^2}{c^2}\right)x = Av + \dot{v} \qquad 9.4\text{-}61$$

Before we can proceed any farther, however, we must completely specify the disturbance $v(t)$ for all future times of interest. Unfortunately this step is not possible for practical problems, and the best we can do is solve the equation for some simple inputs.

For example, for the case of a step input, $v = v_s$, entering a system that originally is at the optimum steady state conditions

$$x(0) = 0 \qquad 9.4\text{-}62$$

the solution is

$$x(t) = C_1 \exp - \left(A\sqrt{1 + \frac{B^2}{c^2 A^2}}\right)$$

$$+ C_2 \exp \left(A\sqrt{1 + \frac{B^2}{c^2 A^2}}\right)t - \left(\frac{v_s}{1 + (B^2/c^2 A^2)}\right) \qquad 9.4\text{-}63$$

It is clear that the state variable will become unbounded unless we set $C_2 = 0$. The value of C_1 is determined so that the solution matches the initial condition given by Eq. 9.4–62. Thus we obtain

$$x = \frac{v_s}{[1 + (B^2/c^2 A^2)]} \left[1 - \exp - \left(A\sqrt{1 + \frac{B^2}{c^2 A^2}}\right) t \right] \qquad 9.4\text{–}64$$

Again we find that as time approaches infinity, the system will not return to the desired steady state conditions. This is identical to the result from servomechanism theory, where we found that step inputs always produce a certain amount of offset. It is clear that if the cost of control is very small, the offset becomes small; whereas if the cost of control is very high, the control system barely reduces the offset (in comparison with the uncontrolled case). This agrees with our intuition and our previous knowledge.

Now that we have determined the optimal path that the system follows, we can use Eq. 9.4–52 to find the optimal control policy for the step input

$$u = \frac{\dot{x} - Ax - v_s}{B} \qquad 9.4\text{–}65$$

This gives the result

$$u = -\frac{A}{B} \left(1 + \sqrt{1 + \frac{B^2}{c^2 A^2}}\right) x - \frac{v_s}{B} \left[1 - \left(\frac{A}{\sqrt{1 + (B^2/c^2 A^2)}}\right) \right] \qquad 9.4\text{–}66$$

Although the gain of the proportional feedback control part of this policy is exactly the same as we had before, we now have to add a constant term that depends on the magnitude of the step input.

We can use a similar procedure to find the optimal control policies for other kinds of disturbances and can follow the procedure described by Denn to extend the results to higher-order systems having only a single control variable.

Example 9.4–4 Proportional plus integral control

Our study of servomechanism theory showed that offset could be removed if an integral mode was included in the controller. Hence it would be useful to find a way to obtain a proportional plus integral controller from optimal control theory. A procedure that leads to this result is described below for the case of a first-order system.[6] The state equation is taken as

$$\dot{x} = Ax + Bu + v \qquad x(0) = 0 \qquad 9.4\text{–}67$$

and the performance index is

$$P = \frac{1}{2} \int_0^{\theta = \infty} (x^2 + c^2 \dot{u}^2)\, dt \qquad 9.4\text{–}68$$

where the rate of change of the control variable is used as a measure of poor performance, rather than the amount of control.

Solution

Our formulation of Pontryagin's minimum principle cannot be applied directly to this kind of a performance index, and therefore we must look for a transformation. For a case where the system is at steady state conditions originally and at time zero a step disturbance enters the plant, $v = v_s$, we can differentiate the state equation to obtain

$$\ddot{x} = A\dot{x} + B\dot{u}, \quad x(0) = 0, \quad \dot{x}(0) = v_s \qquad 9.4\text{-}69$$

Now we can let

$$x_1 = x, \quad x_2 = \dot{x}_1 = \dot{x}, \quad w = \dot{u} \qquad 9.4\text{-}70$$

so that the state equations become

$$\dot{x}_1 = x_2 \qquad x_1(0) = 0 \qquad 9.4\text{-}71$$

$$\dot{x}_2 = Ax_2 + Bw \qquad x_2(0) = v_s \qquad 9.4\text{-}72$$

and the performance index becomes

$$P = \frac{1}{2} \int_0^{\theta = \infty} (x_1^2 + c^2 w^2)\, dt \qquad 9.4\text{-}73$$

This problem is almost the same as that studied in Example 9.4–1. Following the procedure described earlier, we obtain

$$w = -\frac{1}{c} x_1 - \frac{A}{B} \left[1 + \sqrt{1 + \frac{2B}{cA^2}} \right] x_2 \qquad 9.4\text{-}74$$

Substituting the original variables, Eq. 9.4–70, this becomes

$$\dot{u} = -\frac{1}{c} x - \frac{A}{B} \left[1 + \sqrt{1 + \frac{2B}{cA^2}} \right] \dot{x} \qquad 9.4\text{-}75$$

Then integrating this equation, we obtain

$$u(t) = -\frac{A}{B} \left[1 + \sqrt{1 + \frac{2B}{cA^2}} \right] x(t) - \frac{1}{c} \int_0^t x(t)\, dt \qquad 9.4\text{-}76$$

Hence the optimal control law contains both a proportional and an integral control mode.

If we substitute Eq. 9.4–75 into Eq. 9.4–69, we obtain an equation for the optimal control path

$$\ddot{x} + A\sqrt{1 + \frac{2B}{cA^2}}\, \dot{x} + \frac{B}{c} x = 0 \qquad 9.4\text{-}77$$

The characteristic roots of this differential equation are

$$-\frac{A}{2} \sqrt{1 + \frac{2B}{cA^2}} \pm \frac{A}{2} \sqrt{1 - \frac{2B}{cA^2}} \qquad 9.4\text{-}78$$

so that if

$$c < \frac{2B}{A^2} \qquad 9.4\text{-}79$$

We will obtain an underdamped system. By comparing the solution of the optimal control problem with the solution from servomechanism theory that gives a decay ratio of $\frac{1}{4}$, a 45 deg phase margin, and so on, we can develop some "rules of thumb" relating the relative cost of control to the servomechanism design criteria. It is important to note that the solution of Eq. 9.4–77 with the boundary conditions given by Eq. 9.4–69 will always approach zero as time approaches infinity; hence the addition of the integral control mode eliminates offset.

Example 9.4–5 A three-mode controller

If we apply the same approach as described above to a second-order system, we obtain a three-mode controller.[7] Again we consider the linearized CSTR equations

$$\dot{x}_1 = a_{11}x_1 + a_{12}x_2 \qquad\qquad 9.4\text{–}80$$

$$\dot{x}_2 = a_{21}x_1 + a_{22}x_2 + b_{22}u_2 + v_2 \qquad\qquad 9.4\text{–}81$$

where v_2 represents a disturbance in either the feed temperature or the inlet temperature to the heating coil. The performance index is

$$P = \frac{1}{2}\int_0^{\theta=\infty} (c_1 x_1^2 + c_2 x_2^2 + c_3 \dot{u}_2^2)\, dt \qquad\qquad 9.4\text{–}82$$

Solution

Making the transformation

$$x_1 = -y_1 \qquad\qquad 9.4\text{–}4$$

$$x_2 = \frac{a_{11}}{a_{21}}y_1 + \frac{1}{a_{12}}y_2 \qquad\qquad 9.4\text{–}5$$

and letting

$$a_1 = (a_{11}a_{22} - a_{12}a_{21}), \quad a_2 = (a_{11} + a_{22}), \quad u = a_{12}b_{22}u_2$$

$$v = a_{12}v_2, \quad c_{11} = c_1 + c_2\frac{a_{11}^2}{a_{12}^2}, \quad 2c_{12} = \frac{2c_2 a_{11}}{a_{12}^2} \qquad\qquad 9.4\text{–}83$$

$$c_{22} = \frac{c_2}{a_{12}^2} \qquad c_{33} = \frac{c_3}{a_{12}^2 b_{22}^2}$$

the state equations become

$$\dot{y}_1 = y_2 \qquad\qquad 9.4\text{–}84$$

$$\dot{y}_2 = -a_1 y_1 - a_2 y_2 + u + v \qquad\qquad 9.4\text{–}85$$

and the performance index becomes

$$P = \frac{1}{2}\int_0^{\theta} (c_{11}y_1^2 + 2c_{12}y_1 y_2 + c_{22}y_2^2 + c_{33}\dot{u}^2)\, dt \qquad\qquad 9.4\text{–}86$$

It is interesting to note that we can also write these state equations in the

---

[7] Taken from Ch. 8. of reference 7 on p. 261.

form

$$\ddot{y}_1 = \dot{y}_2 = -a_1 y_1 + a_2 y_2 + u + v$$
$$= -a_1 y_1 - a_2 \dot{y}_1 + u + v$$

or

$$\ddot{y}_1 + a_2 \dot{y}_1 + a_1 y_1 = u + v \qquad 9.4\text{-}87$$

We will need this result later in our analysis.

Since the performance index contains $\dot{u}$ rather than u, we need to introduce an additional transformation. Instead of following the procedure described in the previous example, we will introduce a new state variable and let

$$y_3 = u \qquad 9.4\text{-}88$$

$$\dot{y}_3 = \dot{u} = w \qquad 9.4\text{-}89$$

With these definitions, the state equations become

$$\dot{y}_1 = y_2 \qquad 9.4\text{-}90$$

$$\dot{y}_2 = -a_1 y_1 - a_2 y_2 + y_3 + v \qquad 9.4\text{-}91$$

$$\dot{y}_3 = w \qquad 9.4\text{-}92$$

and the performance index becomes

$$P = \frac{1}{2} \int_0^{\theta = \infty} (c_{11} y_1^2 + 2c_{12} y_1 y_2 + c_{22} y_2^2 + c_{33} w^2)\, dt \qquad 9.4\text{-}93$$

Now we can write the Hamiltonian as

$$H = \tfrac{1}{2}(c_{11} y_1^2 + 2c_{12} y_1 y_2 + c_{22} y_2^2 + c_{33} w^2) + \lambda_1 y_2$$
$$+ \lambda_2(-a_1 y_1 - a_2 y_2 + y_3 + v) + \lambda_3 w \qquad 9.4\text{-}94$$

so that the adjoint equations are

$$\dot{\lambda}_1 = -c_{11} y_1 - c_{12} y_2 + a_1 \lambda_2 \qquad 9.4\text{-}95$$

$$\dot{\lambda}_2 = -c_{12} y_1 - c_{22} y_2 - \lambda_1 + a_2 \lambda_2 \qquad 9.4\text{-}96$$

$$\dot{\lambda}_3 = -\lambda_2 \qquad 9.4\text{-}97$$

and the optimal control law is

$$\frac{\partial H}{\partial w} = 0 = c_{33} w + \lambda_3$$

or

$$w = -\frac{\lambda_3}{c_{33}} \qquad 9.4\text{-}98$$

If we assume that we can find a solution having the form

$$\lambda_1 = m_{11} y_1 + m_{12} y_2 + m_{13} y_3 + n_1 v \qquad 9.4\text{-}99$$

$$\lambda_2 = m_{12} y_1 + m_{22} y_2 + m_{23} y_3 + n_2 v \qquad 9.4\text{-}100$$

$$\lambda_3 = m_{13} y_1 + m_{23} y_2 + m_{33} y_3 + n_3 v \qquad 9.4\text{-}101$$

where the terms m_{ij} and n_i are unknown constants, and follow the procedure we have described in most of the previous examples, we find that our assumed solution is valid when

$$c_{11}c_{33} = m_{13}^2 + 2a_1 m_{12} c_{33} \qquad\qquad 9.4\text{-}102$$

$$c_{12}c_{33} = m_{13}m_{23} - m_1 c_{33} + a_2 m_{12} c_{33} + a_1 m_{22} c_{33} \qquad 9.4\text{-}103$$

$$0 = m_{13}m_{33} - m_{12}c_{33} + a_1 m_{23} c_{33} \qquad\qquad 9.4\text{-}104$$

$$c_{22}c_{33} = m_{23}^2 - 2m_{12}c_{33} + 2a_2 m_{22} c_{33} \qquad\qquad 9.4\text{-}105$$

$$0 = m_{23}m_{33} - m_{13}c_{33} + a_2 m_{23} c_{33} - m_{22}c_{33} \qquad 9.4\text{-}106$$

$$0 = m_{33}^2 - 2m_{23}c_{33} \qquad\qquad 9.4\text{-}107$$

$$m_{12}c_{33} = a_1 n_2 + m_{13} n_3 \qquad\qquad 9.4\text{-}108$$

$$m_{22}c_{33} = -n_1 c_{33} + a_2 n_2 c_{33} + m_{23} n_3 \qquad\qquad 9.4\text{-}109$$

$$m_{23}c_{33} = -n_2 c_{33} + m_{33} n_3 \qquad\qquad 9.4\text{-}110$$

which provides nine equations that can be used to evaluate the nine unknown constants. After some manipulation, we find that m_{33} is the root of the quartic equation.

$$m_{33}^4 + 4a_2 c_{33} m_{33}^3 + 4(a_1 + a_2^2)c_{33}^2 m_{33}^2 + 8(a_1 a_2 c_{33} - \sqrt{c_{11}c_{33}})c_{33}^2 m_{33}$$
$$- 4(c_{22} + 2a_2\sqrt{c_{11}c_{33}})c_{33}^3 = 0 \qquad\qquad 9.4\text{-}111$$

Once this root has been determined, the other parameters of interest can be evaluated, using the expressions

$$m_{13} = -a_1 m_{33} + \sqrt{c_{11}c_{33}} \qquad\qquad 9.4\text{-}112$$

$$m_{23} = \frac{m_{33}^2}{2c_{33}} \qquad\qquad 9.4\text{-}113$$

$$n_3 = m_{33} \qquad\qquad 9.4\text{-}114$$

Substitution of Eq. 9.4–101 into Eq. 9.4–98 gives the optimal control law as

$$w = -\frac{m_{13}}{c_{33}} y_1 - \frac{m_{23}}{c_{33}} y_2 - \frac{m_{33}}{c_{33}} y_3 - \frac{n_3}{c_{33}} v \qquad 9.4\text{-}115$$

Then, using Eqs. 9.4–84, 9.4–88, 9.4–89, and 9.4–114, we can write the equation as

$$\dot{u} = -\frac{m_{13}}{c_{33}} y_1 - \frac{m_{23}}{c_{33}} \dot{y}_1 - \frac{m_{33}}{c_{33}}(u + v) \qquad 9.4\text{-}116$$

Finally, substituting Eq. 9.4–87, we obtain

$$\dot{u} = -\left(\frac{m_{33}}{c_{33}}\right)\ddot{y}_1 - \left(\frac{m_{23} + a_2 m_{33}}{c_{33}}\right)\dot{y}_1 - \left(\frac{m_{13} + a_1 m_{33}}{c_{33}}\right)y_1 \qquad 9.4\text{-}117$$

Integrating and using Eqs. 9.4–112 and 9.4–113 to eliminate m_{13} and m_{23}, we find that

$$u(t) = -\left(\frac{m_{33}}{c_{33}}\right)\dot{y}_1 - \left(\frac{m_{33}^2 + 2a_2 c_{33} m_{33}}{c_{33}^2}\right)y_1(t) - \left(\sqrt{\frac{c_{11}}{c_{33}}}\right)\int_0^t y_1(t)\, dt$$

or since

$$y_1 = -x_1 \qquad\qquad\qquad 9.4\text{-}4$$

we obtain the three-mode control law

$$u(t) = +\left(\frac{m_{33}}{c_{33}}\right)\dot{x}_1 + \left(\frac{m_{33}^2 + 2a_2 c_{33} m_{33}}{c_{33}^2}\right)x_1(t) + \sqrt{\frac{c_{11}}{c_{33}}}\int_0^t x_1(t)\, dt \qquad 9.4\text{-}118$$

The standard equation for a three-mode controller is often written

$$u(t) = K_c\left[x(t) + \frac{1}{T_i}\int_0^t x(t)\, dt + T_D \frac{dx(t)}{dt}\right] \qquad 9.4\text{-}119$$

Hence by comparing coefficients we find that

$$K_c = \frac{m_{33}}{c_{33}^2}(m_{33} + 2a_2 c_{33}) \qquad T_i = \frac{m_{33}}{c_{33}^2}\left(\sqrt{\frac{c_{33}}{c_{11}}}\right)(m_{33} + 2a_2 c_{33})$$

$$T_D = \frac{c_{33}}{m_{33} + 2a_2 c_{33}} \qquad\qquad 9.4\text{-}120$$

For sufficiently large values of c_{11} or c_{22}, an approximate solution of Eq. 9.4-111 is

$$m_{33} = [4(c_{22} + 2a_2\sqrt{c_{11}c_{33}})c_{33}^3]^{1/4} \qquad 9.4\text{-}121$$

which is frequently useful for making first estimates of the system parameters.

Now that explicit expressions for the controller gains are available, it would be possible to compare the optimal control system with one designed by means of conventional servomechanism techniques. In this way we could develop a better understanding of how to select reasonable values for the weighting factors in the performance index. It should be recognized, however, that the preceding problem has a very special form, and that, in general, we do not expect to obtain three-mode controllers as the solution to optimization problems.

Example 9.4-6 A multivariable proportional plus integral controller

Most of the foregoing discussion has been limited to a single control variable. We would like to extend the results to multivariable plants, for most chemical processes fall in this category. Our development below is given in terms of matrix notation in order to conserve space, but by now the reader should be able to solve the problem without this notation if he finds it confusing. The equations describing the system are considered to be

$$\dot{\mathbf{x}} = \mathbf{A}\mathbf{x} + \mathbf{B}\mathbf{u} + \mathbf{C}\mathbf{v} \qquad\qquad 9.4\text{-}122$$

where initially the system is at the optimum steady state conditions, $\mathbf{x}(0) = \mathbf{0}$, and at time equal to zero a number of step disturbances, $\mathbf{v}$, enter the system.

The performance index is taken as

$$P = \frac{1}{2} \int_0^{\theta=\infty} (\mathbf{x}^T \mathbf{Q} \mathbf{x} + \dot{\mathbf{u}}^T \mathbf{R} \dot{\mathbf{u}}) \, dt \qquad 9.4\text{-}123$$

where $\mathbf{Q}$ and $\mathbf{R}$ are symmetric, positive definite matrices.

Solution

This functional does not fit our formulation of the maximum principle, so that again we must seek an appropriate transformation. Differentiating the state equation

$$\ddot{\mathbf{x}} = \mathbf{A}\dot{\mathbf{x}} + \mathbf{B}\dot{\mathbf{u}} \qquad 9.4\text{-}124$$

and letting

$$\mathbf{y}_1 = \mathbf{x}, \quad \mathbf{y}_2 = \dot{\mathbf{y}}_1 = \dot{\mathbf{x}}, \quad \dot{\mathbf{u}} = \mathbf{w} \qquad 9.4\text{-}125$$

we obtain the new set of state equations

$$\dot{\mathbf{y}}_1 = \mathbf{y}_2 \qquad 9.4\text{-}126$$

$$\dot{\mathbf{y}}_2 = \mathbf{A}\mathbf{y}_2 + \mathbf{B}\mathbf{w} \qquad 9.4\text{-}127$$

and the corresponding performance index

$$P = \frac{1}{2} \int_0^{\theta=\infty} (\mathbf{y}_1^T \mathbf{Q} \mathbf{y}_1 + \mathbf{w}^T \mathbf{R} \mathbf{w}) \, dt \qquad 9.4\text{-}128$$

The Hamiltonian becomes

$$H = \tfrac{1}{2} (\mathbf{y}_1^T \mathbf{Q} \mathbf{y}_1 + \mathbf{w}^T \mathbf{R} \mathbf{w}) + \boldsymbol{\lambda}_1^T \mathbf{y}_2 + \boldsymbol{\lambda}_2^T (\mathbf{A}\mathbf{y}_2 + \mathbf{B}\mathbf{w}) \qquad 9.4\text{-}129$$

the adjoint equations are

$$\dot{\boldsymbol{\lambda}}_1 = -\mathbf{Q}\mathbf{y}_1 \qquad 9.4\text{-}130$$

$$\dot{\boldsymbol{\lambda}}_2 = -\boldsymbol{\lambda}_1 - \mathbf{A}^T \boldsymbol{\lambda}_2 \qquad 9.4\text{-}131$$

and the optimal control law is

$$\frac{\partial H}{\partial \mathbf{w}} = 0 = \mathbf{R}\mathbf{w} + \mathbf{B}^T \boldsymbol{\lambda}_2$$

or

$$\mathbf{w} = -\mathbf{R}^{-1} \mathbf{B}^T \boldsymbol{\lambda}_2 \qquad 9.4\text{-}132$$

where we are assuming that $\mathbf{R}^{-1}$ exists.

Now we assume a solution having the form

$$\boldsymbol{\lambda}_1 = \mathbf{K}\mathbf{y}_1 + \mathbf{L}\mathbf{y}_2 \qquad \boldsymbol{\lambda}_2 = \mathbf{M}\mathbf{y}_1 + \mathbf{N}\mathbf{y}_2 \qquad 9.4\text{-}133$$

where $\mathbf{K}, \mathbf{L}, \mathbf{M}$, and $\mathbf{N}$ are matrices whose elements must be determined. The assumed solutions imply that

$$\dot{\boldsymbol{\lambda}}_1 = \mathbf{K}\dot{\mathbf{y}}_1 + \mathbf{L}\dot{\mathbf{y}}_2 \qquad \dot{\boldsymbol{\lambda}}_2 = \mathbf{M}\dot{\mathbf{y}}_1 + \mathbf{N}\dot{\mathbf{y}}_2 \qquad 9.4\text{-}134$$

or after substituting the state equations and optimal control law,

$$\dot{\boldsymbol{\lambda}}_1 = -\mathbf{L}\mathbf{B}\mathbf{R}^{-1}\mathbf{B}^T \mathbf{M}\mathbf{y}_1 - (\mathbf{L}\mathbf{B}\mathbf{R}^{-1}\mathbf{B}^T \mathbf{N} - \mathbf{K} - \mathbf{L}\mathbf{A})\mathbf{y}_2 \qquad 9.4\text{-}135$$

$$\dot{\boldsymbol{\lambda}}_2 = -\mathbf{N}\mathbf{B}\mathbf{R}^{-1}\mathbf{B}^T \mathbf{M}\mathbf{y}_1 - (\mathbf{N}\mathbf{B}\mathbf{R}^{-1}\mathbf{B}^T \mathbf{N} - \mathbf{M} - \mathbf{N}\mathbf{A})\mathbf{y}_2 \qquad 9.4\text{-}136$$

Substituting the assumed solution into the adjoint equations gives the expressions

$$\dot{\lambda}_1 = -\mathbf{Q}\mathbf{y}_1 \qquad 9.4\text{--}130$$

$$\dot{\lambda}_2 = -(\mathbf{K} + \mathbf{A}^T\mathbf{M})\mathbf{y}_1 - (\mathbf{L} + \mathbf{A}^T\mathbf{N})\mathbf{y}_2 \qquad 9.4\text{--}137$$

and in order for our assumed solution to be correct, both sets of expressions must give the same results. Hence we find that

$$\mathbf{L}\mathbf{B}\mathbf{R}^{-1}\mathbf{B}^T\mathbf{M} = \mathbf{Q} \qquad 9.4\text{--}138$$

$$-\mathbf{L}\mathbf{B}\mathbf{R}^{-1}\mathbf{B}^T\mathbf{N} + \mathbf{K} + \mathbf{L}\mathbf{A} = 0 \qquad 9.4\text{--}139$$

$$-\mathbf{N}\mathbf{B}\mathbf{R}^{-1}\mathbf{B}^T\mathbf{M} + \mathbf{K} + \mathbf{A}^T\mathbf{M} = 0 \qquad 9.4\text{--}140$$

$$-\mathbf{N}\mathbf{B}\mathbf{R}^{-1}\mathbf{B}^T\mathbf{N} + \mathbf{M} + \mathbf{N}\mathbf{A} + \mathbf{L} + \mathbf{A}^T\mathbf{N} = 0 \qquad 9.4\text{--}141$$

so that we have four nonlinear matrix equations that we can solve to determine $\mathbf{K}, \mathbf{L}, \mathbf{M},$ and $\mathbf{N}$.

Once these values have been established, we can substitute the assumed solution, Eq. 9.4–133, into the optimal control law, Eq. 9.4–132, to obtain

$$\mathbf{w} = -\mathbf{R}^{-1}\mathbf{B}^T\lambda_2 = -\mathbf{R}^{-1}\mathbf{B}^T\mathbf{M}\mathbf{y}_1 - \mathbf{R}^{-1}\mathbf{B}^T\mathbf{N}\mathbf{y}_2 \qquad 9.4\text{--}142$$

which in terms of the original variables, Eq. 9.4–125, becomes

$$\dot{\mathbf{u}} = -\mathbf{R}^{-1}\mathbf{B}^T\mathbf{M}\mathbf{x} - \mathbf{R}^{-1}\mathbf{B}^T\mathbf{N}\dot{\mathbf{x}} \qquad 9.4\text{--}143$$

Integrating this expression then gives us a multivariable proportional plus integral controller[8]

$$\mathbf{u}(t) = -\mathbf{R}^{-1}\mathbf{B}^T\mathbf{N}\mathbf{x}(t) - \mathbf{R}^{-1}\mathbf{B}^T\mathbf{M}\int_0^t \mathbf{x}(t)\,dt \qquad 9.4\text{--}144$$

Example 9.4–7   Extension to nonlinear systems

The great advantages of the control systems just described is that they are simple to implement, they are multivariable systems, and the optimum values of the controller gains only depend on the assumptions in the problem formulation—that is, the weighting factors in the performance index. However, they are based on an analysis of linear system equations; therefore, in some cases, such as Example 9.4–1, they predict that negative compositions or other physically unrealistic values will be obtained. Of course, conventional servomechanism theory suffers from exactly the same difficulty, because it is also limited to linear systems.

Our development of the minimum principle applies equally well to nonlinear plants; thus we should be able to use optimization theory to design multivariable controllers that correspond to the exact nonlinear system equations. We must keep in mind, however, that if we must solve a complicated nonlinear boundary value problem on-line in order to find the optimal control action,

---

[8] Some recent studies indicate that the criterion for complete controllability is not satisfied for this system, and a solution does not exist, unless the number of control variables is equal to the number of state variables.

the cost of the controller will be quite high, whereas an approximate solution of
the optimization equations that is simple to implement might lead to a higher
overall profit. An illustration of the difference between these two approaches is
given below for the very simple problem of an isothermal stirred-tank reactor.
The equation describing the nonlinear system for a second-order reaction is

$$V\frac{dA}{dt} = q(A_f - A) - kVA^2 \qquad 9.4\text{--}145$$

If we consider that the reactor temperature, or rather the rate constant k, is
used as a control variable and let

$$x = \frac{A - A_s}{A_f}, \quad u = k - k_s, \quad \beta_1 = 2k_s A_s + \frac{q}{V}, \quad \beta_2 = \frac{A_s^2}{A_f},$$

$$\beta_3 = k_s A_f, \quad \beta_4 = 2A_s, \quad \beta_5 = A_f \qquad 9.4\text{--}146$$

then the state equation can be put into the form

$$\dot{x} = -\beta_1 x - \beta_2 u - \beta_3 x^2 - \beta_4 xu - \beta_5 x^2 u \qquad 9.4\text{--}147$$

We will consider the performance index

$$P = \frac{1}{2}\int_0^{\theta=\infty} (x^2 + c^2 u^2)\, dt \qquad 9.4\text{--}148$$

and look for the control variable that forces the system to go from an initial
measured derivation, $x(0) = x_0$, back to the optimum steady state design condition, $x(\theta) = 0$, while giving the minimum value of P.
The Hamiltonian is

$$H = \tfrac{1}{2}(x^2 + c^2 u^2) - \lambda(\beta_1 x + \beta_2 u + \beta_3 x^2 + \beta_4 xu + \beta_5 x^2 u) \qquad 9.4\text{--}149$$

the adjoint equation is

$$\dot{\lambda} = -x + \beta_1 \lambda + 2\beta_3 \lambda x + \beta_4 \lambda u + 2\beta_5 \lambda xu \qquad 9.4\text{--}150$$

and the optimal control condition is

$$\frac{\partial H}{\partial u} = 0 = c^2 u - \beta_2 \lambda - \beta_4 \lambda x - \beta_5 \lambda x^2$$

or

$$u = \beta_2 \lambda + \beta_4 \lambda x + \beta_5 \lambda x^2 \qquad 9.4\text{--}151$$

One way of solving this set of nonlinear equations is to use the optimal
control law, Eq. 9.4–151, to eliminate the control variable from the state equation, Eq. 9.4–147, and the adjoint equation, Eq. 9.4–150. We anticipate that it
might take a very long time to return to the final steady state, so we solve the
state and adjoint equations with the boundary conditions with $x(0) = x_0$ and
$\lambda(\theta) = 0$, where θ is much larger than the time constant of the linearized equations (see Eqs. 9.2–6, 9.2–7, and 9.2–11). After we have determined x and λ
as a function of time in this way, we can substitute the results into the optimal
control law, Eq. 9.4–151, to find u as a function of t. Finally, we adjust the

control variable so that this relationship is satisfied or another disturbance is observed.

Unfortunately this procedure does not lead to a feedback control system that is simple to implement. A much more serious drawback, however, is encountered for higher-order systems, for then it becomes very difficult to solve the nonlinear boundary value problem. This means that an on-line digital computer will be required to implement the optimal control policy in a feed-forward manner. Therefore it seems worthwhile to look for approximate solutions to the optimization problem, which are simpler to implement.

It is clear that if we set β_3, β_4, and β_5 in the state equation, Eq. 9.4–147, equal to zero, we will obtain the linearized equation for the plant. From our previous studies of this problem, we know that the optimal controller for this case is simply a proportional feedback system. One method we used to obtain this result was to assume a solution having the form $\lambda = mx$ and then to find the value of m that was consistent with the optimization equations.

Remembering the ideas we developed in the chapter on perturbation theory, suppose we assume that we can find an approximate solution of the nonlinear optimization equations having the form

$$\lambda = m_1 x + m_2 x^2 + m_3 x^3 + \ldots \qquad 9.4\text{–}152$$

where $m_1, m_2, \ldots$ are unknown constants. For small values of x and u, this should give us a solution that is identical to the linear problem. Also by increasing the number of terms, we should be able to get some idea of the importance of the nonlinear terms or at least determine the region where the linear analysis is valid. Another way of looking at the problem is to consider that our assumed solution represents a Taylor series expansion of the unknown exact solution, and we expect that this series expansion will be consistent with the optimization equations.[6]

Following our normal procedure, we know that if the assumed solution is valid at every instant of time, we must also have

$$\dot{\lambda} = m_1 \dot{x} + 2m_2 x \dot{x} + 3m_3 x^2 \dot{x} + \cdots \qquad 9.4\text{–}153$$

Using Eqs. 9.4–147, 9.4–151, and 9.4–152 to eliminate $\dot{x}, u$, and λ from this equation, we find that

$$\dot{\lambda} = -\left(\beta_1 m_1 + \frac{\beta_2^2}{c^2} m_1^2\right) x - \left(\beta_3 m_1 + 2\beta_1 m_2 + 2\frac{\beta_2 \beta_4}{c^2} m_1^2 + 3\frac{\beta_2^2}{c^2} m_1 m_2\right) x^2$$

$$- \left[2\beta_3 m_2 + 3\beta_1 m_3 + \left(2\frac{\beta_2 \beta_5}{c^2} + \frac{\beta_4^2}{c^2}\right) m_1^2 + 2\frac{\beta_2^2}{c^2} m_2^2 + 6\frac{\beta_2 \beta_4}{c^2} m_1 m_2 \right.$$

$$\left. + 4\frac{\beta_2^2}{c^2} m_1 m_3\right] x^3 + \cdots \qquad 9.4\text{–}154$$

Similarly, we can substitute the optimal control law, Eq. 9.4–151, and the assumed solution, Eq. 9.4–152, into the adjoint equation, Eq. 9.4–150, to obtain

$$\dot{\lambda} = (\beta_1 m_1 - 1)x + \left(\beta_1 m_2 + \beta_3 m_1 + \frac{\beta_2 \beta_4}{c^2}\right)x^2$$

$$+ \left[\beta_1 m_3 + 2\beta_3 m_2 + \left(\frac{\beta_4^2}{c^2} + 2\frac{\beta_2 \beta_5}{c^2}\right)m_1^2 + 2\frac{\beta_2 \beta_4}{c^2} m_1 m_2\right]x^3 + \cdots$$

$$9.4\text{–}155$$

In order for our assumed solution to be valid for arbitrary values of x, we must have

$$m_1^2 + 2\frac{\beta_1 c^2}{\beta_2^2} m_1 - \frac{c^2}{\beta_2^2} = 0 \qquad\qquad 9.4\text{–}156$$

$$m_2 = -\left(\frac{\beta_3 c^2 + \beta_2 \beta_4 m_1}{\beta_1 c^2 + \beta_2^2 m_1}\right)m_1 \qquad\qquad 9.4\text{–}157$$

$$m_3 = -\frac{\beta_3 \beta_5 m_1^2 + \beta_2 \beta_4 m_1 m_2 + \frac{1}{2}(\beta_4 m_1 + \beta_2 m_2)^2 + \beta_3 c^2 m_2}{\beta_1 c^2 + \beta_2^2 m_1} \qquad 9.4\text{–}158$$

and it is a simple matter to solve these equations for the unknown constants $m_1, m_2, \ldots$. After these constants have been evaluated, we can substitute the assumed solution into the optimal control law and write

$$u = \left(\frac{\beta_2}{c^2} m_1\right)x + \frac{1}{c^2}(\beta_2 m_2 + \beta_4 m_1)x^2$$

$$+ \frac{1}{c^2}(\beta_2 m_3 + \beta_4 m_2 + \beta_5 m_1)x^3 + \cdots \qquad 9.4\text{–}152$$

Even though this is a nonlinear control law, it is simple to implement in a feedback manner. In addition, we can extend this approach to multivariable control of higher-order systems.

SECTION 9.5 TIME OPTIMAL CONTROL

So far all our examples of optimal control theory have been based on an integral square error performance criterion. Also, we have assumed that the solutions always fall within any constraints on the control variables, and therefore we have neglected any potential difficulties that this might introduce. However, the functional chosen for the performance index can have a major effect on the optimal control policy, and we would like to study some of the other optimal control laws we might encounter. It seems quite reasonable, on an intuitive basis, to design control systems that will force the system to return to the optimum steady state design conditions in a minimum time after a disturbance has been observed. Thus we will let

$$P = \int_0^\theta dt \qquad\qquad 9.5\text{–}1$$

and attempt to solve some simple control problems. Again the presentation will be broken down into a number of examples that illustrate the important points of the theory. In all these problems the time required to reach steady state conditions is finite, so that θ is undetermined.

Example 9.5-1 Time optimal control of an inertial system

As a very simple example of a time optimal control problem, we will consider the state equation

$$\ddot{x} = u \qquad\qquad 9.5\text{-}2$$

Solution

It is probable that no chemical processes are described by this state equation, but it has been studied extensively by electrical and mechanical engineers, for it provides a model of an inertial system—that is, one whose state changes with acceleration. In order to put the equation into the required form, we let

$$x_1 = x \qquad\qquad 9.5\text{-}3$$

$$x_2 = \dot{x}_1 \qquad\qquad 9.5\text{-}4$$

so that the state equations become

$$\dot{x}_1 = x_2 \qquad\qquad 9.5\text{-}5$$

$$\dot{x}_2 = u \qquad\qquad 9.5\text{-}6$$

and the performance index is given by Eq. 9.5-1. The Hamiltonian becomes

$$H = 1 + \lambda_1 x_2 + \lambda_2 u \qquad\qquad 9.5\text{-}7$$

and the adjoint equations are

$$\dot{\lambda}_1 = 0 \qquad\qquad 9.5\text{-}8$$

$$\dot{\lambda}_2 = -\lambda_1 \qquad\qquad 9.5\text{-}9$$

If we attempt to find the optimal control policy by differentiation, we obtain

$$\frac{\partial H}{\partial u} = \lambda_2 = 0$$

However, this result must be true for all time; consequently, we would also expect that $\dot{\lambda}_2 = 0$. Then, from Eq. 9.5-9, we learn that λ_1 is always equal to zero, and from Eq. 9.5-7 we find that $H = 1$. Of course, this contradicts Eq. 9.3-38, which states that $H = 0$ for all time whenever the process time is not specified. This apparent discrepancy is easily resolved when we recognize that the Hamiltonian, Eq. 9.5-7, is a linear function of the control variable, and we know from the elementary theory of calculus that the minimum of a linear function is always at one of the endpoints and is never defined by an equation of the type $\partial H/\partial u = 0$.

After reconsidering the problem of choosing the value of u that minimizes the Hamiltonian, we come to the conclusion that if λ_2 is negative, we should make u infinitely large; whereas if λ_2 is positive, we should make u infinitely small. Although this is indeed correct, it is physically unrealistic because we never have an infinite supply of control action at our disposal. Thus before we can proceed with our discussion, it becomes necessary to define the bounds on the control variable explicitly. It is common practice in the control literature to take these bounds as (± 1), because it is always possible to redefine the variables so that this condition is satisfied. We can denote the bounded nature of the control action by writing

$$u_* \leq u \leq u^* \qquad\qquad 9.5\text{-}10$$

or

$$-1 \leq u \leq 1 \qquad\qquad 9.5\text{-}11$$

The fact that the control variable is bounded in no way changes our reasoning. If λ_2 is negative, we should still make u as large as we possibly can, $u = +1$, in order to minimize the Hamiltonian, and vice versa. We can denote this idea that the control variable should always be at its maximum or minimum value by defining a new function, the signum function, such that

$$\text{sgn}\,\phi = \begin{cases} 1, & \text{if } \phi > 0 \\ -1, & \text{if } \phi < 0 \end{cases} \qquad\qquad 9.5\text{-}12$$

Thus the optimal control policy for our problem becomes

$$u = \text{sgn}\,(-\lambda_2) \qquad\qquad 9.5\text{-}13$$

In order to develop a solution to the foregoing equations, the first thing we notice is that the adjoint equations, Eqs. 9.5-8 and 9.5-9, can be solved immediately

$$\lambda_1 = c_1 \qquad\qquad 9.5\text{-}14$$

$$\lambda_2 = c_2 t + c_3 \qquad\qquad 9.5\text{-}15$$

where c_1, c_2, and c_3 are integration constants. For a problem where we measure an initial deviation

$$\text{at } t = 0, \quad x = x_0 = x_{10}, \quad \dot{x} = \dot{x}_0 = x_{20} \qquad 9.5\text{-}16$$

and attempt to force the system to return to the origin in a minimum time θ,

$$\text{at } t = \theta, \quad x = 0 = x_1, \quad \dot{x} = 0 = x_2 \qquad 9.5\text{-}17$$

we have four boundary conditions on the state variables and none on the adjoint variables. This means that it is not possible to evaluate the integration constants in Eqs. 9.5-14 and 9.5-15, at least at this time. However, we can glean one very important piece of information from these equations—and that is that no matter what values we select for c_2 and c_3, the function $\lambda_2(t)$, given by Eq. 9.5-15, can be equal to zero at most once. Therefore from the optimal control law, Eq. 9.5-13, we learn that the control can switch from its maximum to minimum value, or vice versa, once at most.

Since we know now that the control variable will be either $u = +1$ or $u = -1$, we can solve the state equations for both cases. First considering $u = +1$, we obtain

$$x_2 = t + x_{20} \qquad\qquad 9.5\text{-}18$$

$$x_1 = \tfrac{1}{2} t^2 + x_{20} t + x_{10} = \tfrac{1}{2} (t + x_{20})^2 + (x_{10} - \tfrac{1}{2} x_{20}^2) \qquad 9.5\text{-}19$$

Again we find it convenient to eliminate time from the equations to obtain an expression for the system trajectories in the phase plane

$$x_1 = \tfrac{1}{2} x_2^2 + (x_{10} - \tfrac{1}{2} x_{20}^2) \qquad\qquad 9.5\text{-}20$$

A plot of these parabolic paths for various values of x_{10} and x_{20} is shown in Figure 9.5-1. The first thing to notice about this graph is that only one of the trajectories, $x_1 = \tfrac{1}{2} x_2^2$, passes through the origin—that is, the desired endpoint.

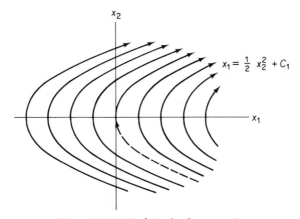

Figure 9.5-1. Trajectories for $u = +1$.

This means that if c_2 and c_3 in Eq. 9.5-15 are both negative, so that $\lambda_2(t)$ is always negative and according to Eq. 9.5-13 we should always use a control $u = +1$, the only valid initial conditions for this solution must lie along the portion of this trajectory that is approaching the origin. In addition, for cases where c_2 is negative and c_3 is positive, so that we expect λ_2 to start off as a positive quantity but become negative at large times and expect u to switch from an initial value of (-1) to a final value of $(+1)$, we must approach the origin along this same curve. In other words, it must also be the switching curve—that is, the locus of points where we must change the control setting from $u = -1$ to $u = +1$.

A similar procedure can be used to find the trajectories for $u = -1$ (see Figure 9.5-2) and to show that the other switching curve must be $x_1 = -\tfrac{1}{2} x_2^2$.

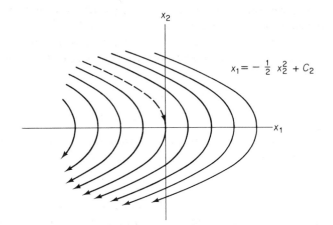

Figure 9.5-2. Trajectories for $u = -1$.

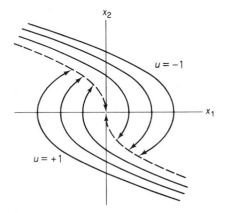

Figure 9.5-3. Optimal trajectories.

Now, if we put the appropriate portions of the two sets of solutions together, as is shown in Figure 9.5–3, the optimal control policy becomes clear. Whenever we measure a deviation that is below the two switching curves, we impose a control $u = +1$ until we reach the switching curve. Next, we make $u = -1$ until we reach the origin, and, finally, we turn off the controller. The procedure for measured deviations that lie above the switching curves is analogous.

This kind of a control, where the control variable is always at its maximum or minimum value, is usually called *bang-bang* control. It is quite simple to implement, as will be discussed later. It is interesting to note, now that we have a complete understanding of the solution of the optimal control problem, that for any particular measured deviation we could use the solution of the state equations, the condition that the Hamiltonian is always equal to zero, and the solutions to the adjoint equations to find the integration constants in the adjoint equations that correspond to that particular initial deviation. A more profound exercise in futility would be hard to imagine.

Example 9.5–2 Time optimal control of a harmonic oscillator

In the very simple example just discussed, only one switch was required for the optimal control system to achieve its objective. As an example that illustrates other kinds of switching behavior, we will consider the time optimal control of a harmonic oscillator

$$\ddot{x} + x = u \qquad\qquad 9.5\text{–}21$$

or the equivalent state equations

$$\dot{x}_1 = x_2 \qquad\qquad 9.5\text{–}22$$

$$\dot{x}_2 = -x_1 + u \qquad\qquad 9.5\text{–}23$$

where u is again bounded

$$-1 \leq u \leq 1 \qquad\qquad 9.5\text{–}24$$

Solution

For this case, the Hamiltonian becomes

$$H = 1 + \lambda_1 x_2 + \lambda_2(-x_1 + u) \qquad\qquad 9.5\text{–}25$$

the adjoint equations are

$$\dot{\lambda}_1 = \lambda_2 \qquad\qquad 9.5\text{–}26$$

$$\dot{\lambda}_2 = -\lambda_1 \qquad\qquad 9.5\text{–}27$$

and we find that the optimal control law is

$$u = \text{sgn}\,(-\lambda_2) \qquad\qquad 9.5\text{–}28$$

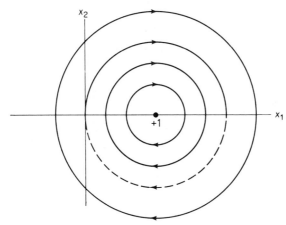

Figure 9.5-4. Trajectories for $u = +1$.

The solution of the adjoint equations can be written as

$$\lambda_2 = -A \sin(t + \phi) \qquad\qquad 9.5\text{-}29$$

where, again, the boundary conditions A and ϕ are unknown. However, when we substitute this result into the optimal control law

$$u = \text{sgn} \left[\sin(t + \phi) \right] \qquad\qquad 9.5\text{-}30$$

we find that the controller must switch from its maximum to minimum value, or vice versa, at time intervals of $t = \pi$. The control still is always a constant during any one interval, and therefore we can solve the state equations for both

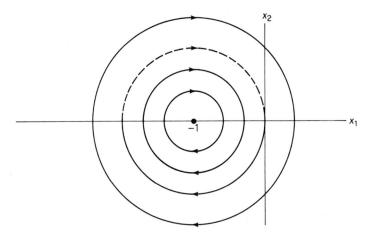

Figure 9.5-5. Trajectories for $u = -1$.

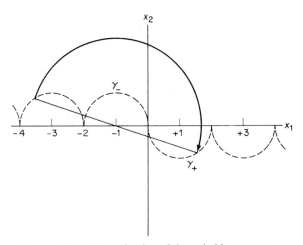

Figure 9.5-6. Determination of the switching curves.

extreme values. When $u = +1$, we find that the trajectories are circles with their centers at $x_1 = 1$ and $x_2 = 0$.

$$(x_1 - 1)^2 + (x_2)^2 = R^2 \qquad\qquad 9.5\text{-}31$$

These are shown in Figure 9.5-4. Actually, only a semicircular portion of these curves can correspond to an optimal path, because we know that the controller must switch after π time units. It is clear that only one curve passes through the origin, and this must be an optimal path, as well as a portion of the switching curve.

Similar results are obtained when $u = -1$ except that the circles have their centers at $x_1 = -1$ and $x_2 = 0$ (see Figure 9.5-5). The semicircle passing through the origin provides another portion of the switching curve. Also, we see that any trajectory that ends on this portion of the switching curve must have a value of $u = +1$. Hence it must lie on a portion of a circle around $x_1 = 1$ and $x_2 = 0$. Since this kind of a path is valid for at most a time interval $t = \pi$, before the control setting would again have to be set at $u = -1$, we can determine another portion of the switching curve by finding the earliest possible

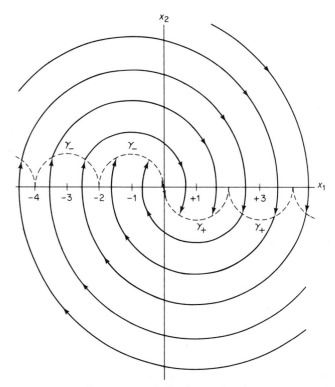

Figure 9.5-7. Optimal control paths.

starting points for all trajectories ending on the $u = -1$ switching curve (see Figure 9.5–6.) Using this backward tracing procedure, we can establish all the optimal control paths, some of which are shown in Figure 9.5–7. The optimal control of an underdamped second-order system is similar, and this problem is presented as an exercise at the end of the chapter.

Example 9.5–3 Multivariable time optimal control of a stirred-tank reactor: the linearized case

Our major purpose in pursuing optimization theory was to develop an approach that could be used to design multivariable control systems. An example we can use to study multivariable, time optimal controllers is the linearized model of a nonisothermal stirred-tank reactor. The state equations are

$$\dot{x}_1 = a_{11}x_1 + a_{12}x_2 + b_{11}u_1 \qquad\qquad 9.5\text{–}32$$

$$\dot{x}_2 = a_{21}x_1 + a_{22}x_2 + b_{21}u_1 + b_{22}u_2 \qquad\qquad 9.5\text{–}33$$

where the control variables must lie in the range

$$u_{1*} \leq u_1 \leq u_1^* \qquad\qquad 9.5\text{–}34$$

$$u_{2*} \leq u_2 \leq u_2^*$$

We assume that we measure an initial deviation and want to return to the optimum steady state design conditions, the origin, in a minimum time.

$$\text{at } t = 0, \qquad x_1 = x_{10}, \quad x_2 = x_{20}$$

$$\text{at } t = \theta, \qquad x_1 = x_2 = 0 \qquad\qquad 9.5\text{–}35$$

Solution

It turns out that we can simplify the algebra involved if we make a canonical transformation and put the equations in diagonal form. For a case of distinct real roots, we let

$$x_1 = y_1 + y_2 \qquad\qquad 9.5\text{–}36$$

$$x_2 = \frac{a_{11} + s_1}{a_{12}} y_1 - \frac{a_{11} + s_2}{a_{12}} y_2 \qquad\qquad 9.5\text{–}37$$

where s_1 and s_2 are the characteristic roots, and follow the procedure described in Example 3.3–4 to obtain

$$\dot{y}_1 = s_1 y_1 - M_{11}u_1 - M_{12}u_2 \qquad\qquad 9.5\text{–}38$$

$$\dot{y}_2 = s_2 y_2 + M_{21}u_1 + M_{12}u_2 \qquad\qquad 9.5\text{–}39$$

where

$$M_{11} = \frac{a_{11}b_{11} + a_{12}b_{21} + s_2 b_{11}}{s_1 - s_2}, \quad M_{12} = \frac{a_{12}b_{22}}{s_1 - s_2}, \quad M_{21} = M_{11} + b_{11} \qquad 9.5\text{–}40$$

The performance index is not affected by this transformation, so the Hamiltonian becomes

$$H = 1 + \lambda_1(s_1 y_1 - M_{11}u_1 - M_{12}u_2) + \lambda_2(s_2 y_2 + M_{21}u_1 + M_{12}u_2) \qquad 9.5\text{–}41$$

and the adjoint equations are

$$\dot{\lambda}_1 = -s_1 y_1 \qquad\qquad\qquad 9.5\text{-}42$$

$$\dot{\lambda}_2 = -s_2 y_2 \qquad\qquad\qquad 9.5\text{-}43$$

Again the Hamiltonian is linear in the control variables and the minimum of the Hamiltanian is obtained when the control variables are at their extreme values. Thus

$$u_1 = \mathrm{sgn}\,(M_{11}\lambda_1 - M_{21}\lambda_2) \qquad\qquad 9.5\text{-}44$$

$$u_2 = \mathrm{sgn}\,M_{12}(\lambda_1 - \lambda_2) \qquad\qquad 9.5\text{-}45$$

where now we are considering the signum function to be

$$\mathrm{sgn}\,\phi = \begin{cases} u_1^*, & \text{if } \phi > 0 \\ u_{1*}, & \text{if } \phi < 0 \end{cases} \qquad\qquad 9.5\text{-}46$$

Following the approach used in the two previous examples, we first solve the adjoint equations

$$\lambda_1 = \lambda_{10} e^{-s_1 t} \qquad\qquad\qquad 9.5\text{-}47$$

$$\lambda_2 = \lambda_{20} e^{-s_2 t} \qquad\qquad\qquad 9.5\text{-}48$$

where λ_{10} and λ_{20} are unknown initial conditions (which cannot be evaluated because there are four boundary conditions on the state variables, Eq. 9.5-35). It is interesting to note that if the original system is stable, so that the characteristic roots s_1 and s_2 are less than zero, the adjoint equations are unstable, and the adjoint variables become unbounded as time approaches infinity. Of course, the same kind of behavior was obtained in the earlier example (see Eq. 9.5-15) and, in fact, is characteristic of all optimization problems.

Substituting the preceding solutions into the optimal control laws, we find that

$$u_1 = \mathrm{sgn}\left[M_{21}\lambda_{20}\,e^{-s_2 t}\left(\frac{M_{11}\lambda_{10}}{M_{21}\lambda_{20}}\,e^{(s_2 - s_1)t} - 1\right) \right] \qquad 9.5\text{-}49$$

$$u_2 = \mathrm{sgn}\left[M_{12}\lambda_{20}\,e^{-s_2 t}\left(\frac{\lambda_{10}}{\lambda_{20}}\,e^{(s_2 - s_1)t} - 1\right) \right] \qquad 9.5\text{-}50$$

We are interested only in the sign of the argument of this function; consequently, we can drop the $e^{-s_2 t}$ term because it is always positive. Also, if we consider a case where

$$s_2 < s_1 < 0, \quad M_{21} > M_{11} > 0, \quad M_{12} < 0 \qquad 9.5\text{-}51$$

which correspond to the parameters given in Table 9.4-1,[1] we can eliminate some of the other constants and write

$$u_1 = \mathrm{sgn}\left[\lambda_{20}\left(r\frac{\lambda_{10}}{\lambda_{20}}\,e^{(s_2 - s_1)t} - 1\right) \right] \qquad 9.5\text{-}52$$

$$u_1 = \mathrm{sgn}\left[-\lambda_{20}\left(\frac{\lambda_{10}}{\lambda_{20}}\,e^{(s_2 - s_1)t} - 1\right) \right] \qquad 9.5\text{-}53$$

[1] See also J. M. Douglas, *Chem. Eng. Sci.*, **21**, 519 (1966).

where

$$r = \frac{M_{11}}{M_{21}} \quad \text{and} \quad 0 < r < 1 \qquad 9.5\text{--}54$$

The initial values of λ_1 and λ_2 must be retained in these expressions because the signs of these quantities are unknown. However, if we consider all possible combinations of signs for these terms, we will be able to determine all of the possible control settings. We know that for the case under consideration, $s_2 < s_1 < 0$, the function $e^{(s_2-s_1)t}$ starts at unity and approaches zero as time goes from zero to infinity. Therefore, whenever λ_{10} and λ_{20} have opposite signs, the signs of the argument of the signum functions can never change; hence the optimal control policy will be either (u_1^*, u_{2*}) or (u_{1*}, u_2^*) and no switching is necessary. If λ_{10} and λ_{20} are both positive, then when t becomes sufficiently large, the final control action must be (u_{1*}, u_2^*); but this may be preceded by (u_{1*}, u_{2*}) and (u_1^*, u_{2*}) settings, depending on the magnitude of the parameters. Similarly, if λ_{10} and λ_{20} are both negative, we find that the final control must be (u_1^*, u_{2*}) for sufficiently large t, but this may be preceded by (u_1^*, u_2^*) and (u_{1*}, u_2^*) settings.

In situations where both the control variables change sign, the time that the system can take on the intermediate control settings, (u_1^*, u_2^*) or (u_{1*}, u_{2*}), only depends on the system parameters. For example, considering a case where both λ_{10} and λ_{20} are positive and $\lambda_{10}/\lambda_{20} > 1/r$, then at some time t_1 the controller will switch from (u_1^*, u_{2*}) to (u_{1*}, u_{2*}), and from the argument of the signum function in Eq. 9.5–52 this must satisfy the equation

$$r \frac{\lambda_{10}}{\lambda_{20}} e^{(s_2-s_1)t_1} = 1$$

or

$$t_1 = \left(\frac{1}{s_2 - s_1}\right) \ln \left(\frac{\lambda_{20}}{\lambda_{10}} \frac{1}{r}\right) \qquad 9.5\text{--}55$$

Similarly, the switching from (u_{1*}, u_{2*}) to (u_{1*}, u_2^*) will occur at some time t_2, and we find from Eq. 9.5–53 that

$$\frac{\lambda_{10}}{\lambda_{20}} e^{(s_2-s_1)t_2} = 1$$

or

$$t_2 = \left(\frac{1}{s_2 - s_1}\right) \ln \left(\frac{\lambda_{20}}{\lambda_{10}}\right) \qquad 9.5\text{--}56$$

Thus the time between the two switchings must be

$$t_2 - t_1 = \left(\frac{1}{s_2 - s_1}\right) \ln r \qquad 9.5\text{--}57$$

and it is clear that the characteristic roots, s_1 and s_2, as well as the parameter r, defined by Eqs. 9.5-54 and 9.5-40, are completely specified by the system parameters.

The foregoing results indicate that the control variables are always constant between switching periods. Therefore we can solve the state equations, Eqs. 9.5-38, and 9.5-39, because u_1 and u_2 are constant. Writing the equations in the form

$$\dot{y}_1 = s_1(y_1 + N_1) \qquad\qquad 9.5\text{-}58$$

$$\dot{y}_2 = s_2(y_2 + N_2) \qquad\qquad 9.5\text{-}59$$

where

$$N_1 = -\left(\frac{M_{11}}{s_1}u_1 + \frac{M_{12}}{s_1}u_2\right) \qquad N_2 = \left(\frac{M_{21}}{s_2}u_1 + \frac{M_{12}}{s_2}u_2\right) \qquad 9.5\text{-}60$$

and calling the transformed initial conditions, given in Eq. 9.5-35, y_{10} and y_{20}, the solutions are

$$t = \frac{1}{s_1}\ln\left(\frac{y_1 + N_1}{y_{10} + N_1}\right) \qquad\qquad 9.5\text{-}61$$

and

$$t = \frac{1}{s_2}\ln\left(\frac{y_2 + N_2}{y_{20} + N_2}\right) \qquad\qquad 9.5\text{-}62$$

After eliminating time from these expressions and manipulating the results somewhat, we obtain an expression for the trajectories in the phase plane

$$y_1 = -N_1 + (y_{10} + N_1)\left(\frac{y_2 + N_2}{y_{20} + N_2}\right)^{s_1/s_2} \qquad\qquad 9.5\text{-}63$$

From our analysis of the possible optimal control policies, we learned that the final control action must be either (u_1^*, u_{2*}) or (u_{1*}, u_2^*). Hence we can find one of the optimal trajectories by letting $u_1 = u_1^*$, $u_2 = u_{2*}$, and $y_{10} = y_{20} = 0$ in the preceding equations. Also, it is apparent that this curve must be a switching curve, which we will call $(\gamma + -)$, for paths where the control is (u_1^*, u_2^*), for it is the only trajectory that leads to the desired endpoint—that is, the origin according to Eqs. 9.5-35, 9.5-36, and 9.5-37. Similarly, if we let $u_1 = u_{1*}$, $u_2 = u_2^*$, and $y_{10} = y_{20} = 0$, we can calculate another optimal trajectory, which will also be a switching curve $(\gamma - +)$.

Now, if we arbitrarily select points on the $(\gamma + -)$ switching curve, so that we have new values for y_{10} and y_{20}; let $u_1 = u_2^*$ and $u_2 = u_2^*$, which we know is the only control that can precede the (u_1^*, u_{2*}) control; and integrate the state equations backward in time, using Eqs. 9.5-63 and either 9.5-61 or 9.5-62, for a length of time $t_2 - t_1 = [1/(s_2 - s_1)]\ln r$, we can calculate various other optimal paths and the switching curve $(\gamma + +)$. Next, we can repeat this procedure for initial values on the $(\gamma - +)$ curve, and $u_1 = u_{1*}$ and $u_2 = u_{2*}$, to obtain other optimal paths and the $(\gamma - -)$ switching curve. It should be noted, however, that it might be possible for the system to reach the origin with a (u_1^*, u_2^*) control before a switching to (u_1^*, u_{2*}) is required, and therefore the $(\gamma + +)$ curve might have a discontinuous slope. Similar results can be expected for the $(\gamma - -)$ case.

Once the $(\gamma + +)$ and $(\gamma - -)$ curves have been calculated. we can select various starting points on these curves, new values for y_{10} and y_{20} in Eq. 9.5–63, set the control variables to (u_{1*}, u_2^*) or (u_1^*, u_{2*}), respectively, and calculate the remaining portions of all the possible optimal control paths. In other words, with this backward tracing procedure, we can fill the whole phase plane with optimal trajectories. The results corresponding to the parameters given in Table 9.5–1 are shown in Figure 9.5–8.

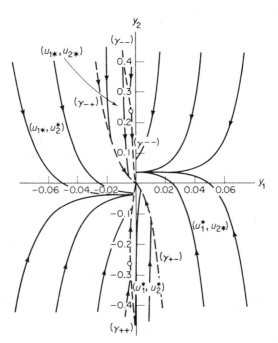

Figure 9.5-8. Optimal trajectories. [Reproduced from J. M. Douglas, *Chem. Eng. Sci.*, **21**, 519 (1966), by permission of Pergamon Press.]

Now that we have found the optimal paths in terms of the canonical variables, we can use Eqs. 9.5–36 and 9.5–37 to transform the trajectories back into the x_1 and x_2 variables, which are the dimensionless deviations of composition and temperature from the optimum steady state design equations. Also, using the definitions given in Eqs. 3.3–8 and 3.3–11, we can transform the results into the dimensionless composition versus temperature phase plane; that is,

$$z_1 = \frac{A}{A_f} = \frac{A_s}{A_f} + x_1 \qquad z_2 = \frac{TC_p\rho}{(-\Delta H)A_f} = \frac{T_sC_p\rho}{(-\Delta H)A_f} + x_2 \qquad 9.5\text{-}64$$

The results in this coordinate system are presented in Figure 9.5-9. The graph makes it clear that cases where it is necessary to switch both control variables fall outside the region of interest. In addition, it also shows that some of the optimal paths move into the region of negative composition before they return to the desired steady state design conditions. Obviously the real system can never follow these paths, and we again find that the linearized system equations sometime give misleading results.

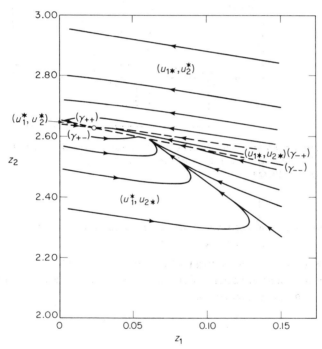

Figure 9.5-9. Optimal trajectories. [Reproduced from J. M. Douglas, *Chem. Eng. Sci.*, **21**, 519 (1966), by permission of Pergamon Press.]

It is a simple matter to implement the optimal control policy in a feedback manner, as is shown in Figure 9.5–10. Only a small number of standard analog-computing components are required. However, in the neighborhood of the steady state operating point, where the inputs to the relays approach zero, the relays will tend to chatter. This behavior is undesirable because of wear considerations, but it is easy to prevent by modifying the relay characteristics either to include a certain amount of dead zone or by using a proportional controller in this region.

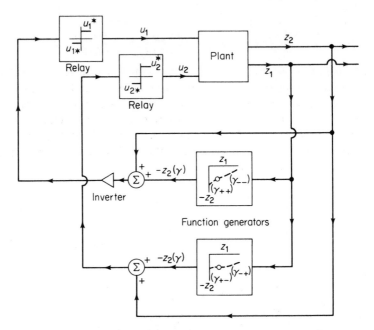

Figure 9.5-10. Time optimal control system. [Reproduced from J. M. Douglas, *Chem. Eng. Sci.*, **21**, 519 (1966), by permission of Pergamon Press.]

Example 9.5–4 *Multivariable time optimal control of a stirred-tank reactor: the nonlinear case*

The previous example illustrates the importance of extending our results to the nonlinear system equations. For a problem where there is a cooling coil in the reactor, the system equations can be written

$$\dot{z}_1 = \frac{q}{V}(1 - z_1) - kz_1 \qquad\qquad 9.5\text{--}65$$

$$\dot{z}_2 = \frac{q}{V}(z_f - z_2) - \frac{UA_cKq_c(z_2 - z_c)}{VC_p\rho(1 + Kq_c)} + kz_1 \qquad 9.5\text{--}66$$

where the definitions used are almost the same as those given in Example 3.3–1.

Solution

The Hamiltonian for the nonlinear system is

$$H = 1 + \lambda_1\left[\frac{q}{V}(1 - z_1) - kz_1\right]$$

$$+ \lambda_2\left[\frac{q}{V}(z_f - z_2) - \frac{UA_cKq_c(z_2 - z_c)}{VC_p\rho(1 + Kq_c)} + kz_1\right] \qquad 9.5\text{--}67$$

and the adjoint equations become

$$\dot{\lambda}_1 = \left(\frac{q}{V} + k\right)\lambda_1 - k\lambda_2 \qquad 9.5\text{-}68$$

$$\dot{\lambda}_2 = \left[\frac{EC_p\rho k z_1}{R(-\Delta H)A_f z_2^2}\right]\lambda_1 + \left[\frac{q}{V} + \frac{UA_cKq_c}{VC_p\rho(1 + Kq_c)} - \frac{EC_p\rho k z_1}{R(-\Delta H)A_f z_2^2}\right]\lambda_2$$

$$9.5\text{-}69$$

The Hamiltonian is a linear function of the feed rate, so that the value of q that minimizes H is given by

$$q = \text{sgn} - \left[\left(\frac{1 - z_1}{V}\right)\lambda_1 + \left(\frac{z_f - z_2}{V}\right)\lambda_2\right] \qquad 9.5\text{-}70$$

The coolant flow rate, q_c, appears in a nonlinear way in the Hamiltonian, but this nonlinear function can never possess an internal minimum. Thus the value of q_c that minimizes H is given by the expression

$$q_c = \text{sgn} \left[\frac{UA_c(z_2 - z_c)}{VC_p\rho}\lambda_2\right] \qquad 9.5\text{-}71$$

The bounds on the control variables are defined by the expressions

$$q_s + u_{1*} \leq q \leq q_s + u_1^* \qquad q_{cs} + u_{2*} \leq q_c \leq q_{cs} + u_2^*$$

and the boundary conditions for the problem are taken as

$$\begin{aligned} \text{at } t = 0, \quad & z_1 = z_{10}, \quad z_2 = z_{20} \\ \text{at } t = \theta, \quad & z_1 = z_{1s}, \quad z_2 = z_{2s} \end{aligned} \qquad 9.5\text{-}72$$

In addition, since the time θ is not specified, we know that the Hamiltonian must be equal to zero at every instant of time

$$H(t) = 0 \qquad 9.5\text{-}73$$

These equations resemble the ones in the previous example except that they are nonlinear. Therefore we will not be able to find analytical solutions, to determine the maximum number of switches implied by the equations, or the time between switchings. Also, at first glance it appears as if it will be a difficult job to solve the equations numerically, for the problem has mixed boundary conditions.

However, in a very small region around the equilibrium point, we would expect that the solution of the nonlinear problem reduces to the solution of the linearized equations (because the form of the optimization equations is the same in both cases and the nonlinearities must become negligible). Therefore we can use the result of the previous example as a first guess of the optimal control policies in this region. In other words, we can set $q = q_s + u_1^*$ and $q_c = q_{cs} + u_{2*}$, numerically integrate the state equations backward in time (by letting $t = -\tau$) starting from the equilibrium point, and in this way we obtain a first estimate of the nonlinear switching curve $(\gamma + -)$. This procedure can be repeated to estimate the other switching curves of interest, $(\gamma + +), (\gamma - -)$, and $(\gamma - +)$. Now if we assume that the nonlinear solution has the same sequence of optimal control policies as the linearized problem, which again we expect to be valid in some sufficiently small neighborhood of the equilibrium point, we can select starting points on the switching curves, set the control variables to the appropriate values given in the previous example, and continue to integrate backward in time. In this way we can fill the phase plane with optimal trajectories, providing that our assumed solution is correct. The results of this procedure are given in Figure 9.5-11. The computed switching curves

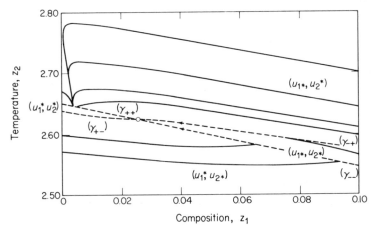

Figure 9.5-11. Optimal trajectories. [Reproduced from J. M. Douglas, *Chem. Eng. Sci.*, **21**, 519 (1966), by permission of Pergamon Press.]

certainly resemble those obtained for the linearized problem. However, the optimal paths are somewhat different, and we note that the problem of observing negative compositions has disappeared.

Of course, it is necessary to verify this assumed solution by solving the complete set of nonlinear optimization equations. Again we start at the equilibrium point and attempt to develop a backward tracing procedure. This process can be accomplished by selecting an arbitrary value for λ_2, using Eq. 9.5-71 to find the value of q_c, substituting these results and the equilibrium values z_{1s} and z_{2s} into the Hamiltonian, Eq. 9.5-67 (which must be equal to zero according to Eq. 9.5-73), and using the Hamiltonian to calculate values for λ_1 which correspond to $q_s + u_1^*$ and $q_s + u_{1*}$.

Now we can use Eq. 9.5-70 to determine whether or not these values for λ_1 and q are correct. This procedure will give us a consistent set of initial conditions for the state and adjoint equations. We numerically integrate these backward in time until the arguments of one of the signum functions in Eqs. 9.5-70 or 9.5-71 becomes equal to zero, and at this point we switch the appropriate control variable to its other extreme and continue the integration. This procedure generates one complete optimal trajectory, as well as a portion of one of the switching curves. By selecting a number of values for λ_2, we can fill the whole phase plane with optimal paths.

This technique was used to verify that the assumed solutions described above satisfy the optimization equations. The implementation of the optimal controller is identical to that shown in Figure 9.5–10 except that we use the nonlinear switching curves shown in Figure 9.5–11.

An experimental investigation of a controller of this type was undertaken by Javinsky and Kadlec.[2] The saponification of ethyl acetate in a laboratory, nonisothermal stirred-tank reactor was controlled using the "bang-bang" policy. Excellent agreement between the theory and experiment was observed.

Example 9.5–5 Time optimal control of a second-order
overdamped system with dead time

Koppel and Latour[3] studied the time optimal control of a second-order overdamped, system with dead time. The system transfer function they considered was

$$\frac{\tilde{x}(s)}{\tilde{u}(s)} = \frac{-e^{-a\tau s}}{(\tau s + 1)(b\tau s + 1)} \qquad 9.5\text{–}74$$

where $b\tau$ is the smallest of the two time constants, $0 < b < 1$, $a\tau$ is the dead time, and the negative sign arises because the plant was part of a feedback loop. As we mentioned in Section 5.3, this type of model is often used to approximate the dynamic characteristics of distributed parameter plants or high-order lumped parameter processes. Techniques for finding the constants in the model from a simple step response experiment also were discussed. It should be noted, however, that the overall system gain has been taken as unity, which implies that the system variables have been normalized.

The differential equation corresponding to this transfer function can be written as

$$b\tau^2 \ddot{x}(t) + (b + 1)\tau \dot{x}(t) + x(t) = -u(t - a\tau) \qquad 9.5\text{–}75$$

We assume that up until time equal zero everything is at the desired steady state design conditions (all of the deviations from steady state are zero), but that at time zero we measure an initial deviation in the plant effluent. Then we want the system to return to steady state conditions in a minimum time. Thus the boundary conditions are

$$\begin{array}{ll} \text{at } t = 0, & x = x_0, \quad \dot{x} = 0 \\ \text{at } t = \theta, & x = \dot{x} = 0 \end{array} \qquad 9.5\text{–}76$$

Solution

In order to put the equations into the appropriate form for applying the

[2] M. A. Javinsky and R. H. Kadlec, *AIChE Journal*, **16**, 916 (1970).
[3] L. B. Koppel and P. R. Latour, *Ind. Eng. Chem. Fundamentals*, **4**, 463 (1965).

minimum principle, we let

$$x_1 = x \qquad x_2 = \dot{x}_1 \qquad\qquad 9.5\text{-}77$$

so that the state equations become

$$\dot{x}_1 = x_2 \qquad x_1(0) = x_0 \qquad\qquad 9.5\text{-}78$$

$$\dot{x}_2 = -\frac{1}{b\tau^2} x_1 - \frac{b+1}{b\tau} x_2 - \frac{u(t-a\tau)}{b\tau^2} \qquad x_2(0) = 0 \qquad 9.5\text{-}79$$

Hence the Hamiltonian becomes

$$H = 1 + \lambda_1 x_2 - \lambda_2 \left[\left(\frac{1}{b\tau^2} \right) x_1 + \left(\frac{b+1}{b\tau} \right) x_2 + \frac{u(t-a\tau)}{b\tau^2} \right] \qquad 9.5\text{-}80$$

the adjoint equations are

$$\dot{\lambda}_1(t) = \left(\frac{1}{b\tau^2} \right) \lambda_2(t) \qquad\qquad 9.5\text{-}81$$

$$\dot{\lambda}_2(t) = -\lambda_1(t) + \left(\frac{b+1}{b\tau} \right) \lambda_2(t) \qquad\qquad 9.5\text{-}82$$

and the optimal control policy is

$$u(t - \tau) = \operatorname{sgn}(-\lambda_2) \qquad\qquad 9.5\text{-}83$$

where the bounds on the control variable are taken as

$$u_* \le u \le u^* \qquad\qquad 9.5\text{-}84$$

    In order to solve the problem, we initially ignore the dead time and develop the solution for the overdamped system. After solving the adjoint equations, in terms of unknown initial conditions, and substituting the expression for $\lambda_2(t)$ into the optimal control law, we find that there can be at most one switch of the control variable between the maximum and minimum value. For this constant control and no dead time, we can solve the state equations for the extreme values of the control variables. By piecing together the trajectories, we can determine the switching curves and the optimal paths in the phase plane. Since

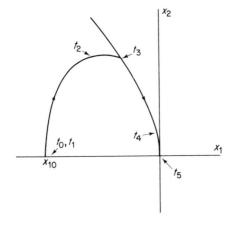

Figure 9.5-12. An optimal control path. [Reproduced from L. B. Koppel and P. R. Latour, *Ind. Eng. Chem. Fundamentals*, **4**, 463 (1965), by permission of the ACS.]

the results are essentially the same as those obtained in Example 9.5-3, except that we now have only a single control variable, we will omit the details and only show one optimal path and a portion of the $(\lambda +)$ switching curve (see Figure 9.5-12).

If we now consider the effect of the dead time, we know that we will not be able to see any change in the output until a time $a\tau$ has elapsed after a

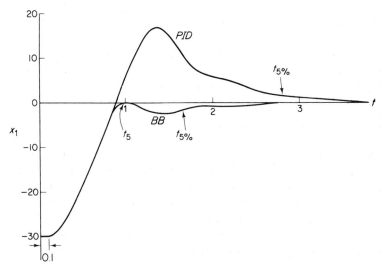

Figure 9.5-13. System response. [Reproduced from L. B. Koppel and P. R. Latour, *Ind. Eng. Chem. Fundamentals,* **4,** 463 (1965), by permission of the ACS.]

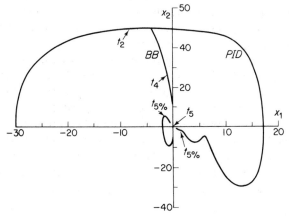

Figure 9.5-14. Phase plane response. [Reproduced from L. B. Koppel and P. R. Latour, *Ind. Eng. Chem. Fundamentals,* **4,** 463 (1965), by permission of the ACS.]

change has been made in u. Hence if we observe a deviation $(-x_0)$ in the output and immediately set $u = u_*$, the output will not respond until $t_1 = a\tau$. Also, we must recognize that it will be necessary to switch the control variable to $u = u^*$ at a time t_2, which is $(a\tau)$ time units before the switching curve $(\gamma+)$ is reached, and that we will have to shut off the bang-bang controller at time t_4, again $(a\tau)$ time units before the system would arrive at the origin with a $u = u^*$ control (see Figure 9.5-12). Koppel and Latour proved that this procedure satisfied the optimal control equations, and they noted that the presence of the dead time made it difficult to implement the optimal control law. In a numerical study, they compared the time optimal controller with a well-tuned three-mode controller, for a particular set of system parameters, and obtained a significant improvement in the settling time, $t_{5\%}$. Some of their results are shown in Figures 9.5-13 and 9.5-14.

Extension to Higher-Order Systems

Although conceptually it is a straightforward task to extend the results of time optimal control theory to higher-order systems, the practical difficulties associated with the implementation of the optimal control system can become overwhelming. For a general third-order system, the switching curves will become surfaces in the three-dimensional phase space, and for fourth-order systems they will become hypersurfaces. This means that we will no longer be able to use analog-computing components as function generators, and instead will have to use a digital computer to determine the optimal control policy.

However, even this approach is not always practical. For example, Lapidus and Luus[4] attempted to solve the time optimal control problem for a gas obsorber containing six plates (similar to Example 4.6–3) on an IBM 7094. After an hour of running time, their algorithm still had not converged, and therefore it is apparent that we could not use this approach as an on-line control system. They also found that this difficulty can be overcome for linear dynamic systems by writing the equations in discrete form and by using a linear-programming routine to solve the time optimal problem. With this approach, the running times for the absorber problem were of the order of 10 sec, but for some runs they encountered overflow of the 32,000-word-storage capacity of the computer. Their book, as well as that by Denn,[5] provides a great deal of useful information on numerical techniques for solving optimization problems.

Despite the fact that it seems as if it will not be feasible to develop bang-bang control systems for large-scale chemical process units, an interesting study by McNeill and Sacks[6] indicates that in some cases the qualitative ideas can be implemented successfully. They were concerned with the control of a large

[4] See footnote 2, p. 257.
[5] Footnote 5, p. 261.
[6] G. A. McNeill and J. D. Sacks, *Chem. Eng. Prog.*, **65**, No. 3, 33 (1969).

distillation column that was used to separate ethylbenzene and xylene. Originally the product quality was very poor. Experimental tests showed that the column dynamics were approximately first-order but that the column time constant was of the order of 20 hours. By installing a bang-bang controller, this time constant could be reduced to about 1 hour. Moreover, the improvement in product quality resulted in a savings of over $ 25,000 per year and the additional 10 percent throughput that could be obtained from the column was worth $ 100,000 per year. When savings in this range can be realized, it does not really matter if the control policy is suboptimal.

SECTION 9.6 OTHER APPROACHES TO OPTIMAL CONTROL PROBLEMS

The field of optimal control theory is relatively new; consequently, most of the research effort has been spent on trying to find solutions to problems where the performance index seems to be a physically realistic quantity. The hope was that the control systems developed in this way would be far superior to systems designed using conventional servomechanism theory. Although a number of published examples certainly show that this effort has been justified, there is no real reason to restrict the application of optimization theory to this class of problems. For example, even though it might not be realistic to include a $\dot{u}^2$ term in an integral square error performance index, rather than u^2, the fact that this leads to a proportional plus integral controller, or a three-mode controller, as an optimal solution for certain problems is certainly an interesting result and one we probably would not hesitate to implement. Another class of problems of great interest is to find multivariable control systems that are simple to implement, even for high-order nonlinear plants. Although we have obtained some systems that fall into this class, it would be nice to have a more general procedure for design purposes. Also, in some cases it might be worthwhile to attempt to formulate optimal control problems that are very different from the optimal regulator problem we have considered in our previous work. These are the kinds of problems we will consider in this section.

An Instantaneously Optimal Controller

As we found from our previous studies, a number of optimal control problems have such complex solutions that the optimal control policy cannot be implemented in a feedback manner. Hence some investigators have looked for suboptimal policies that are simpler to implement. For example, it has been suggested that if we have a performance criterion of the form

$$P = \int_0^\theta F(\mathbf{x}, \mathbf{u}) \, dt \qquad 9.6\text{--}1$$

that instead of looking for the control action, $\mathbf{u}(t)$, which minimizes P, we could simply choose the control that instantaneously minimizes F or that drives F to zero as rapidly as possible. Of course, if we choose to follow this approach, we are completely ignoring the future consequences of our present action, a sort of "Damn the torpedoes, full speed ahead" philosophy. Nevertheless, we might obtain some interesting resulting results in this way.

As a particular example, we will choose F to be a positive, quadratic form

$$F = \tfrac{1}{2}\mathbf{x}^T\mathbf{C}\mathbf{x} \qquad\qquad 9.6\text{-}2$$

or

$$F = \tfrac{1}{2}(c_{11}x_1^2 + 2c_{12}x_1x_2 + c_{22}x_2^2 + \cdots) \qquad\qquad 9.6\text{-}3$$

where $\mathbf{C}$ is a symmetric, positive definite matrix. With this choice, we are guaranteed that F will always be positive for all deviations of the x's, both positive and negative. Now we want to find the control variables such that we force this function to approach zero as rapidly as possible, or, in other words, we look for the control variables that minimize the time derivative of F

$$\dot{F} = \mathbf{x}^T\mathbf{C}\dot{\mathbf{x}} \qquad\qquad 9.6\text{-}4$$

If we consider state equations having the form

$$\dot{\mathbf{x}} = \mathbf{f}(\mathbf{x}) + \mathbf{B}(\mathbf{x})\mathbf{u} \qquad\qquad 9.6\text{-}5$$

where $\mathbf{f}(\mathbf{x})$ is a nonlinear vector function of $\mathbf{x}$ and $\mathbf{B}(\mathbf{x})$ is a nonlinear matrix function of $\mathbf{x}$, we can write Eq. 9.6-4 as

$$\dot{F} = \mathbf{x}^T\mathbf{C}\mathbf{f} + \mathbf{x}^T\mathbf{C}\mathbf{B}\mathbf{u} \qquad\qquad 9.6\text{-}6$$

Since this expression is a linear function of the control variables, which always will be the case if the state equations are linear in $\mathbf{u}$, the optimal control can exist only at the extreme values of $\mathbf{u}$. For a case where the bounds on $\mathbf{u}$ are given by

$$\mathbf{u}_* \leq \mathbf{u} \leq \mathbf{u}^* \qquad\qquad 9.6\text{-}7$$

we must have

$$\mathbf{u} = \mathrm{sgn}\,(-\mathbf{B}^T\mathbf{C}\mathbf{x}) \qquad\qquad 9.6\text{-}8$$

where $\mathrm{sgn}\,\boldsymbol{\phi}$ is a vector function whose elements are defined by the previous signum function. Thus we immediately obtain a feedback control law, and if $\mathbf{B}$ is independent of $\mathbf{x}$, it is simply a linear switching law. Although it is somewhat surprising to obtain this kind of a result even for a case where the state equations can be nonlinear, it is an extremely useful result because it is simple to implement, even for high-order systems.

Paradis and Perlmutter[1] have applied this approach to a stirred-tank reactor

---

[1] W. O. Paradis and D. D. Perlmutter, *AIChE Journal*, **12**, 876, 883 (1966).

with a cooling coil

$$\frac{d(A - A_s)}{dt} = \frac{q}{V}(A_f - A) - kA \qquad 9.6\text{-}9$$

$$\frac{d(T - T_s)}{dt} = \frac{q}{V}(T_f - T) + \frac{(-\Delta H)}{C_p \rho}kA - \frac{T - T_c}{VC_p \rho}u_2 \qquad 9.6\text{-}10$$

where the control variable is taken as the monotonic function

$$u_2 = \frac{UA_c Kq_c}{(1 + Kq_c)} \qquad 9.6\text{-}11$$

Letting

$$x_1 = A - A_s, \quad x_2 = T - T_s, \quad \mathbf{B} = \begin{pmatrix} 0 & 0 \\ 0 & -\dfrac{T - T_c}{VC_p \rho} \end{pmatrix} \qquad 9.6\text{-}12$$

we find that

$$\mathbf{B}^T \mathbf{C}^T \mathbf{x} = \begin{pmatrix} 0 & 0 \\ 0 & -\dfrac{T - T_c}{VC_p \rho} \end{pmatrix}\begin{pmatrix} c_{11} & c_{12} \\ c_{12} & c_{22} \end{pmatrix}\begin{pmatrix} A - A_s \\ T - T_s \end{pmatrix}$$

$$= \begin{pmatrix} 0 \\ -\left(\dfrac{T - T_c}{VC_p \rho}\right)[c_{12}(A - A_s) + c_{22}(T - T_s)] \end{pmatrix} \qquad 9.6\text{-}13$$

From Eq. 9.6-8

$$\mathbf{u} = \begin{pmatrix} u_1 \\ u_2 \end{pmatrix} = \text{sgn}\,(-\mathbf{B}^T \mathbf{C}^T \mathbf{x})$$

or

$$u_2 = \text{sgn}\,[(T - T_s) + \alpha(A - A_s)] \qquad 9.6\text{-}14$$

where $(T - T_c)/VC_p\rho$ has been removed from the argument of the signum function because it is always positive and

$$\alpha = \frac{c_{12}}{c_{22}} \qquad 9.6\text{-}15$$

Thus we obtain a linear switching law for our suboptimal feedback controller. Also, for a case where $c_{12} = 0$, which is the value used in most of our previous problems, then

$$u = \text{sgn}\,(T - T_s) \qquad 9.6\text{-}16$$

which is an on-off controller; that is, when the temperature exceeds the steady state value, we increase the coolant flow rate to the maximum allowable value and vice versa. This is the kind of control used for the heating systems in most homes.

Paradis and Perlmutter computed the reactor response for the control law given by Eq. 9.6-16 and a particular set of process parameters. Some of their

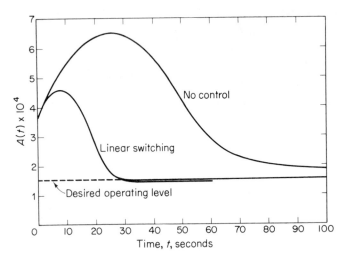

Figure 9.6-1. Composition response. [Reproduced from W. O. Paradis and D. D. Perlmutter, *AIChE Journal*, **12**, 876, 883 (1966), by permission.]

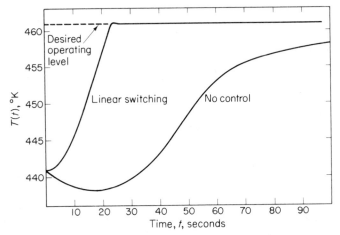

Figure 9.6-2. Temperature response. [Reproduced from W. O. Paradis and D. D. Perlmutter, *AIChE Journal*, **12**, 876, 883 (1966), by permission.]

results are shown in Figures 9.6–1 and 9.6–2. We can see from the graph that the on-off controller gives a better response than the uncontrolled system, although the controller, a relay, will chatter as the desired steady state is approached. This difficulty can be avoided either by putting a dead zone in the relay or by using a proportional controller in this neighborhood, as was mentioned in the time optimal control study.

A somewhat obvious extension of the procedure presented above is to select the weighting matrix C in Eq. 9.6–2 so that it is a Liapunov function. In other words, we not only want Eq. 9.6–2 to be a positive, quadratic form but we also require that the derivative of F in Eq. 9.6–4 is a negative, quadratic form. With this choice we can guarantee that the controlled plant will be asymptotically stable. An on-line application of this technique to a 15-tray pilot-scale rectifying column separating ethylene glycol monomethylether (methylcellosolve) and water was published by Brosilow and Handley.[2] Their results indicate that excellent control is achieved despite large changes in the feed flow rate or the controller set point.

An Inverse Problem

Since the suboptimal control policy we obtained in the preceding example is so easy to implement and since for simple problems the policy reduces to one commonly encountered in practice, it seems reasonable to wonder whether or not it can ever correspond to an optimal control law for some performance index that we have not considered thus far. This idea of starting out with an arbitrary control law and looking for the performance criterion that will make the control policy optimal is called *an inverse problem*. Kalman[3] first suggested the approach, and Denn[4] has applied it to a number of examples, including the one below.

We will consider the linear multivariable system

$$\mathbf{x} = \mathbf{Ax} + \mathbf{Bu} \qquad\qquad 9.6–17$$

and fix the control policy as

$$\mathbf{u} = \text{sgn}\,(-B^T C\mathbf{x}) \qquad\qquad 9.6–18$$

where C is chosen as a symmetric, positive definite matrix. For simplicity, we consider the case of symmetric bounds on the control variables, so that

$$-\mathbf{u}^* \leq \mathbf{u} \leq \mathbf{u}^* \qquad\qquad 9.6–19$$

The general expression for the performance criterion is

$$P = \int_0^\theta F\,dt \qquad\qquad 9.6–20$$

where the functional F is unknown for our problem. Nevertheless, we can write the Hamiltonian as

$$H = F + \boldsymbol{\lambda}^T \mathbf{Ax} + \boldsymbol{\lambda}^T \mathbf{Bu} \qquad\qquad 9.6–21$$

[2] C. B. Brosilow and K. R. Handley, *AIChE Journal*, **14**, 467 (1968).
[3] R. E., Kalman, *J. Basic Engr.*, **86**, 51 (1964).
[4] See Ch. 8. of reference 7, on p. 261.

so that the adjoint equations become

$$\dot{\boldsymbol{\lambda}} = -\frac{\partial F}{\partial \mathbf{x}} - \mathbf{A}^T \boldsymbol{\lambda} \qquad\qquad 9.6\text{-}22$$

and the control that minimizes the Hamiltonian is

$$\mathbf{u} = \text{sgn}\,(-\mathbf{B}^T \boldsymbol{\lambda}) \qquad\qquad 9.6\text{-}23$$

In order to obtain this last expression, we have assumed that F is independent of $\mathbf{u}$. Now if we compare this expression for the optimal control law with the result we desire to obtain, we find that we must have

$$\boldsymbol{\lambda} = \mathbf{C}\mathbf{x} \qquad\qquad 9.6\text{-}24$$

Substituting this relationship into the adjoint equations, Eqs. 9.6–22, we obtain

$$\dot{\boldsymbol{\lambda}} = -\frac{\partial F}{\partial \mathbf{x}} - \mathbf{A}^T \mathbf{C}\mathbf{x} \qquad\qquad 9.6\text{-}25$$

However, Eq. 9.6–24 also implies that

$$\dot{\boldsymbol{\lambda}} = \mathbf{C}\dot{\mathbf{x}} \qquad\qquad 9.6\text{-}26$$

Eliminating $\dot{\mathbf{x}}$, $\mathbf{u}$, and $\boldsymbol{\lambda}$, using Eqs. 9.6–17, 9.6–23, and 9.6–24, we find that

$$\dot{\boldsymbol{\lambda}} = \mathbf{C}\mathbf{A}\mathbf{x} + \mathbf{C}\mathbf{B}\,\text{sgn}\,(-\mathbf{B}^T \mathbf{C}\mathbf{x}) \qquad\qquad 9.6\text{-}27$$

Since both equations for $\dot{\boldsymbol{\lambda}}$, Eqs. 9.6–25 and 9.6–27, must be identical, we obtain a set of partial differential equations that define F.

$$\frac{\partial F}{\partial \mathbf{x}} = -(\mathbf{C}\mathbf{A} + \mathbf{A}^T \mathbf{C})\mathbf{x} - \mathbf{C}\mathbf{B}\,\text{sgn}\,(-\mathbf{B}^T \mathbf{C}\mathbf{x}) \qquad\qquad 9.6\text{-}28$$

Integration of these equations leads to the result

$$F = -\tfrac{1}{2}\,\mathbf{x}^T(\mathbf{C}\mathbf{A} + \mathbf{A}^T \mathbf{C})\mathbf{x} + |\mathbf{u}^{*T}\mathbf{B}^T \mathbf{C}\mathbf{x}| \qquad\qquad 9.6\text{-}29$$

The integration constant that normally would appear in this expression has been set equal to zero, so that the condition $H = 0$ for all time will be satisfied for the case where θ is unspecified. Now if we let

$$\mathbf{Q} = -(\mathbf{C}\mathbf{A} + \mathbf{A}^T \mathbf{C}) \qquad\qquad 9.6\text{-}30$$

we obtain

$$F = \tfrac{1}{2}\,\mathbf{x}^T \mathbf{Q}\mathbf{x} + |\mathbf{u}^{*T}\mathbf{B}^T \mathbf{C}\mathbf{x}| \qquad\qquad 9.6\text{-}31$$

and

$$P = \int_0^\theta (\tfrac{1}{2}\,\mathbf{x}^T \mathbf{Q}\mathbf{x} + |\mathbf{u}^{*T}\mathbf{B}^T \mathbf{C}\mathbf{x}|)\,dt \qquad\qquad 9.6\text{-}32$$

Denn has discussed a number of the features of the solution for a case where there is a single control variable. He was able to prove, using Liapunov functions, that the controlled system will be asymptotically stable, providing that the uncontrolled system is asymptotically stable. Also, he noted that it is a straightforward task to extend the analysis to nonlinear systems described by the state equations

$$\dot{x} = f(x) + B(x)u \qquad 9.6\text{-}33$$

However, as Denn points out,[4] Thau was able to show that the same control law is obtained from any sign preserving relation between λ and x, in addition to that given by Eq. 9.6-24. Thus for a simple first-order system

$$\dot{x} = Ax + Bu \qquad 9.6\text{-}34$$

the control law

$$u = \text{sgn}\,(-QBx) \qquad 9.6\text{-}35$$

is optimal not only for the performance index

$$P = \int_0^{\theta} [AQx^2 + |QBx|]\,dt \qquad 9.6\text{-}36$$

which would result from the analysis above, but also for the functionals

$$P = \int_0^{\theta} (AB^2Q^3x^4 + |B^3Q^3x^3|)\,dt \qquad 9.6\text{-}37$$

and

$$P = \int_0^{\theta} \left(\frac{A}{B}x\sinh(BQx) + |\sinh(BQx)|\right)dt \qquad 9.6\text{-}38$$

Anticipatory Control

Thus far we have restricted our attention to optimal control problems where we measure a deviation from some desired steady state operating condition and look for the control settings that will force the system to return to the desired steady state in an optimal way. However, it is a simple matter to think of some problems where we can predict that the disturbance will enter the system sometime in the future; for example, if we are going to have to change feed tanks supplying a catalytic cracking unit and the feed compositions in the two tanks are somewhat different. On an intuitive basis, we would expect that for these problems it might be preferable to start changing our control variables before the disturbance entered the system, rather than wait until we measure that the output is deviating from the desired value, so that we will already have made some compensation for the disturbance. A problem of this type was solved by Oldenburger and Chang.[5] Their analysis is reviewed below.

A very simple first-order system is considered

$$\dot{x} = u + v \qquad 9.6\text{-}39$$

where x is the state variable, u is the control variable, and v is a disturbance. Only the case of step disturbances, $v = $ constant, is considered here, but the results for pulse inputs are also discussed in the original paper. If the system

[5] R. Oldenburger and R. C. C. Chang, *Proceedings of the Joint Automatic Control Conference,* 441, Stanford University, Stanford, Calif., June 1964.

initially is at steady state, $x = \dot{x} = u = v = 0$, and it is known that at some time, $t = 0$, a step disturbance enters the plant, all we would have to do is to set $u = -v$ at $t = 0$ and we could keep the plant output constant at the steady state conditions, $x = \dot{x} = 0$. This is a simple feedforward control policy. However, this can be accomplished only if it is possible to switch u from zero to $u = -v$ instantaneously. Since this step is never possible for any physical system, we consider that the rate of change of the control variable is bounded and let

$$-1 \leq \dot{u} \leq 1 \qquad\qquad 9.6\text{-}40$$

If we look for the control function that gives us the minimum time to return to steady state, we take the performance index as

$$P = \int_0^\theta dt \qquad\qquad 9.6\text{-}41$$

Differentiating Eq. 9.6-39 gives

$$\ddot{x} = \dot{u}$$

or letting

$$x_1 = x, \quad \dot{x}_1 = x_2, \quad \dot{u} = w \qquad\qquad 9.6\text{-}43$$

the system equations become

$$\dot{x}_1 = x_2 \qquad\qquad 9.6\text{-}44$$
$$\dot{x}_2 = w \qquad\qquad 9.6\text{-}45$$

and the bounds on the control variable are

$$-1 \leq w \leq 1 \qquad\qquad 9.6\text{-}46$$

Now if we measure the disturbance at time zero, the problem is identical with the first example discussed in Section 9.5, and therefore the solutions are known.

When $u = +1$,

$$x_1 = \tfrac{1}{2}(t + x_{20})^2 + (x_1 - \tfrac{1}{2}x_{20}^2) = \tfrac{1}{2}x_2^2 + (x_{10} - \tfrac{1}{2}x_{20}^2) \qquad 9.6\text{-}47$$

$$x_2 = t + x_{20} \qquad\qquad 9.6\text{-}48$$

$$\text{Switching curve } x_1 = \tfrac{1}{2}x_2^2 \qquad\qquad 9.6\text{-}49$$

When $u = -1$,

$$x_1 = -\tfrac{1}{2}(t - x_{20})^2 + (x_{10} + \tfrac{1}{2}x_{20}^2) = -\tfrac{1}{2}x_2^2 + (x_{10} + \tfrac{1}{2}x_{20}^2) \qquad 9.6\text{-}50$$

$$x_2 = -t + x_{20} \qquad\qquad 9.6\text{-}51$$

$$\text{Switching curve } x_1 = -\tfrac{1}{2}x_2^2 \qquad\qquad 9.6\text{-}52$$

For our problem, at $t = 0$,

$$x = x_1 = 0 \qquad \dot{x} = +v = x_2 \qquad\qquad 9.6\text{-}53$$

Thus, from Figure 9.6–3, we see that we must start off with $w = -1$ and follow the path given by Eqs. 9.6–50 and 9.6–51. After some time T_1, when we reach the switching curve given by Eq. 9.6–49, we switch the controller to $w = +1$ and follow the paths given by Eqs. 9.6–47 and 9.6–48. We can write equations for the condition when switching occurs

$$x_{1s} = -\tfrac{1}{2}(T_1 - v)^2 + \tfrac{1}{2}v^2 \qquad x_{2s} = -T_1 + v \qquad \text{9.6–54}$$

and

$$x_{1s} = \tfrac{1}{2}x_{2s}^2 \qquad \text{9.6–55}$$

so that the switching time is

$$T_1 = (1 + \tfrac{1}{2}\sqrt{2})v \qquad \text{9.6–56}$$

and the state variables at this time are

$$x_{1s} = \tfrac{1}{4}v \qquad x_{2s} = \tfrac{1}{2}\sqrt{2}\ v \qquad \text{9.6–57}$$

Starting from this point, we can now determine the time required to reach the origin, T_2, with $w = +1$,

$$x_2 = -t + x_{2s}$$

or

$$0 = -T_2 + \tfrac{1}{2}\sqrt{2}\ v \qquad \text{9.6–58}$$

Hence the minimum time we obtain for the normal problem is

$$\theta = T_1 + T_2 = (1 + \sqrt{2})\,v \qquad \text{9.6–59}$$

The optimal trajectory for $x_1(t) = x(t)$ is shown in Figure 9.6–3.

To extend the results, we would like to develop a solution for a case where we know that the step disturbance will occur at $t = 0$. Both $\dot{x}$ and $\dot{u}$ can be discontinuous at this point, so that we have need of one-sided limits.

$$\lim_{t \to 0} \dot{x} = \dot{x}_{0^-} \ \text{for } t < 0 \qquad \lim_{t \to 0} \dot{x} = \dot{x}_{0^+} \ \text{for } t > 0 \qquad \text{9.6–60}$$

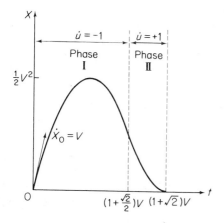

Figure 9.6-3. Optimal system response. (Reproduced from R. Oldenburger and R. C. C. Chang, *Proceedings of the Joint Automatic Control Conference*, p. 441, Stanford Univ., June, 1964, by permission.)

From the original state equation, Eq. 9.6–39, we have

$$\dot{x}_{0^-} = u_0 \qquad \dot{x}_{0^+} = u_0 + v \qquad\qquad 9.6\text{–}61$$

where u_0 might be some initial control setting. Eliminating u_0 from these expressions, we obtain

$$\dot{x}_{0^+} = \dot{x}_{0^-} - v \qquad\qquad 9.6\text{–}62$$

Now we consider an arbitrary point P_0 on the x axis where x_0 is positive (see Figure 9.6–4). It is apparent from Eq. 9.6–62 that we can always choose a value for $\dot{x}_{0^+}$ that is sufficiently negative that $\dot{x}_{0^-}$ will be negative and at the same time it satisfies the inequality

$$x_0 + \tfrac{1}{2}\,\dot{x}_{0^+}|\dot{x}_{0^+}| < 0 \quad \text{or} \quad x_1 + \tfrac{1}{2}\,x_{2^+}|x_{2^+}| < 0 \qquad\qquad 9.6\text{–}63$$

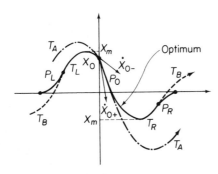

Figure 9.6-4. System response. (Reproduced from R. Oldenburger and R.C.C. Chang, *Proceedings of the Joint Automatic Control Conference*, p. 441, Stanford Univ., June, 1964, by permission.)

so that the trajectory will pass below the time axis. We can apply our previous solution for the initial conditions $(x_0, \dot{x}_{0^+})$ or (x_{10}, x_{20^+}) and find the time optimal path to return to the equilibrium point $x_0 = \dot{x}_0 = 0$ or $x_1 = x_2 = 0$. It is simple to show that the optimal control settings are $w = +1$ and $w = -1$, respectively. The optimal trajectory is the curve T_R on Figure 9.6–4 and the switch point is labeled P_R.

In order to find out what control we should have used up until $t = 0$, we introduce a new time variable $t = -\tau$ and look for the time optimal solution going backward in time, starting from the point $(x_0, \dot{x}_{0^-})$ or (x_{10}, x_{20^-}). The solution is again straightforward. We find that the optimal control settings should be $w = -1$ and $w = +1$, respectively. The optimal path is shown as T_L on Figure 9.6–4 and the switch point is P_L.

Thus for any given initial state $(x_0, \dot{x}_{0^+})$, we can find the time optimal response of the system. However, we must still find a way to determine the values of x_0 and $\dot{x}_{0^+}$ corresponding to the best of these optimal paths. This can be accomplished by studying the effect of changes in the initial values of $\dot{x}_{0^+}$. It is clear that this value is the initial slope of the trajectory. Also, as we make $\dot{x}_{0^+}$ more negative, the optimal path must pass farther below the time axis, and it will require a longer time to return to the endpoint, $x = \dot{x} = 0$ at $t = \theta$

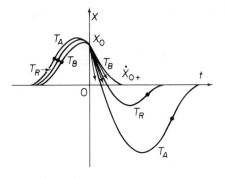

Figure 9.6-5. Variations in the optimal path. (Reproduced from R. Oldenburger and R. C. C. Chang, *Proceedings of the Joint Automotic Control Conference*, p. 441, Stanford Univ., June, 1964, by permission.)

(see curve T_A on Figure 9.6–5). Conversely, if we make $\dot{x}_{0^+}$ less negative, the amount of overshoot decreases; also, the curves on the opposite sides of $t = 0$ become more symmetrical. The best response is obtained when the curves are exactly symmetrical (see Figure 9.6–6), which occurs when

$$\dot{x}_{0^+} = -\dot{x}_{0^-} = -\tfrac{1}{2}v \qquad 9.6\text{–}64$$

After some additional manipulations, it can be shown that the minimum time occurs when

$$x_0 = \tfrac{1}{8}v^2 \qquad 9.6\text{–}65$$

and for this value, the total response time is $\theta = v$ (see Figure 9.6–6). Thus we find that the anticipatory control system leads to a 59 percent reduction in the response time, from $(1 + \sqrt{2})v$ to v, and a 75 percent decrease in the

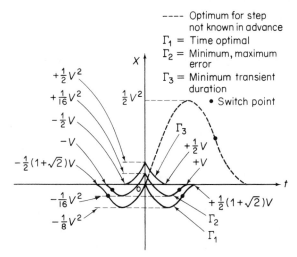

Figure 9.6-6. Time optimal path. (Reproduced from R. Oldenburger and R. C. C. Chang, *Proceedings of the Joint Automatic Control Conference*, p. 441, Stanford Univ., June., 1964, by permission.)

maximum observed error, from $\frac{1}{2} v^2$ to $\frac{1}{8} v^2$, as compared to the normal-time optimal solution.

Oldenburger also presents results for the optimum response for $x \le 0$ and for the minimal maximum error, and he has extended the analysis to square pulse disturbances.

9.7 THE EFFECT OF THE PERFORMANCE CRITERION ON THE OPTIMAL CONTROL LAW; SINGULAR SOLUTIONS

The previous examples make it clear that the form of the optimal control law we obtain from the maximum principle depends to a large extent on the performance index we have chosen. We can gain an even better insight into this phenomenon if we study the results for a single state equation and a number of performance functions. The system we choose to consider is

$$\ddot{x} = u \qquad\qquad 9.7\text{–}1$$

or its equivalent

$$\dot{x}_1 = x_2 \qquad\qquad 9.7\text{–}2$$

$$\dot{x}_2 = u \qquad\qquad 9.7\text{–}3$$

where the bounds on the control are

$$-1 \le u \le +1$$

This plant is selected because it is very simple and because an extensive number of results are available, which turn out not to be so simple in certain situations.

For the case of a minimum-time performance index

$$P = \int_0^\theta dt \qquad\qquad 9.7\text{–}4$$

The solution of this problem is given in detail in Section 9.5. The optimal control policy is "bang-bang," and the switching curves are given by the expression

$$x_1 + \frac{1}{2} x_2 |x_2| = 0 \qquad\qquad 9.7\text{–}5$$

The optimal trajectories are repeated in Figure 9.7–1 for the purpose of comparison.

Fuller[1] studied the same system for a case where

$$P = \frac{1}{2} \int_0^\theta x_1^2 \, dt \qquad\qquad 9.7\text{–}6$$

He used very special methods to show that a bang-bang solution was again optimal but that the switching curve was given by the expression

$$x_1 + 0.4446 x_2 |x_2| = 0 \qquad\qquad 9.7\text{–}7$$

[1] A. T. Fuller, *Proceedings of the First International Congress of the IEAC*, p. 510, Butterworths, London, 1961.

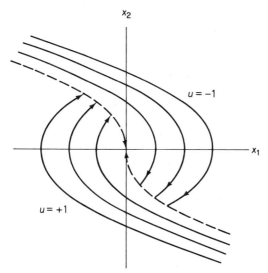

Figure 9.7-1. Time optimal control paths.

With this result, it requires an infinite number of switchings to drive the system to the origin, but it reaches this desired endpoint in a finite time. In a later paper Fuller[2] showed that this solution satisfies the minimum principle, although it is difficult to use the optimization equations to develop the solution because it is necessary to find the smallest root of a fourth-order polynomial. However, Wonham[3] demonstrated that the Hamilton-Jacobi equation could be used for this purpose. The details are too lengthy and complex to be included here. Some optimal trajectories obtained using an analog computer solution of the optimization equations[4] are shown in Figure 9.7-2.

Wonham and Johnson[5] studied the problem for the performance index

$$P = \frac{1}{2} \int_0^\theta (x_1^2 + x_2)^2 \, dt \qquad 9.7\text{-}8$$

Actually, they developed the theory for a general nth-order equation with a single control variable (see Eq. 9.4–13), but their numerical study was limited to the problem we are discussing. Since this example involves an important feature of optimization problems that we have not mentioned before, we will develop some of the details of the solution. We write the Hamiltonian as

$$H = \tfrac{1}{2} (x_1^2 + x_2^2) + \lambda_1 x_2 + \lambda_2 u \qquad 9.7\text{-}9$$

so that the adjoint equations are

$$\dot{\lambda}_1 = -x_1 \qquad 9.7\text{-}10$$

[2] A. T. Fuller, *J. Electron. Contr.*, **17,** 301 (1964).
[3] W. M. Wonham, *J. Electron. Contr.*, **15,** 59 (1963).
[4] P. J. Brennan, and A. P. Roberts, *J. Electron. Contr.*, **12,** 345 (1962).
[5] W. M. Wonham, and C. O. Johnson, *J. Basic Engr.*, **86,** 107 (1964).

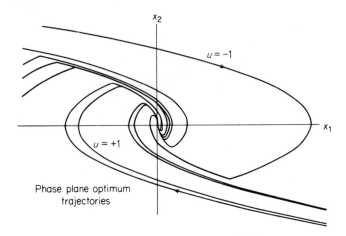

Figure 9.7-2. Optimal trajectories for $P = \dfrac{1}{2} \int_0^\theta x_1^2 \, dt$. [Reproduced from P. J. Brennan and A. P. Roberts, *J. Electron Contr.*, **12**, 345 (1962), by permission.]

$$\dot{\lambda}_2 = -x_2 - \lambda_1 \qquad\qquad 9.7\text{-}11$$

and the optimal control law is

$$u = \text{sgn}\,(-\lambda_2) \qquad\qquad 9.7\text{-}12$$

However, the factor we have ignored up to this point is that there might be situations where

$$\lambda_2(t) = 0 \qquad\qquad 9.7\text{-}13$$

over some finite time interval. If this is the case, the Hamiltonian will be independent of the control variable over that interval and we cannot use it to find the optimal control policy. This does not necessarily mean that the problem is unsolvable for this case. Certainly, if $\lambda_2(t) = 0$, we must have that

$$\dot{\lambda}_2(t) = 0 \qquad\qquad 9.7\text{-}14$$

over that interval. Then from Eq. 9.7–11 we see that

$$\lambda_1 = -x_2$$

or

$$\dot{\lambda}_1 = -\dot{x}_2 \qquad\qquad 9.7\text{-}15$$

Eliminating λ_1 between Eq. 9.7–15 and 9.7–10, we obtain

$$\dot{x}_2 = x_1 \qquad\qquad 9.7\text{-}16$$

Solving this equation simultaneously with the first state equation, Eq. 9.7–2, we find that

$$\ddot{x}_2 - x_2 = 0$$

or

$$x_2 = c_1 e^{-t} + c_2 e^t \qquad\qquad 9.7\text{-}17$$

The output will become unbounded unless we set $c_2 = 0$. The integration constant c_1 is evaluated from the initial condition $x_2(0) = x_{20}$. Thus

$$x_2 = x_{20} e^{-t} \qquad\qquad 9.7\text{-}18$$

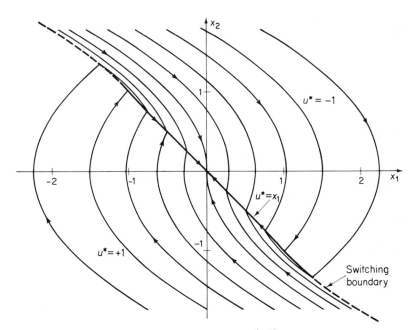

Figure 9.7-3. Optimal trajectories for $P = \dfrac{1}{2}\displaystyle\int_2^\theta (x_1^2 + x_2^2)\,dt$. [Reproduced from W. M. Wonham and C. D. Johnson, *J. Basic Engr.*, **86**, 107 (1964), by permission.]

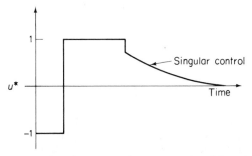

Figure 9.7-4. Optimal control settings. [Reproduced from W. M. Wonham and C. D. Johnson, *J. Basic Engr.*, **86**, 107 (1964), by permission.]

Now we can use the second state equation, Eq. 9.7–4, to evaluate the control variable that satisfies the optimization equations

$$u = \dot{x}_2 = -x_{20}e^{-t} = -x_2 \qquad\qquad 9.7\text{–}19$$

Alternately, using Eq. 9.7–16, we could write

$$u = x_1 \qquad\qquad 9.7\text{–}20$$

This is called a *singular control* because it satisfies the optimization equations when the Hamiltonian is independent of the control variable. From Eqs. 9.7–19 and 9.7–20, we find that the phase plane trajectories for the singular control are given by

$$x_1 + x_2 = 0 \qquad\qquad 9.7\text{–}21$$

which is just the equation for a straight line through the origin. From the constraints on the control variables and Eqs. 9.7–19 and 9.7–20, we see that this path can only be optimal in the region where

$$-1 \leq x_1 \leq 1 \qquad -1 \leq x_2 \leq 1 \qquad\qquad 9.7\text{–}22$$

Wonham and Johnson proved that the singular trajectory above was optimal in the vicinity of the origin and that a bang-bang controller should precede the singular control. They developed equations for the switching curves and presented some numerical examples. A portion of their results are reproduced in Figures 9.7–3 and 9.7–4.

In a subsequent investigation, Wonham and Johnson[6] studied the optimal

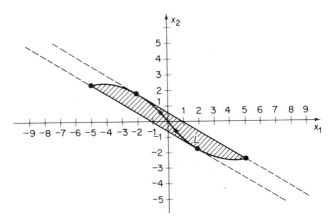

Figure 9.7-5. Region for the linear feedback controller. [Reproduced from C. D. Johnson and W. M. Wonham, *Proceedings of the Joint Automatic Control Conference*, p. 317, Stanford Univ., June, 1964, by permission.]

[6] C. D. Johnson and W. M. Wonham, *Proceedings of the Joint Automatic Control Conference*, p. 317, Stanford University, Stanford, Calif., June 1964.

control of the system, using the performance index

$$P = \frac{1}{2} \int_0^\theta (x_1^2 + x_2^2 + u^2)\, dt \qquad \qquad 9.7\text{-}23$$

Again, they present the theory for the nth-order plant with a single control variable. No singular solution is obtained for this case. Providing that the constraints on the control variables are not violated, the optimal controller is just a linear feedback system similar to that obtained in Example 9.4-1. The region where this solution is valid is bounded by two straight lines corresponding to $u = \pm 1$ and the two system trajectories that are tangent to these lines (see Figure 9.7-5). The remainder of the optimal paths are obtained by a backward-tracing procedure, starting at various points on the boundary of this region (see Figure 9.7-6). As is shown in the graph, the optimal control is initially saturated at $u = +1$, but it passes through a region where u again falls inside the constraints before it becomes saturated at $u = -1$ (see Figure 9.7-7).

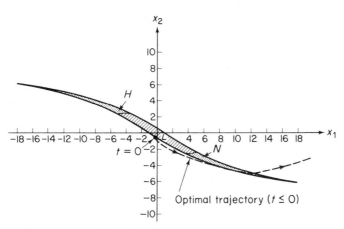

Figure 9.7-6. Backward tracing. [Reproduced from C. D. Johnson and W. M. Wonham, *Proceedings of the Joint Automatic Control Conference*, p. 317, Stanford Univ., June, 1964, by permission.]

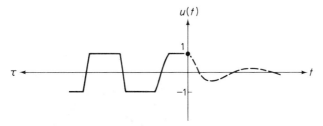

Figure 9.7-7. Optimal control settings. [Reproduced from C. D. Johnson and W. M. Wonham, *Proceedings of the Joint Automatic Control Conference*, p. 317, Stanford Univ., June, 1964, by permission.]

R T HATCH

Singular Solutions

    The previous example illustrates the fact that in some cases the Hamiltonian may become independent of one or more of the control variables over a certain interval of time, and therefore our normal approach of selecting the controls that minimize the Hamiltonian breaks down. For these cases, special methods are needed to find the control variables that satisfy the optimization equations, and the results are called singular controls. Actually, every problem in optimization theory should be checked for the possibility of singular solutions by setting the coefficient of the control variable in the Hamiltonian equal to zero and looking for solutions of the equations. We have not done this in our previous discussions only because it detracts from the presentation by adding an extra degree of complexity. However, the reader should go back and convince himself that no singular solutions exist for the examples discussed earlier.

    As another illustration of singular control, we will consider a reversible reaction in a nonisothermal CSTR. This problem has been discussed in detail by Siebenthal and Aris.[7] The state equations are written in the form

$$\frac{dA}{dt} = \frac{q}{V}(A_f - A) + r(A, T) \qquad\qquad 9.7\text{-}24$$

$$\frac{dT}{dt} = \frac{q}{V}(T_f - T) + \frac{(-\Delta H)}{C_p \rho} r(A, T) - h(T - T_c) \qquad 9.7\text{-}25$$

where $r(A, T)$ represents the reaction rate of the reversible reaction and

$$h = \frac{U A_c K q_c}{V C_p \rho (1 + K q_c)} \qquad\qquad 9.7\text{-}26$$

The equations are made dimensionless by letting

$$z_1 = \frac{A}{A_f}, \quad z_2 = \frac{T C_p \rho}{(-\Delta H) A_f}, \quad \tau = \frac{qt}{V}$$

$$h_d = \frac{qh}{V A_f} \qquad r_d = \frac{qr}{V A_f} \qquad\qquad 9.7\text{-}27$$

and they are written in terms of deviations from the desired steady state operating point by letting

$$x_1 = z_1 - z_{1s} = \frac{A - A_s}{A_f} \qquad x_2 = z_2 - z_{2s} = \frac{(T - T_s)C_p \rho}{(-\Delta H) A_f} \quad 9.7\text{-}28$$

$$u = h_d - h_{ds} \qquad z_{cs} = \frac{(T_s - T_c)C_p}{(-\Delta H) A_f}.$$

$$R(x_1, x_2) = r_d - r_{ds} \qquad \beta = 1 + h_{ds}$$

[7] C. D. Siebenthal and R. Aris, *Chem. Eng. Sci.*, **19**, 729, 747 (1964).

Hence the state equations become

$$\dot{x} = -x_1 + R(x_1, x_2) \qquad 9.7\text{-}29$$

$$\dot{x}_2 = R(x_1, x_2) - \beta x_2 - (x_2 + z_{cs})u \qquad 9.7\text{-}30$$

where we are considering that the heat-transfer coefficient is the control variable.

For a minimum-time performance index, we can write the Hamiltonian as

$$H = 1 + \lambda_1[-x_1 + R(x_1, x_2)] + \lambda_2[R(x_1, x_2) - \beta x_2 - (x_2 + z_{cs})u] \qquad 9.7\text{-}31$$

The adjoint equations are

$$\dot{\lambda}_1 = \lambda_1\left(1 - \frac{\partial R}{\partial x_1}\right) - \lambda_2 \frac{\partial R}{\partial x_2} \qquad 9.7\text{-}32$$

$$\dot{\lambda}_2 = -\lambda_1 \frac{\partial R}{\partial x_2} + \lambda_2\left(\beta + u - \frac{\partial R}{\partial x_2}\right) \qquad 9.7\text{-}33$$

and the optimal control policy is given by

$$u = \text{sgn}\,(-\lambda_2) \qquad 9.7\text{-}34$$

unless $\lambda_2(t) = 0$ over some finite time interval.

In order to check the possibility of obtaining a singular solution, we let

$$\lambda_2(t) = 0 \quad \text{and} \quad \dot{\lambda}_2(t) = 0 \qquad 9.7\text{-}35$$

From Eq. 9.7–33 we find that

$$\lambda_1 \frac{\partial R}{\partial x_2} = 0 \qquad 9.7\text{-}36$$

If $\lambda_1(t)$ in this expression is set equal to zero, we must also have that $\dot{\lambda}_1(t) = 0$. Then since λ_1 and λ_2 are both equal to zero, from Eq. 9.7–31 we see that $H = 1$ for all time. However, this contradicts the "stopping condition," $H(t) = 0$, which must be satisfied because the final time has not been specified. Thus we learn that $\lambda_1(t)$ cannot be equal to zero. For the case of a reversible endothermic reaction, we know from the elementary theory of reaction kinetics that $\partial R/\partial x_2$, which is proportional to the derivative of the reaction rate with respect to temperature, is always positive. Therefore Eq. 9.7–36 can never be satisfied and a singular solution cannot exist. Alternatively, for a reversible exothermic reaction, $\partial R/\partial x_2$ can be equal to zero, and a singular solution might exist. For this case, the singular control can be found by solving the set of simultaneous equations

$$\dot{\lambda}_1 = \lambda_1\left(1 - \frac{\partial R}{\partial x_1}\right) \qquad 9.7\text{-}37$$

$$\dot{x}_1 = -x_1 + R(x_1, x_2) \qquad 9.7\text{-}38$$

$$\dot{x}_2 = R(x_1, x_2) - \beta x_2 - (x_2 + z_{cs})u \qquad 9.7\text{-}39$$

for $u(t)$, so that the system boundary conditions are satisfied

$$x_1(0) = x_{10}, \quad x_2(0) = x_{20}(0), \quad x_1(\theta) = 0, \quad x_2(\theta) = 0 \qquad 9.7\text{-}40$$

and the Hamiltonian, Eq. 9.7-31, is equal to zero for all time.

Siebenthal and Aris presented a numerical study for a reversible reaction that was first-order in each direction. They found that a singular solution was never optimal unless the desired steady state corresponded to a case where the reaction rate was a maximum, a case of particular interest, for this condition gives the highest yield per unit volume of reactor. Under these circumstances, they not only found that singular solutions were optimal for certain regions of the phase plane but they also showed that for some starting conditions an infinite number of "bang-bang" paths satisfied the optimization equations. An interested reader should refer to their paper for additional details.

9.8 DISTRIBUTED PARAMETER SYSTEMS

Even though the results given previously in this chapter are extensive, they are limited to lumped parameter systems. However, many of the plants of interest to chemical engineers are described by partial differential equations, and so it is necessary to extend our analysis to distributed parameter systems. Rather than attempt to develop a very general derivation of the necessary conditions for the optimal control of any distributed parameter process, we will present the details of the derivation for a pair of coupled, second-order, partial differential equations and then apply the method to a steam-heated exchanger.

Derivation of the Minimum Principle

In Example 3.2-4 we discussed a dynamic model for a packed catalytic reactor. For a simple case, where we can neglect the effects inside the catalyst particle and radial gradients but include a homogeneous reaction term and heat transfer from a jacket maintained at a temperature T_H, we can write the model as

$$\frac{\partial c}{\partial t} = D \frac{\partial^2 c}{\partial z^2} - v \frac{\partial c}{\partial z} - kc \qquad 9.8\text{-}1$$

$$C_p \rho \frac{\partial T}{\partial t} = k_H \frac{\partial^2 T}{\partial z^2} - v C_p \rho \frac{\partial T}{\partial z} + (-\Delta H)kc + UA_H(T_H - T) \qquad 9.8\text{-}2$$

An appropriate set of boundary conditions might be (see Eqs. 5.2-71 and 5.2-72)

$$\text{at } z = 0, \quad v c_f(t) = vc - D \frac{\partial c}{\partial z} \quad v c_p \rho T_f(t) = v c_p \rho T - k_H \frac{\partial T}{\partial z} \qquad 9.8\text{-}3$$

$$\text{at } z = Z, \quad \frac{dc}{dz} = 0, \quad \frac{dT}{dz} = 0 \qquad 9.8\text{-}4$$

and at time $t = 0$ we would have some steady state composition and temperature profile in the reactor, which would be determined by an optimum steady state design procedure.

During normal operation of the reactor, the feed composition c_f and the feed temperature T_f might vary somewhat, so that we are interested in designing control systems that would maintain the reactor effluent close to the optimum steady state design conditions. It might be possible to accomplish this by changing the feed rate or velocity v to the reactor or by manipulating the jacket temperature T_H. Both of these control variables are distributed throughout space, and the velocity appears in the boundary conditions as well. We want the control policy to be optimal in some sense. Therefore we must define a performance index similar to those discussed previously, although for some problems we might also want to include an integral over the spatial dimension—for example, if we want to minimize the square of the error between the actual composition and temperature profiles and the desired values of these quantities over the process time θ.

Instead of treating the problem described above directly, we will consider the somewhat more general formulation

$$\frac{\partial x_1}{\partial t} = f_1\left(x_1, x_2, \frac{\partial x_1}{\partial z}, \frac{\partial x_2}{\partial z}, \frac{\partial^2 x_1}{\partial z^2}, \frac{\partial^2 x_2}{\partial z^2}, u_1, u_2\right) \qquad 9.8\text{-}5$$

$$\frac{\partial x_2}{\partial t} = f_2\left(x_1, x_2, \frac{\partial x_1}{\partial z}, \frac{\partial x_2}{\partial z}, \frac{\partial^2 x_1}{\partial z^2}, \frac{\partial^2 x_2}{\partial z^2}, u_1, u_2\right) \qquad 9.8\text{-}6$$

with the boundary conditions

$$\text{at } z = 0, \qquad g_1\left(x_1, x_2, \frac{\partial x_1}{\partial z}, \frac{\partial x_2}{\partial z}, v_1, v_2\right) = 0$$

$$g_2\left(x_1, x_2, \frac{\partial x_1}{\partial z}, \frac{\partial x_2}{\partial z}, v_1, v_2\right) = 0 \qquad 9.8\text{-}7$$

$$\text{at } z = 1, \qquad h_1\left(x_1, x_2, \frac{\partial x_1}{\partial z}, \frac{\partial x_2}{\partial z}, w_1, w_2\right) = 0$$

$$h_2\left(x_1, x_2, \frac{\partial x_1}{\partial z}, \frac{\partial x_2}{\partial z}, w_1, w_2\right) = 0 \qquad 9.8\text{-}8$$

where we have assumed that all the variables have been normalized and that

$$0 < t < \theta \qquad 0 < z < 1$$

In addition, we have included the possibility that the control variables appearing in the system equations, u_1 and u_2, are different from those appearing in the boundary conditions, v_1, v_2, w_1, and w_2. The performance index is taken as

$$P = \int_0^\theta \int_0^1 F\left(x_1, x_2, \frac{\partial x_1}{\partial z}, \frac{\partial x_2}{\partial z}, u_1, u_2\right) dt\, dz \qquad 9.8\text{-}9$$

We proceed in exactly the same way we did for lumped parameter systems, by selecting a set of control variables $\bar{u}_1$, $\bar{u}_2$, $\bar{v}_1$, $\bar{v}_2$, $\bar{w}_1$, and $\bar{w}_2$. Next, we solve the system equations with the appropriate boundary conditions, calling the results $\bar{x}_1, \bar{x}_2, \partial \bar{x}_1/\partial z_1$, and $\partial \bar{x}_2/\partial z$, and, finally, we use this information to evaluate the performance index. Then we make small changes in the control variables, so that we have $\bar{u}_1 + \delta u_1, \bar{u}_2 + \delta u_2, \bar{v}_1 + \delta v_1, \bar{v}_2 + \delta v_2, \bar{w}_1 + \delta w_1,$ and $\bar{w}_2 + \delta w_2$, solve the system equations again, and evaluate the performance index with the new functions, to obtain $\bar{x}_1 + \delta x_1, \bar{x}_2 + \delta x_2, \partial(\bar{x}_1 + \delta x_1)/\partial z,$ $\partial(\bar{x}_2 + \delta x_2)/\partial z$, and δP.

The difference between these solutions is

$$\frac{\partial(\bar{x}_1 + \delta x_1)}{\partial t} - \frac{\partial \bar{x}_1}{\partial t} = \frac{\partial \delta x_1}{\partial t} = f_1 \left[\bar{x}_1 + \delta x_1, \bar{x}_2 + \delta x_2, \frac{\partial(\bar{x}_1 + \delta x_1)}{\partial z}, \frac{\partial(\bar{x}_2 + \delta x_2)}{\partial z}, \right.$$

$$\left. \frac{\partial^2(\bar{x}_1 + \delta x_1)}{\partial z^2}, \frac{\partial^2(\bar{x}_2 + \delta x_2)}{\partial z^2}, \bar{u}_1 + \delta u_1, \bar{u}_2 + \delta u_2 \right]$$

$$- f_1 \left(\bar{x}_1, \bar{x}_2, \frac{\partial \bar{x}_1}{\partial z}, \frac{\partial \bar{x}_2}{\partial z}, \frac{\partial^2 \bar{x}_1}{\partial z^2}, \frac{\partial^2 \bar{x}_2}{\partial z^2}, \bar{u}_1, \bar{u}_2 \right) \qquad \text{9.8-10}$$

If the changes are small enough, we can represent the function on the right-hand side of the equation above by a Taylor series expansion, where we retain only the linear terms

$$\frac{\partial \delta x_1}{\partial t} = \frac{\partial f_1}{\partial x_1} \delta x_1 + \frac{\partial f_1}{\partial x_2} \delta x_2 + \frac{\partial f_1}{\partial(\partial x_1/\partial z)} \frac{\partial \delta x_1}{\partial z} + \frac{\partial f_1}{\partial(\partial x_2/\partial z)} \frac{\partial \delta x_2}{\partial z}$$

$$+ \frac{\partial f_1}{\partial(\partial^2 x_1/\partial z^2)} \frac{\partial^2 \delta x_1}{\partial z^2} + \frac{\partial f_1}{\partial(\partial^2 x_2/\partial z^2)} \frac{\partial^2 \delta x_2}{\partial z^2} + \frac{\partial f_1}{\partial u_1} \delta u_1$$

$$+ \frac{\partial f_1}{\partial u_2} \delta u_2 + \cdots \qquad \text{9.8-11}$$

All the partial derivatives in this expression are evaluated along the initial trajectory. A similar result is obtained for $\partial \delta x_2/\partial t$.

It immediately becomes apparent that we must introduce some new notation in order to prevent the number of terms from becoming unmanageable. Consequently, at this point we will change to summation notation, where a repeated index indicates summation over the number of independent variables, which is two for our problem. Then Eq. 9.8-11 becomes

$$\frac{\partial \delta x_i}{\partial t} = \frac{\partial f_i}{\partial x_j} \delta x_j + \frac{\partial f_i}{\partial(\partial x_j/\partial z)} \frac{\partial \delta x_j}{\partial z} + \frac{\partial f_i}{\partial(\partial^2 x_j/\partial z^2)} \frac{\partial^2 \delta x_j}{\partial z^2}$$

$$+ \frac{\partial f_i}{\partial u_j} \delta u_j + \cdots \qquad \text{9.8-12}$$

for the case where $i = 1$, and we obtain the second equation by letting $i = 2$.

In a similar manner, the difference between the boundary conditions for the two sets of control variables becomes

$$g_1 \left(\bar{x}_1 + \delta x_1, \bar{x}_2 + \delta x_2, \frac{\partial(\bar{x}_1 + \delta x_2)}{\partial z}, \frac{\partial(\bar{x}_2 + \delta x_2)}{\partial z}, \bar{v}_1 + \delta v_1, \bar{v}_2 + \delta v_2 \right)$$

$$- g_1 \left(\bar{x}_1, \bar{x}_2, \frac{\partial \bar{x}_1}{\partial z}, \frac{\partial \bar{x}_2}{\partial z}, \bar{v}_1, \bar{v}_2 \right) = 0 \qquad \text{9.8-13}$$

or after expanding in a Taylor series

$$\frac{\partial g_1}{\partial x_1} \delta x_1 + \frac{\partial g_1}{\partial x_2} \delta x_2 + \frac{\partial g_1}{\partial(\partial x_1/\partial z)} \frac{\partial \delta x_1}{\partial z} + \frac{\partial g_1}{\partial(\partial x_2/\partial z)} \frac{\partial \delta x_2}{\partial z}$$

$$+ \frac{\partial g_1}{\partial v_1} \delta v_1 + \frac{\partial g_1}{\partial v_2} \delta v_2 = 0 \qquad\qquad 9.8\text{-}14$$

Similar results are obtained for g_2, h_1, and h_2. Summarizing the results, in summation notation we have

$$\text{at } z = 0, \qquad \frac{\partial g_i}{\partial x_j} \delta x_j + \frac{\partial g_i}{\partial(\partial x_j/\partial z)} \frac{\partial \delta x_j}{\partial z} + \frac{\partial g_i}{\partial v_j} \delta v_j = 0 \qquad 9.8\text{-}15$$

$$\text{at } z = 1, \qquad \frac{\partial h_i}{\partial x_j} \delta x_j + \frac{\partial h_i}{\partial(\partial x_j/\partial z)} \frac{\partial \delta x_j}{\partial z} + \frac{\partial h_i}{\partial w_j} \delta w_j = 0 \qquad 9.8\text{-}16$$

The change in the performance index becomes

$$(\bar{P} + \delta P) - \bar{P} = \delta P = \int_0^{\theta + \delta\theta} \int_0^1 F\left[\bar{x}_1 + \delta x_1, \bar{x}_2 + \delta x_2, \right.$$

$$\left. \frac{\partial(\bar{x}_1 + \delta x_1)}{\partial z}, \frac{\partial(\bar{x}_2 + \delta x_2)}{\partial z}, \bar{u}_1 + \delta u_1, \bar{u}_2 + \delta u_2 \right] dt\, dz$$

$$- \int_0^{\theta} \int_0^1 F\left(\bar{x}_1, \bar{x}_2, \frac{\partial \bar{x}_1}{\partial z}, \frac{\partial \bar{x}_2}{\partial z}, \bar{u}_1, \bar{u}_2 \right) dt\, dz \qquad 9.8\text{-}17$$

or after expanding in a series and introducing summation notation

$$\delta P = \int_0^{\theta} \int_0^1 \left[\frac{\partial F}{\partial x_j} \delta x_j + \frac{\partial F}{\partial(\partial x_j/\partial z)} \frac{\partial \delta x_j}{\partial z} + \frac{\partial F}{\partial u_j} \delta u_j \right] dt\, dz$$

$$+ \left(\int_0^1 F\, dz \right) \delta\theta \qquad\qquad 9.8\text{-}18$$

Now, if we multiply each of the system equations by a Lagrange multiplier, integrate them with respect to t from zero to θ and with respect to z from zero to unity, and add them to the expression above, we obtain

$$\delta P = \int_0^{\theta} \int_0^1 \left[\left(\frac{\partial F}{\partial x_j} + \lambda_i \frac{\partial f_i}{\partial x_j} \right) \delta x_j - \lambda_i \frac{\partial \delta x_i}{\partial t} + \left(\frac{\partial F}{\partial(\partial x_j/\partial z)} + \lambda_i \frac{\partial f_i}{\partial(\partial x_j/\partial z)} \right) \frac{\partial \delta x_j}{\partial z} \right.$$

$$\left. + \lambda_i \frac{\partial f_i}{\partial(\partial^2 x_j/\partial z^2)} \frac{\partial^2 \delta x_j}{\partial z^2} + \left(\frac{\partial F}{\partial u_j} + \lambda_i \frac{\partial f_i}{\partial u_j} \right) \delta u_j \right] dt\, dz$$

$$+ \left(\int_0^1 F\, dz \right) d\theta \qquad\qquad 9.8\text{-}19$$

From integration by parts, we find that

$$\int_0^{\theta} \int_0^1 \left(\lambda_i \frac{\partial \delta x_i}{\partial t} dt \right) dz = \int_0^1 [\lambda_i(\theta)\, \delta x_i(\theta) - \lambda_i(0)\, \delta x_i(0)]\, dz$$

$$- \int_0^{\theta} \int_0^1 \delta x_1 \frac{\partial \lambda_i}{\partial t} dt\, dz \qquad\qquad 9.8\text{-}20$$

$$\int_0^\theta \int_0^1 \left\{ \left[\frac{\partial F}{\partial (\partial x_j/\partial z)} + \frac{\lambda_i \, \partial f_i}{\partial (\partial x_j/\partial z)} \right] \frac{\partial \delta x_j}{\partial z} dz \right\} dt$$

$$= \int_0^\theta \left\{ \left[\frac{\partial F}{\partial (\partial x_j/\partial z)} + \frac{\lambda_i \, \partial f_i}{\partial (\partial x_j/\partial z)} \right] \delta x_j \Big|_0^1 \right\} dt$$

$$- \int_0^\theta \int_0^1 \left\{ \frac{\partial}{\partial z} \left[\frac{\partial F}{\partial (\partial x_j/\partial z)} + \frac{\lambda_i \, \partial f_i}{\partial (\partial x_j/\partial z)} \right] \delta x_j \right\} dz \, dt \qquad 9.8\text{-}21$$

$$\int_0^\theta \int_0^1 \left[\lambda_i \, \frac{\partial f_i}{\partial (\partial^2 x_j/\partial z^2)} \frac{\partial}{\partial z} \left(\frac{\partial \delta x_j}{\partial z} \right) dz \right] dt$$

$$= \int_0^\theta \left\{ \left[\lambda_i \, \frac{\partial f_i}{\partial (\partial^2 x_j/\partial z^2)} \frac{\partial \delta x_j}{\partial z} \right] \Big|_0^1 \right\} dt$$

$$- \int_0^\theta \int_0^1 \frac{\partial}{\partial z} \left[\lambda_i \, \frac{\partial f_i}{\partial (\partial^2 x_j/\partial z^2)} \right] \frac{\partial \delta x_j}{\partial z} dz \, dt \qquad 9.8\text{-}22$$

The final term in this last expression also can be integrated by parts, so that Eq. 9.8–22 becomes

$$\int_0^\theta \int_0^1 \lambda_i \, \frac{\partial f_i}{\partial (\partial^2 x_j/\partial z^2)} \frac{\partial^2 \delta x_j}{\partial z^2} dz \, dt$$

$$= \int_0^\theta \left\{ \left[\lambda_i \, \frac{\partial f_i}{\partial (\partial^2 x_j/\partial z^2)} \frac{\partial \delta x_j}{\partial z} \right] \Big|_0^1 \right\} dt$$

$$- \int_0^\theta \left\{ \left[\frac{\partial}{\partial z} \left[\lambda_i \, \frac{\partial f_i}{\partial (\partial^2 x_j/\partial z^2)} \right] \delta x_j \Big|_0^1 \right] \right\} dt$$

$$+ \int_0^\theta \int_0^1 \frac{\partial^2}{\partial z^2} \left[\lambda_i \, \frac{\partial f_i}{\partial (\partial^2 x_j/\partial z^2)} \right] \delta x_j \, dz \, dt \qquad 9.8\text{-}23$$

Substituting Eqs. 9.8–20, 9.8–21, and 9.8–23 into Eq. 9.8–19 gives the expression

$$\delta P = \int_0^\theta \int_0^1 \left(\left\{ \frac{\partial \lambda_i}{\partial t} + \frac{\partial F}{\partial x_j} + \lambda_i \, \frac{\partial f_i}{\partial x_j} - \frac{\partial}{\partial z} \left[\frac{\partial F}{\partial (\partial x_j/\partial z)} + \lambda_i \, \frac{\partial f_i}{\partial (\partial x_j/\partial z)} \right] \right. \right.$$

$$+ \left. \left. \frac{\partial^2}{\partial z^2} \left[\lambda_i \, \frac{\partial f_i}{\partial (\partial^2 x_j/\partial z^2)} \right] \right\} \delta x_j + \left(\frac{\partial F}{\partial u_j} + \lambda_i \, \frac{\partial f_i}{\partial u_j} \right) \delta u_j \right) dz \, dt$$

$$+ \left(\int_0^1 F \, dz \right) \delta \theta - \int_0^1 [\lambda_i(\theta) \, \delta x_i(\theta) - \lambda_i(0) \, \delta x_i(0)] \, dz$$

$$+ \int_0^\theta \left\{ \left[\frac{\partial F}{\partial (\partial x_j/\partial z)} + \lambda_i \, \frac{\partial f_i}{\partial (\partial x_j/\partial z)} \right] \delta x_j \Big|_0^1 \right\} dt$$

$$+ \int_0^\theta \left\{ \left[\lambda_i \, \frac{\partial f_i}{\partial (\partial^2 x_j/\partial z^2)} \frac{\partial \delta x_j}{\partial z} \right] \Big|_0^1 \right\} dt$$

$$- \int_0^\theta \left\{ \frac{\partial}{\partial z} \left[\lambda_i \, \frac{\partial f_i}{\partial (\partial^2 x_j/\partial z^2)} \right] \delta x_j \Big|_0^1 \right\} dt \qquad 9.8\text{-}24$$

Now, if we define the adjoint variables, except for their boundary conditions, by the partial differential equations

$$\frac{\partial \lambda_i}{\partial t} = -\left\{\frac{\partial F}{\partial x_j} + \lambda_i \frac{\partial f_i}{\partial x_j} - \frac{\partial}{\partial z}\left[\frac{\partial F}{\partial(\partial x_j/\partial z)} + \lambda_i \frac{\partial f_i}{\partial(\partial x_j/\partial z)}\right] \right.$$
$$\left. + \frac{\partial^2}{\partial z^2}\left[\lambda_i \frac{\partial f_i}{\partial(\partial^2 x_j/\partial z^2)}\right]\right\}$$
$$9.8\text{-}25$$

we can significantly simplify the expression above. Also, we multiply Eqs. 9.8-15 by a new set of Lagrange multipliers v_i and Eqs. 9.8-16 by another set μ_i, integrate both sets of equations from zero to θ, and add them to Eq. 9.8-24.

$$\delta P = \int_0^\theta \int_0^1 \left(\frac{\partial F}{\partial u_j} + \lambda_i \frac{\partial f_i}{\partial u_j}\right) \delta u_j \, dz \, dt + \left(\int_0^1 F \, dz\right) \delta\theta - \int_0^1 [\lambda_i(\theta) \, \delta x_i(\theta)$$

$$- \lambda_i(0) \delta x_i(0)] \, dz + \int_0^\theta \left\{\frac{\partial F}{\partial(\partial x_j/\partial z)} + \lambda_i \frac{\partial f_i}{\partial(\partial x_j/\partial z)}\right.$$

$$\left. - \frac{\partial}{\partial z}\left[\lambda_i \frac{\partial f_i}{\partial(\partial^2 x_j/\partial z^2)}\right] + \mu_i \frac{\partial h_i}{\partial x_j}\right\} \delta x_j \bigg|_{z=1} dt - \int_0^\theta \left\{\frac{\partial F}{\partial(\partial x_j/\partial z)}\right.$$

$$\left. + \lambda_i \frac{\partial f_i}{\partial(\partial x_j/\partial z)} - \frac{\partial}{\partial z}\left[\lambda_i \frac{\partial f_i}{\partial(\partial^2 x_j/\partial z^2)}\right] - v_i \frac{\partial g_i}{\partial x_j}\right\} \delta x_j \bigg|_{z=0} dt$$

$$+ \int_0^\theta \left\{\left[\lambda_i \frac{\partial f_i}{\partial(\partial^2 x_j/\partial z^2)} + \mu_i \frac{\partial h_i}{\partial(\partial x_j/\partial z)}\right]\frac{\partial \delta x_j}{\partial z}\right\}\bigg|_{z=1} dt$$

$$+ \int_0^\theta \mu_i \frac{\partial h_i}{\partial w_j} \delta w_j \, dt - \int_0^\theta \left\{\left[\lambda_i \frac{\partial f_i}{\partial(\partial^2 x_j/\partial z^2)} - v_i \frac{\partial g_i}{\partial(\partial x_j/\partial z)}\right]\frac{\partial \delta x_j}{\partial z}\right\}\bigg|_{z=0}$$

$$+ \int_0^\theta v_i \frac{\partial g_i}{\partial v_j} \delta v_j \, dt$$
$$9.8\text{-}26$$

By defining the boundary conditions on the adjoint variables so that at $z = 0$,

$$\lambda_i \frac{\partial f_i}{\partial(\partial^2 x_j/\partial z^2)} - v_i \frac{\partial g_i}{\partial(\partial x_j/\partial z)} = 0$$

$$\frac{\partial F}{\partial(\partial x_j/\partial z)} + \lambda_i \frac{\partial f_i}{\partial(\partial x_j/\partial z)} - \frac{\partial}{\partial z}\left(\lambda_i \frac{\partial f_i}{\partial(\partial^2 x_j/\partial z^2)}\right) - v_i \frac{\partial g_i}{\partial x_j} = 0 \quad 9.8\text{-}27$$

at $z = 1$,

$$\lambda_i \frac{\partial f_i}{\partial(\partial^2 x_j/\partial z^2)} + \mu_i \frac{\partial h_i}{\partial(\partial x_j/\partial z)} = 0$$

$$\frac{\partial F}{\partial(\partial x_j/\partial z)} + \lambda_i \frac{\partial f_i}{\partial(\partial x_j/\partial z)} - \frac{\partial}{\partial z}\left[\lambda_i \frac{\partial f_i}{\partial(\partial^2 x_j/\partial z^2)}\right] + \mu_i \frac{\partial h_i}{\partial x_j} = 0 \quad 9.8\text{-}28$$

the equation reduces to

$$\delta P = \int_0^\theta \int_0^1 \left(\frac{\partial F}{\partial u_j} + \lambda_i \frac{\partial f_i}{\partial u_j}\right) \delta u_j \, dt \, dz + \int_0^\theta \mu_i \frac{\partial h_i}{\partial w_j} \delta w_j \, dt$$

$$+ \int_0^\theta v_i \frac{\partial g_i}{\partial v_j} \delta v_j \, dt + \left(\int_0^1 F \, dz\right) \delta\theta - \int_0^1 [\lambda_i(\theta) \, \delta x_i(\theta) - \lambda_i(0) \, \delta x_i(0)] \, dz$$

$$9.8\text{-}29$$

If the initial values of x_1 and x_2 at $t = 0$ are fixed, then $\delta x_i(0) = 0$. Since we can also write

$$x_1(\theta + \delta\theta) \simeq x_i(\theta) + f_i(\theta)\,\delta\theta$$

$$\simeq \bar{x}_i(\theta) + \delta x_i(\theta) + f_i(\theta)\,\delta\theta \qquad \text{9.8-30}$$

for a problem where the value of θ is unknown but the final value of x_i is specified—that is, $x_1(\theta + \delta\theta) = \bar{x}_i(\theta)$—then

$$\delta x_i(\theta) = -f_i(\theta)\,\delta\theta \qquad \text{9.8-31}$$

so that Eq. 9.8–29 becomes

$$\delta P = \int_0^\theta \int_0^1 \left(\frac{\partial F}{\partial u_j} + \lambda_i \frac{\partial f_i}{\partial u_j} \right) \delta u_j \, dz \, dt + \int_0^\theta \mu_i \frac{\partial h_i}{\partial w_j} \delta w_j \, dt$$

$$+ \int_0^\theta \nu_i \frac{\partial g_i}{\partial v_j} \delta v_j \, dt + \left[\int_0^1 (F + \lambda_i f_i)|_{t=\theta} \, dz \right] \delta\theta \qquad \text{9.8-32}$$

Thus, we have obtained an expression that contains only the changes in the control variables and the variation of the process time. It is clear that no matter what sign the coefficient of the $\delta\theta$ term has, as we change $\delta\theta$ from a positive to a negative value, the sign of δP must change. However, if our initial choices of the controls and process time were those that minimized the performance criterion, we know that any change must make P increase in the neighborhood of the minimum. The only way it is possible to reconcile these two facts is to require that

$$\int_0^1 (F + \lambda_i f_i)|_{t=0} \, dz = 0 \qquad \text{9.8-33}$$

Similarly, if we let

$$\delta u = -\epsilon \int_0^1 \left(\frac{\partial F}{\partial u_j} + \lambda_i \frac{\partial f_i}{\partial u_j} \right) dz \qquad \text{9.8-34}$$

the dependence of δP on δu becomes

$$\delta P = -\epsilon \int_0^\theta \left[\int_0^1 \left(\frac{\partial F}{\partial u_j} + \lambda_i \frac{\partial f_i}{\partial u_j} \right) dz \right]^2 dt \qquad \text{9.8-35}$$

so that the only way we can be certain that δP will not decrease in the neighborhood of the minimum is to require that

$$\int_0^1 \left(\frac{\partial F}{\partial u_j} + \lambda_i \frac{\partial f_i}{\partial u_j} \right) dz = 0 \qquad \text{9.8-36}$$

Exactly the same procedure leads to the results

$$\mu_i \frac{\partial h_i}{\partial w_j} = 0 \qquad \nu_i \frac{\partial g_i}{\partial v_j} = 0 \qquad \text{9.8-37}$$

It is possible to define the three Hamiltonian functions

$$H = F + \lambda_i f_i, \quad K_0 = v_i g_i \quad \text{at } z = 0$$

$$K_1 = \mu_i h_i \quad \text{at } z = 1 \qquad \qquad 9.8\text{-}38$$

and to use these to develop a set of averaged canonical equations, as well as a simpler necessary condition. In this way the optimal control problem for distributed parameter plants can be made analogous to the lumped parameter formulation.[1]

The preceding analysis makes it painfully clear that the optimal control of a distributed parameter process is much more difficult than the optimal control of a lumped parameter system. For all but the simplest problems, we will have to solve coupled sets of nonlinear, partial differential equations with mixed boundary conditions. It is not surprising, therefore, that few applications of the method have been discussed in the literature. However, the analytical expressions for the necessary conditions are fairly recent, so that significant extensions of the results can be expected.

Example 9.8-1 Optimal control of a steam-heated exchanger

The equation describing a steam-heated exchanger for a case where the wall heat capacitance is negligible was presented in Example 6.2-1.

$$\rho C_p A \frac{\partial T}{\partial t} + \rho C_p A v \frac{\partial T}{\partial z} = h A_H (T_c - T) \qquad \qquad 6.2\text{-}1$$

We are interested in synthesizing a control system that will manipulate the condensate temperature, T_c, so that we have a case where the control variable is distributed throughout space. To simplify the algebra, we let

$$x = \frac{T - T_s}{T_{cs} - T_f}, \quad u = \frac{T_c - T_{cs}}{T_{cs} - T_f}, \quad \tau = \frac{h A_H t}{\rho C_p A}, \quad Z = \frac{h A_H z}{\rho C_p A v} \qquad 9.8\text{-}39$$

where T_s = the steady state temperature within the exchanger, T_{cs} = steady state design value of the condensate temperature, T_f = feed temperature to the exchanger, and L = value of Z at the end of the exchanger. After we substitute these definitions, the system equation becomes

$$\frac{\partial x}{\partial \tau} + \frac{\partial x}{\partial Z} = u - x \qquad \qquad 9.8\text{-}40$$

If we measure some initial deviation in the steady state temperature profile

$$x(0, Z) = x_i \qquad \qquad 9.8\text{-}41$$

we want to adjust the control variable in such a way that we return to the steady state design condition while minimizing the quadratic performance index

$$P = \frac{1}{2} \int_0^{\theta = \infty} \int_0^L [C(Z) x^2(\tau, Z) \, dZ + u^2(\tau)] \, d\tau \qquad 9.8\text{-}42$$

---

[1] See Ch. 11 of reference 7 on p. 264.

With this formulation, we can consider the special case where we only want to control the temperature at the outlet from the exchanger simply by choosing the weighting function to be

$$C(Z) = C_s \delta(L - Z) \qquad\qquad 9.8\text{-}43$$

where C_s is a constant and $\delta(L - Z)$ is the Dirac delta function. In this instance we obtain

$$P = \frac{1}{2} \int_0^{\theta = \infty} [C_s x^2(\tau, L) + u^2(\tau)]\, d\tau \qquad\qquad 9.8\text{-}44$$

Solution

We first note that there are no control variables at the system boundaries, so that the functions g_1, g_2, h_1. and h_2, given by Eqs. 9.8–7 and 9.8–8, are all equal to zero. This means that we can set the Lagrange multipliers v_i and μ_i equal to zero and neglect the Hamiltonian functions K_0 and K_1 given by Eqs. 9.8–38. Rewriting the system equation, Eq. 9.8–40, in the form

$$\frac{\partial x}{\partial \tau} = f = -x - \frac{\partial x}{\partial Z} + u \qquad\qquad 9.8\text{-}45$$

we find that the remaining Hamiltonian function in Eqs. 9.8–38 becomes

$$H = F + \lambda_i f_i = \frac{1}{2}(Cx^2 + u^2) + \lambda\left(-x - \frac{\partial x}{\partial Z} + u\right) \qquad 9.8\text{-}46$$

The adjoint variable appearing in this expression is defined by Eq. 9.8–25

$$\frac{\partial \lambda}{\partial \tau} = -\left\{\frac{\partial F}{\partial x} + \lambda\frac{\partial f}{\partial x} - \frac{\partial}{\partial Z}\left[\lambda\frac{\partial f}{\partial(\partial x/\partial Z)}\right]\right\}$$

$$= -Cx + \lambda - \frac{\partial \lambda}{\partial Z} \qquad\qquad 9.8\text{-}47$$

and the boundary conditions for the adjoint variable are given by Eqs. 9.8–27 and 9.8–28

$$\lambda(\tau, 0) = \lambda(\tau, L) = 0 \qquad\qquad 9.8\text{-}48$$

From Eq. 9.8–36 we see that the necessary condition for the optimal control policy is

$$\int_0^L \frac{\partial H}{\partial u}\, dZ = 0 = \int_0^L (u + \lambda)\, dZ \qquad\qquad 9.8\text{-}49$$

Since the control variable, u, depends only on time and not position, this last result can be written

$$u = -\int_0^L \lambda(\tau, Z)\, dZ \qquad\qquad 9.8\text{-}50$$

In order to avoid confusion later, we write this last expression in terms of the "dummy variable" ϕ

$$u = -\int_0^L \lambda(\tau, \phi)\, d\phi \qquad\qquad 9.8\text{-}51$$

The initial and final values of the temperature are specified at every point within the exchanger (i.e., Eq. 9.8–41) and

$$x(\theta, Z) = 0 \qquad\qquad 9.8\text{-}52$$

We might attempt to use the stopping condition, Eq. 9.8–33, to determine the optimal value of θ. However, from our earlier analysis of the optimal control of linear, lumped parameter processes with quadratic performance criteria, we anticipate that the optimal profile will approach the desired steady state condition in an asymptotic manner. For this kind of a situation, we must let θ approach infinity, as we indicated in Eqs. 9.8–42 and 9.8–44.

We also use our experience with lumped parameter systems to develop a solution to the optimization equations. If we were to replace the spatial derivatives in the system equation and adjoint expression by finite-difference approximations, we would obtain sets of linear, ordinary differential equations. Moreover, the formulation of the optimization equations would look exactly as they did in our earlier lumped parameter study. Hence we would know the solution for that case. To go from the solution of this discretized problem back to the continuous system, it would be necessary to sum (or to integrate) all the adjoint equations.[2] In addition, we expect that the adjoint variable will explicitly depend on τ and Z. Thus we assume a solution having the form[1]

$$\lambda(\tau, Z) = \int_0^L M(\tau, Z, \phi)x(\tau, \phi)\, d\phi \qquad\qquad 9.8\text{-}53$$

By analogy with the lumped parameter analysis, we assume that the unknown quantity M will be symmetric in its spatial arguments

$$M(\tau, Z, \phi) = M(\tau, \phi, Z) \qquad\qquad 9.8\text{-}54$$

Also, from Eq. 9.8–48 we find that the boundary conditions on M must be

$$M(\tau, 0, \phi) = M(\tau, L, \phi) = 0 \qquad\qquad 9.8\text{-}55$$

In order to see if our assumed solution will indeed satisfy the optimization equations, we first substitute Eq. 9.8–46 for λ in the adjoint expression, Eq. 9.8–47,

$$\frac{\partial \lambda}{\partial \tau} + \frac{\partial \lambda}{\partial Z} = -Cx + \int_0^L M(\tau, Z, \phi)x(\tau, \phi)\, d\phi \qquad\qquad 9.8\text{-}56$$

From this point on we will not bother to indicate the explicit dependence of M on τ, but we will keep track of the spatial variables. Also, for convenience later, we choose to write the preceding expression in the equivalent form

$$\frac{\partial \lambda}{\partial \tau} + \frac{\partial \lambda}{\partial Z} = -\int_0^L [C(\phi)\delta(Z - \phi) - M(Z, \phi)]x(\phi)\, d\phi \qquad\qquad 9.8\text{-}57$$

[2] A discussion of the limiting procedure has been published by T. M. Pell, Jr. and R. Aris, *Ind. Eng. Chem. Fundamentals*, **9**, 15 (1970).

A second relationship for the adjoint variable can be developed by differentiating our assumed solution. Thus we find from Eq. 9.8–53 that

$$\frac{\partial \lambda}{\partial \tau} + \frac{\partial \lambda}{\partial Z} = \int_0^L \frac{\partial M(Z, \phi)}{\partial \tau} x(\phi)\, d\phi + \int_0^L M(Z, \phi) \frac{\partial x(\phi)}{\partial \tau}\, d\phi$$

$$+ \int_0^L \frac{\partial M(Z, \phi)}{\partial Z} x(\phi)\, d\phi \qquad\qquad 9.8\text{–}58$$

From the system equation, Eq. 9.8–45, we see that

$$\frac{\partial x(\phi)}{\partial \tau} = -x(\phi) - \frac{\partial x(\phi)}{\partial \phi} + u \qquad\qquad 9.8\text{–}59$$

Using the expression for the optimal control action, Eq. 9.8–51, and our assumed solution, Eq. 9.8–53, we find that

$$u = -\int_0^L \lambda(\phi)\, d\phi = -\int_0^L \left[\int_0^L M(\phi, \psi) x(\psi)\, d\psi \right] d\phi \qquad 9.8\text{–}60$$

Substituting these relationships into Eq. 9.8–58, we obtain

$$\frac{\partial \lambda}{\partial \tau} + \frac{\partial \lambda}{\partial Z} = \int \frac{\partial M(Z, \phi)}{\partial \tau} x(\phi)\, d\phi - \int_0^L M(Z, \phi) x(\phi)\, d\phi$$

$$- \int_0^L M(Z, \phi) \frac{\partial x(\phi)}{\partial \phi}\, d\phi$$

$$- \int_0^L M(Z, \phi) \left[\int_0^L \left\{ \int_0^L M(\sigma, \psi) x(\psi)\, d\psi \right\} d\sigma \right] d\phi$$

$$+ \int_0^L \frac{\partial M(Z, \phi)}{\partial Z} x(\phi)\, d\phi \qquad\qquad 9.8\text{–}61$$

Integrating the third term on the right-hand side of this equation by parts, we get

$$\int_0^L M(Z, \phi) \frac{\partial x(\phi)}{\partial \phi}\, d\phi = \left[M(Z, \phi) x(\phi) \right]\Big|_0^L + \int_0^L \frac{\partial M(Z, \phi)}{\partial \phi} x(\phi)\, d\phi$$

$$= \int_0^L \frac{\partial M(Z, \phi)}{\partial \phi} x(\phi)\, d\phi \qquad\qquad 9.8\text{–}62$$

where Eq. 9.8–55 has been used to simplify the expression. Replacing the third term on the right-hand side of Eq.9.8–61 by this result, changing the order of integration in the last term, and rearranging the equation, we obtain

$$\frac{\partial \lambda}{\partial \tau} + \frac{\partial \lambda}{\partial Z} = \int_0^L \left\{ \frac{\partial M(Z, \phi)}{\partial \tau} + \frac{\partial M(Z, \phi)}{\partial Z} + \frac{\partial M(Z, \phi)}{\partial \phi} - M(Z, \phi) \right.$$

$$\left. - \left[\int_0^L M(Z, \sigma)\, d\sigma \right] \left[\int_0^L M(\psi, \phi)\, d\psi \right] \right\} x(\phi)\, d\phi \qquad 9.8\text{–}63$$

This expression has the same form as our earlier result, Eq. 9.8–57; therefore our assumed solution will be valid only if the two equations are identical. Thus we must have

$$\frac{\partial M}{\partial \tau} + \left(\frac{\partial M}{\partial Z} + \frac{\partial M}{\partial \phi}\right) - 2M - \left[\int_0^L M(Z, \sigma)\, d\sigma\right]\left[\int_0^L M(\psi, \phi)\, d\psi\right]$$

$$+ C(\phi)\delta(Z - \phi) = 0 \qquad\qquad 9.8\text{-}64$$

Note that any solution of this equation will satisfy the symmetry condition we mentioned earlier, Eq. 9.8-54. Also, in the special case where the flow velocity is equal to zero, it can be shown[3] that after two integrations this relationship is equivalent to the Riccati equation we obtained for lumped parameter processes. For the case where we let θ approach infinity, we obtain a solution that is independent of time.

To keep the analysis as simple as possible, we restrict our attention to the control of the effluent temperature, so that $C(\phi) = C_s\delta(L - \phi)$, and we let $\theta = \infty$. The optimal control action is given by Eqs. 9.8-51 and 9.8-53, which we write as

$$u(\tau) = \int_0^L K(Z)x(\tau, Z)\, dZ \qquad\qquad 9.8\text{-}65$$

where the controller gain is

$$K(Z) = -\int_0^L M(Z, \phi)\, d\phi \qquad\qquad 9.8\text{-}66$$

The presence of the $\delta(L - \phi)$ term in Eq. 9.8-64 means that the steady state solution of this expression is discontinuous at the values $Z = L$, $\phi = L$. However, the method of characteristics can be used to develop an implicit solution of the equation, which is valid everywhere except at those points.

$$M(Z, \phi) = -\int_Z^{L - (\phi - Z)\, LU\, (\phi - Z)} e^{2(Z - \psi)} K(\psi)K(\psi - Z + \phi)\, d\psi$$

$$+ C_s e^{-2(L - \phi)}\delta(\phi - Z) \qquad\qquad 9.8\text{-}67$$

where the Heaviside step function $U(\phi - Z)$ is defined as

$$U(\phi - Z) = \begin{cases} 0, & \text{if } \phi < Z \\ 1, & \text{if } \phi > Z \end{cases} \qquad\qquad 9.8\text{-}68$$

Integrating this result gives an expression for the feedback gain

$$K(Z) = -C_s e^{-2(L - Z)} + \int_Z^L K(\psi)e^{2(Z - \psi)}\, d\psi \int_{\psi - Z}^L K(\phi)\, d\phi \qquad 9.8\text{-}69$$

Then, using an iterative approach, we can solve this equation for the optimal gain.

From Eq. 9.8-65 we see that the implementation of the optimal control action will require measured values of the temperature at every point within the exchanger. In other words, the controller acts on measured values of the optimal

[3] See Ch. 11 of reference 7 on p. 264.

profile. Obviously it would be impractical to install an infinite number of thermocouples along the length of the exchanger. Thus we need to look for the optimal approximation of the optimal profile; we leave this an "overwhelming" exercise for the reader.

SECTION 9.9 SUMMARY

The use of classical servomechanism theory to design control systems is essentially limited to linear plants with a single input and a single output. Since most chemical plants are multivariable, nonlinear systems, it became necessary to look for a new basis for the design of process controllers. This chapter showed that Pontryagin's minimum principle could be used for this purpose. This approach involved a dynamic optimization, which seems to be a natural extension of the normal, optimum steady state design analysis.

The necessary conditions that an optimal control system must satisfy were derived, and the results were used to design control systems for a number of different kind of plants. Both multivariable and nonlinear processes were considered, but most of the results were for linear systems. Implementation of the controller was stressed as being an important factor, and therefore the multivariable proportional controller, the multivariable proportional plus integral controller, and the controller with linear switching appear to be the most significant results. Unfortunately the optimum values of the controller gains obtained from the theory depend on the weighting factors in the performance index, and, at present, no quantitative procedure exists for selecting these. Problems where profit or operating costs are used as the measure of plant performance are conspicuously absent from the discussion. This kind of a study is needed in order to develop "rules of thumb" for selecting the weighting factors.

In addtion, problems associated with the computation of the optimal control policy and trajectories have not been considered. However, the literature on this subject is voluminous, and an interested reader is encouraged to refer to the book by Denn and that by Lapidus and Luus for a detailed treatment.

EXERCISES

1. (A†) The optimum steady state design of an isothermal CSTR was described in Example 2.1-1. If we assume that the value of the product, component B, is \$5/lb mole, we can estimate the profit produced by the plant as it returns to the optimum steady state operating point from an arbitrary initial condition representing the effect of a disturbance. Similarly, we can calculate the profit produced when we install an optimal, linear feedback controller on the plant (see Eq. 9.2-15), where the controller manipulates the feed rate to compensate for measured deviations in the reactor composition. It might be possible to use these calculations as a basis for selecting the weighting factor, c^2, which appears in the performance index (see Eq. 9.2-4). Try to assess the value of this procedure by considering cases where the initial value of A is 0.30 and 0.20 lb mole/ft³.

2. (B†) Compare the response of the optimal proportional plus integral controller described in Example 9.4-4 to the results obtained in the previous exercise. Study the sensitivity of your solution to variations in the weighting parameter c.

3. (B) In Example 9.4-1 we developed a solution of the optimal control equations by assuming that the adjoint variables were linear combinations of the state variables; see Eqs. 9.4-19 and 9.4-20. As an alternate approach, see if you can obtain the same solution by eliminating the adjoint variables from the set of equations describing the necessary conditions.

4. (A) The equation describing a first-order system can usually be linearized to give

$$\dot{x} = Ax + Bu$$

Find the optimum control policy $u(t)$ that forces the plant from some initial condition x_0 to the origin while minimizing the performance criterion[1]

$$P = \int_0^\theta (C^2 + u^2)\, dt$$

Discuss the meaning of your solution.

5. (A) According to Foust et al.,[2] the equation describing a filter press can be written as

$$\frac{dV}{dt} = \frac{(-\Delta P)g_c A^2}{\mu(wV\alpha + R_m A)}$$

where V = volume of filtrate collected, t = time, A = filtration area, μ = viscosity, w = concentration of solids in the feed slurry, R_m = resistance of filter medium and piping, and α = specific resistance. For many materials, α can be described by the equation

$$\alpha = \alpha_0 + b(-\Delta P)^s$$

Assuming that these relationships are correct, how should we manipulate the pressure drop across the filter cake in order to maximize the amount of filtrate we can collect in a specified time? Compare your optimal solution to the conventional operating procedure and discuss the results, noting that s can never be greater than unity. Use the "Inverse problem" formulation to see if a constant rate filtration can ever correspond to an optimal policy for some unknown relationship between α and ΔP.

6. (B†) Optimal control theory can be applied to a wide variety of problems. For example, consider a case where we attempt to establish an optimal production schedule for a particular plant. We let I = inventory, P = production rate, and s = sales rate, and we write the equation describing the system as

$$\frac{dI}{dt} = P - s$$

[1] Taken from p. 97 of reference 7 on p. 264.

[2] A. S. Foust, L. A. Wenzel, C. W. Clump, L. Maus, and L. B. Anderson, *Principles of Unit Operations*, p. 492, Wiley, N.Y., 1962.

Now, if we assume that the incremental costs associated with deviations of the production rate and inventory from the design values can be described by a quadratic function, we can write the performance index as

$$J = \int_0^\theta [C_i(I - I_s)^2 + C_p(P - P_s)^2] \, dt$$

where I_s and P_s are the design values of I and P, C_i and C_p are constant cost factors, and θ is a fixed operating interval. Once the sales rate $s(t)$ has been specified, show how you can determine the optimal production rate.[3] Would it ever be desirable to implement the optimal policy in a feedback manner?[4]

7. (B†) A number of chemical engineers have developed an interest in bioengineering problems in recent years because there are significant areas of overlap between the disciplines. As an illustration, Denn[4] presents a problem which relates the CO_2 concentration in the body tissue x to the pulmonary ventilation u,

$$\alpha_0 \ddot{x} + (\alpha_1 + \alpha_2 u)\dot{x} + \alpha_3 u x = \alpha_4 + \alpha_5 u$$

where the coefficients α_i are constants. He assumes that the "control mechanism" in the body manipulates the pulmonary ventilation to minimize the performance measure

$$P = \frac{1}{2} \int_0^\theta (x^2 + C^2 u^2) \, dt$$

and asks you to determine the equations describing the optimal control law. Also, you are supposed to find an approximate solution of the equations and compare it to the commonly used expression

$$u = a + bx$$

8. (B) Material balances for the batch distillation of a binary mixture can be written as

$$\frac{ds}{dt} = -D$$

$$\frac{dx_s}{dt} = \frac{D}{s}(x_s - y)$$

where $s = $ amount of material in the still, $x_s = $ composition of the more volatile component in the still, $D = $ distillate rate, and $y = $ overhead composition of the more volatile component. Suppose that we want to maximize the distillate rate

$$P = \int_0^\theta D \, dt$$

while maintaining a specified average purity of the overhead

$$y_{\mathrm{av}} = \frac{\displaystyle\int_0^\theta y D \, dt}{\displaystyle\int_0^\theta D \, dt}$$

What are the necessary conditions for the optimal control policy?[5]

[3] Taken from p. 134 of reference 7 on p. 264 based on a paper by Holt, et al.
[4] Taken from p. 134 of reference 7 on p. 264 based on a paper by Grodins, et al.
[5] Taken from p. 132 of reference 7 on p. 264 based on a paper by Converse and Gross.

9. (B) Use the method described in Example 9.4-1 to determine a nonlinear, optimal control policy for the plant described in Exercise 1 above.

10. (C†) Johnson[6] studied the optimal control of a first-order nonlinear system described by the equation

$$\frac{dx}{dt} = f_1(x) + uf_2(x) \qquad x(0) = x_0$$

and showed that the control policy which would force the plant to return to the origin while minimizing the performance index

$$P = \int_0^\theta [g(x) + C^2 u^2] \, dt$$

where C = constant and f_1, f_2, and g are continuous, differentiable functions of x, was

$$u = -\frac{f_1(x)}{f_2(x)} - \frac{x}{f_2(x)} \left\{ \left[\frac{f_1(x)}{x} \right]^2 + \frac{[f_2(x)]^2[g(x) + \alpha]}{C^2 x^2} \right\}^{1/2}$$

In this expression, α is a function of x_0 and θ. Show that this solution satisfies the appropriate optimization equations.

Apply this method to the plant described in Exercise 1. Also, compare your result to the solution of the previous exercise.

11. (C†) The derivation of the optimal linear feedback controller can be extended to nonautonomous systems of the form

$$\dot{\mathbf{x}} = \mathbf{A}(t)\mathbf{x} + \mathbf{B}(t)\mathbf{u}$$

or to cases of finite operating intervals θ, by replacing the assumed solution, Eq. 9.4-42, by the more general expression,

$$\boldsymbol{\lambda} = \mathbf{M}(t)\mathbf{x}$$

With this assumption, the matrix Riccati equation becomes

$$\dot{\mathbf{M}} - \mathbf{M}\mathbf{B}\mathbf{R}^{-1}\mathbf{B}^T\mathbf{M} + \mathbf{M}\mathbf{A} + \mathbf{A}^T\mathbf{M} + \mathbf{C} = 0$$

which is identical to our earlier result except that it includes time derivative terms. When the operating interval is specified as some finite value, it will not be possible to return to the origin, so that we must change the system boundary conditions to $\lambda_1(\theta) = \lambda_2(\theta) = 0$. However, as θ approaches infinity, it has been shown that the two solutions become identical.

Weber and Lapidus[7] suggested using the more general procedure as a way of estimating the opimal control of a nonlinear autonomous system. Since flow rates are normally used as control variables, the state equations can be put into the form

$$\dot{\mathbf{x}} = [\mathbf{f}_1(\mathbf{x})]\mathbf{x} + [\mathbf{f}_2(\mathbf{x})]\mathbf{u}$$

Although at first glance it might appear that functions $\mathbf{f}_1(\mathbf{x})$ and $\mathbf{f}_2(\mathbf{x})$ correspond to the matrices $\mathbf{A}$ and $\mathbf{B}$ after we apply the linearization procedure, this is not always the case. Nevertheless, Lapidus proposes that we develop an approximate solution by initially evaluating these functions at steady state conditions, so that the state equations have constant coefficients, and then evaluating the normal linear feedback solu-

---

[6] Taken from p. 133 of reference 7 on p. 264 based on a paper by Johnson.

[7] A. P. J. Weber and L. Lapidus, *AIChE Journal*, **17**, 641, 649 (1971).

tion. In this way we obtain a first estimate of $\mathbf{x}(t)$. Now we can substitute these time functions into $\mathbf{f}_1(\mathbf{x})$ and $\mathbf{f}_2(\mathbf{x})$, and we see that we have replaced a nonlinear problem by one that is linear but has variable coefficients. However, we can again solve this linear problem to obtain an improved estimate of $\mathbf{x}$. By iterating in this way, we will eventually arrive at an approximate solution to the optimal control problem.

Consider the simple isothermal reactor system described in Exercise 1 above and see if you can use the Lapidus procedure to find the optimal control policy corresponding to a quadratic performance index.

12. (B) The optimum design conditions and the linearized equations describing a single reaction in a nonisothermal CSTR were presented in Exercise 33, p. 220, Volume 1. Establish the time optimal control policy corresponding to these linearized equations. Consider cases where the flow rate of heating fluid is the only available control variable and where both the feed rate and the flow of heat-transfer fluid can be manipulated. Assume that there is sufficient flexibility in the design that the steady state values of the control variables are one-half their maximum values.

13. (C) See if you can find the time optimal control policy corresponding to the complete set of nonlinear state equations for the plant described in the previous exercise. Again, first consider a case where $u_1 = 0$, and then try to solve the problem with two control variables. If your numerical method gives you strange answers, how can you ascertain the source of the difficulty? Also, suggest a way of avoiding this kind of a problem.

14. (B) Determine the time optimal control policy for the linearized system equations presented in Exercise 34, p. 220, Volume 1. Initially, solve the problem with just a single control variable, u_2, and then repeat the analysis for the multivariable case. Assume that the maximum allowable control settings are twice the steady state values.

15. (B) Design a multivariable, linear feedback control system based on the linearized equations approximating the behavior of the reactor described in Exercise 33, p. 220, Volume 1. Let the weighting matrix on the control variables in the quadratic performance index be

$$\mathbf{R} = \begin{pmatrix} 0.1 & 0 \\ 0 & 0.025 \end{pmatrix}$$

and consider two weighting matrices for state variables

$$\mathbf{C} = \begin{pmatrix} 0.05 & 0 \\ 0 & 20 \end{pmatrix} \quad \text{and} \quad \mathbf{C} = \begin{pmatrix} 0.5 & 0 \\ 0 & 200 \end{pmatrix}$$

Estimate the profit produced by the reactor, as well as the integral square error, as the reactor approaches the optimum steady state operating point from at least two different initial conditions.

16. (B) Use the technique described by Paridis and Perlmutter[8] to design a suboptimal, multivariable feedback controller for the plant described in the exercise above. Compare the profit produced by the plant for the two kinds of control system.

17. (C†) The parameters corresponding to an optimum steady state design of a single irreversible reaction in a nonisothermal CSTR were presented in Exercise

[8] Footnote 1, p. 318.

33, p. 220, Volume 1, and discussed in Exercise 15 above. Find an optimal start-up policy for the reactor. Carefully list your assumptions.

18. (C) Consider the linearized equations for a reactor presented in Exercise 35, p. 220, Volume 1, along with the two weighting matrices,

$$C = \begin{pmatrix} 1.0 & 0 & 0 \\ 0 & 10.0 & 0 \\ 0 & 0 & 1.0 \end{pmatrix} \quad \text{and} \quad C = \begin{pmatrix} 0.5 & 0 & 0 \\ 0 & 50.0 & 0 \\ 0 & 0 & 5.0 \end{pmatrix}$$

Apply the method of Paridis and Perlmutter to develop a multivariable feedback control system for the plant. How would you implement this policy? Find the profit produced by the plant as it returns to steady state after several initial disturbances. Consider controllers based on both of the weighting matrices given above.

19. (C) Establish the time optimal control policy for the linearized equations given in Exercise 35, p. 220, Volume 1. Compare your results to the profit obtained in the previous exercise. How would you implement the optimal policy?

Deliberate Unsteady State Operation

Periodic Processing

10

Throughout the text we have implicitly assumed that the optimum steady state design always gives the best plant performance and, therefore, that we should design control systems to make a unit operate as close as possible to these conditions. However, we also made several somewhat vague references to the fact that this assumption might not always be valid, and in this chapter we will explore some of the exceptions to the general rule. Actually, the fact that periodic processing is sometimes superior to steady state operation has been widely known since the advent of pulsed extraction in 1935. The reason for the improvement was attributed to the generation of local turbulence in the vicinity of the plates by the pulsations, so that there was a higher overall mass-transfer coefficient. Since this is a very specific phenomenon and since the process economics indicated that the cost required to supply the pulsations was higher than the incremental return of the additional product obtained, the study of periodic processing was not pursued.

However, in the last few years it has been shown that the ideas of periodic operation can be applied to a wide variety of plant units, and, in some cases, the time average performance of the oscillating system can be superior to the optimum steady state design. Of course, in these cases, we have to abandon our previous "rules of thumb" and look for new approaches to the design and control problem. For example, we find that sometimes it is advantageous to use a positive feedback control system to make a stable plant become unstable. This chapter provides an introduction to this idea of deliberate unsteady state processing. It considers pulsed operation (sinusoidal inputs to stable plants), controlled cycling (square wave variations in flow rate, where the flow rate is zero during a portion of the operating cycle), chemical oscillators (locally unstable systems that generate periodic outputs even when the inputs are maintained constant), and the optimal periodic operation of both lumped and distributed parameter plants.

SECTION 10.1 PULSED OPERATION

Most chemical processes are designed to operate at a steady state condition, and preferably this is chosen as the optimum steady state. It is well known that some of the inputs will vary with time, but the steady state design is based on the time average value of these fluctuating quantities. Also, attempts are made to remove or damp these disturbances by installing surge tanks or control systems, so that the controlled plant is forced to have a relatively constant output close to the optimum steady state value. In order to show that this approach does not always correspond to the best performance of the system, we will present a number of counterexamples. The case of sinusoidal inputs to stable plants is the simplest to handle mathematically and therefore will be considered in this section. Although some terminology confusion exists in the literature, this case is commonly referred to as pulsed operation. The periodic operation of simple lumped parameter systems will be considered first, and then the results will be extended to distributed parameter plants.

Single Reaction in an Isothermal Stirred-Tank Reactor

Perhaps the simplest system we could consider is a single, second-order, isothermal reaction in a CSTR. The optimum steady state design of the unit was described in Example 2.1-1. There we outlined a procedure for obtaining the values of q, A, and V that minimized the total cost

$$C_T = C_V V + C_f q A_f \qquad 10.1\text{-}1$$

and satisfied the production equation

$$G = q(A_f - A) \qquad 10.1\text{-}2$$

as well as the reactor material balance

$$q(A_f - A) - kVA^2 = 0 \qquad 10.1\text{-}3$$

assuming that the system parameters C_V, C_f, G, A_f, and k had been specified. Once the optimum steady state design has been established, we would use a surge tank and/or a conventional control system to compensate for fluctuations in the feed composition A_f. However, before we install one of these damping devices, we would like to see how detrimental sinusoidal variations in this quantity might be.

The simplest procedure is to consider the case of very low frequency fluctuations,[1] where the system dynamics are negligible. Then the steady state material balance can be used to find the effect of fluctuations on the output. Letting

$$x = \frac{A}{A_{fs}}, \quad x_f = \frac{A_f}{A_{fs}}, \quad V_R = \frac{kVA_{fs}}{q} \qquad 10.1\text{-}4$$

and solving Eq. 10.1-3 for x, gives the expression

$$x = -\tfrac{1}{2} V_R [1 - (1 + 4V_R x_f)^{1/2}] \qquad 10.1\text{-}5$$

For a sinusoidal input

$$x_f = 1 + a \sin \omega t \qquad 10.1\text{-}6$$

Substituting this relationship into Eq. 10.1-5 and expanding the result in a Taylor series about $x_f = 1$, we obtain

$$x \approx x_s + a(1 + 4V_R)^{-1/2} \sin \omega t - a^2 V_R (1 + 4V_R)^{-3/2} \sin^2 \omega t + \cdots$$

$$10.1\text{-}7$$

where x_s is merely the steady state composition

$$x_s = -\tfrac{1}{2} V_R [1 - (1 + 4V_R)^{1/2}] \qquad 10.1\text{-}8$$

Now if we calculate the time average of the output, we find that

$$x_{\text{av}} = x_s - \frac{a^2 V_R}{2} (1 + 4V_R)^{-3/2} \qquad 10.1\text{-}9$$

so the time-average effluent composition of reactant is lower than the normal steady state value. This behavior is due to the presence of the nonlinear reaction rate term; that is, more is gained when the feed composition is high than is lost when it is low, and it means that the periodic process gives a higher time average conversion or that the same production rate can be achieved with a lower flow rate. In fact, by working out the remaining details, it can be shown that the fluctuations always improve the system performance. Therefore, it is never worthwhile to spend money to install a control system or surge tank upstream of the reactor. Instead, it would be better to put the surge tank on the downstream side.

In order to determine the effect of the system dynamics,[2] we include the accumulation term in the material balance

$$V \frac{dA}{dt} = q(A_f - A) - kVA^2 \qquad 10.1\text{-}10$$

let

---

[1] This problem is taken from J. M. Douglas and D. W. T. Rippen, *Chem. Eng. Sci.*, **21**, 305 (1966).

[2] J. M. Douglas, *Ind. Eng. Chem. Proc. Design Develop.*, **6**, 34 (1967).

$$x = x_s + y, \quad \alpha = 2V_R x_s, \quad \tau = \frac{qt}{V}, \quad \omega_0 = \frac{V}{q}\omega \qquad \text{10.1-11}$$

and insert an artificial parameter μ before the nonlinear term

$$\frac{dy}{d\tau} + (1 + \alpha)\, y = a \sin \omega_0 \tau - \mu V_R y^2 \qquad \text{10.1-12}$$

A perturbation solution is developed by assuming a solution

$$y = y_0 + \mu y_1 + \mu^2 y_2 + \cdots \qquad \text{10.1-13}$$

substituting this expression into Eq. 10.1-12, and equating terms having like coefficients of powers of μ. In this way we obtain the set of equations

$$\frac{dy_0}{d\tau} + (1 + \alpha)y_0 = a \sin \omega_0 \tau \qquad \text{10.1-14}$$

$$\frac{dy_1}{d\tau} + (1 + \alpha)y_1 = -V_R y_0^2 \qquad \text{10.1-15}$$

$$\frac{dy_2}{d\tau} + (1 + \alpha)y_2 = -2V_R y_0 y_1 \qquad \text{10.1-16}$$

The pseudo-steady state solution of Eq. 10.1-14 is

$$y_0 = \frac{a}{[(1 + \alpha)^2 + \omega_0^2]}[(1 + \alpha)\sin \omega_0 \tau - \omega_0 \cos \omega_0 \tau] \qquad \text{10.1-17}$$

Using this expression to eliminate y_0 from Eq. 10.1-15 and then solving the equation, we obtain

$$y_1 = \frac{-a^2 V_R}{2(1 + \alpha)[(1 + \alpha)^2 + \omega_0^2]} + \frac{a^2 V_R}{[(1 + \alpha)^2 + \omega_0^2]}\left\{\left[\frac{2(1 + \alpha)^2 - \omega_0^2}{(1 + \alpha)^2 + 4\omega_0^2}\right]\right.$$

$$\left. \times \sin 2\omega\tau + \frac{\tfrac{1}{2}(1 + \alpha)[(1 + \alpha)^2 - 5\omega_0^2]}{[(1 + \alpha)^2 + 4\omega_0^2]} \cos 2\omega_0 \tau\right\} \qquad \text{10.1-18}$$

Although it is possible to evaluate a second-order correction function from Eqs. 10.1-16, 10.1-17, and 10.1-18, we will hope that two terms are adequate for our purpose. We write our complete solution, Eq. 10.1-13 with $\mu = 1$, as

$$y = \left[\frac{a}{(1 + \alpha)^2 + \omega_0^2}\right][(1 + \alpha)\sin \omega_0 \tau - \omega_0 \cos \omega_0 \tau]$$

$$- \frac{a^2 V_R}{2(1 + \alpha)[(1 + \alpha)^2 + \omega_0^2]}$$

$$+ \frac{a^2 V_R}{[(1 + \alpha)^2 + \omega_0^2]}\left\{\left[\frac{2(1 + \alpha)^2 - \omega_0^2}{(1 + \alpha)^2 + 4\omega_0^2}\right]\sin 2\omega_0 \tau\right.$$

$$\left. + \frac{(1 + \alpha)}{2}\left[\frac{(1 + \alpha)^2 - 5\omega_0^2}{(1 + \alpha)^2 + 4\omega_0^2}\right]\cos 2\omega_0 \tau\right\} \qquad \text{10.1-19}$$

Hence the frequency response of the nonlinear system can be written, approximately, as the linear frequency response, plus a d-c component, plus higher harmonics. The time average value of the output is simply the constant term, which can be written as

$$x_{av} = x_s - \frac{a^2 V}{2(1 + 4V)^{1/2}(1 + 4V + \omega_0^2)} \qquad 10.1\text{-}20$$

It is clear that at low-forcing frequencies this result becomes identical to Eq. 10.1-9, and at very high frequencies the change in the average operating level approaches zero. This is in complete accord with our intuition, for we expect that low-frequency signals will not be damped as they pass through a first-order dynamic system but that high-frequency inputs will be almost completely damped.

The fact that we obtain an improvement for feed composition fluctuations should also have been expected. This variable was considered constant in the optimum steady state design procedure, and certainly we would expect the design to depend on the particular value specified. If we plot qA versus A_f, using Eq. 10.1-3 (see Figure 10.1-1), we find that the curve is concave downward, so that varying A_f around some average value will always give us a lower time average

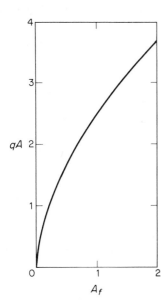

Figure 10.1-1. Reponse surface for feed composition.

value of A than that corresponding to the mean feed composition. Thus allowing the fluctuations around the specified mean value will always give a better performance than the steady state design based on the mean.

A similar analysis can be used to study the effect of flow rate fluctuations. However, this variable was selected in such a way as to minimize the system cost; therefore our intuition leads us to believe that the performance of the

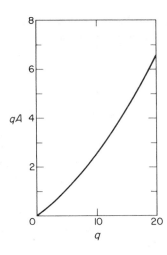

Figure 10.1-2. Response surface for flow rate.

periodic system would be poorer than conventional operation. Figure 10.1–2 shows that a plot of qA versus q, from Eq. 10.1–3, is concave upward, so that fluctuations would lead to higher time average values of the fraction unreacted or lower conversions. The steady state and dynamic analytical results lead to the same conclusion.

The question now remains: What happens if we simultaneously introduce sinusoidal variations in feed composition and flow rate? At first thought we might expect that the reduced performance caused by flow oscillations at least partially compensates for the improvements obtained with composition fluctuations. However, the response surface plots, the steady state, and dynamic analyses show that this is not correct. Instead we find that we can obtain improvements that are higher than with composition fluctuations alone, providing that we have the correct phase angle between the two signals. Of course, this means that we should introduce flow fluctuations in an attempt to amplify the effect of feed composition fluctuations, which certainly is a policy in contradiction with our conventional control concepts. It is quite reasonable in terms of physical considerations, however, since we decrease the flow rate when the feed composition is high and increase it when the feed composition is low; and because of the nonlinear reaction rate, we gain more in the first case than we lose in the second.

Unfortunately, quantitative calculations show that the potential improvement by periodic operation of this simple process is very small. The reason is that a quadratic nonlinearity actually does not have a large effect on the system for input fluctuations of the order of 10 percent. The example is also somewhat misleading because the implication can be drawn that dynamic considerations are always insignificant and that a careful steady state investigation always leads to the correct results. In the next example we will consider a case where the total effect is due to the system dynamics.

Complex Reactions in an Isothermal Stirred-Tank Reactor

One of the most critical problems in reactor design in how to obtain the maximum yield, or selectivity, in the reactor when the process follows a complex reaction mechanism. As a simple illustration of this kind of problem, we will consider the parallel reaction scheme $2A \rightarrow B$ and $A \rightarrow C$, which was studied in some detail by Horn.[3] The dynamic system equations are

$$V\frac{dA}{dt} = q(A_f - A) - k_1 V A^2 - k_2 V A$$

$$V\frac{dB}{dt} = q(B_f - B) + k_1 V A^2$$

10.1–21

For simplicity, we consider the steady state design problem where we choose the reactor temperature that maximizes the yield of component B, although we recognize that this normally will not correspond to the optimum economic design. Letting

$$x_1 = \frac{A}{A_f}, \quad x_2 = \frac{B}{A_f}, \quad u = \frac{Vk_{10}}{q}(e^{-E_1/RT})A_f, \quad \rho = \frac{E_2}{E_1}$$

$$B_f = 0, \quad a = \left(\frac{Vk_{20}}{q}\right)\left(\frac{V}{q}k_{10}A_f\right)^{-\rho}, \quad \tau = \frac{qt}{V}$$

10.1–22

the dimensionless system equations become

$$\frac{dx_1}{d\tau} = 1 - x_1 - ux_1^2 - au^\rho x_1$$

10.1–23

$$\frac{dx_2}{d\tau} = -x_2 + ux_1^2$$

10.1–24

The optimum steady state design can be determined by setting the time derivatives in these equations equal to zero, eliminating x_1 to obtain the yield of component B (or x_2) as an explicit function of u, and then setting the partial derivative of x_2 with respect to u equal to zero. With this procedure, we find that

$$x_2 = \frac{1}{2u}[(1 + au^\rho)^2 + 2u - (1 + au^\rho)\sqrt{(1 + au^\rho)^2 + 4u}] \quad 10.1\text{–}25$$

and

$$u = \left[\frac{1}{(2\rho - 1)a}\right]^{1/\rho} \quad 10.1\text{–}26$$

We want to compare this optimum steady state operation with the behavior we obtain if we vary u sinusoidally. The results will be somewhat different if we consider fluctuations in the reactor temperature, rather than the rate constant u, but certainly our intuition and past prejudice would lead us to believe that periodic operation is inferior to the steady state performance in either case.

[3] F. J. M. Horn and R. C. Lin, *Ind. Eng. Chem. Proc. Design Develop.*, **6**, 21 (1967).

Before using our perturbation technique to develop an approximate solution for the nonlinear dynamic equations, we let

$$y_1 = x_1 - x_{1s}, \quad y_2 = x_2 - x_{2s}, \quad v = u - u_s, \quad a_{11} = + (1 + au_s^\rho + 2u_s x_{1s})$$

$$a_{21} = 2x_{1s}u_s, \quad b_{11} = + (x_{1s}^2 + a\rho u_s^{\sigma-1}x_{1s}), \quad b_{22} = x_{1s}^2, \quad c_{11} = -u_s$$

$$c_{13} = -\tfrac{1}{2} a\rho(\rho - 1)u_s^{\rho-2}x_{1s}, \quad c_{21} = u_s, \quad c_{22} = 2x_{1s}, \quad c_{12} = -(a\rho u_s^{\rho-1} + 2x_{1s})$$

10.1–27

expand the nonlinear functions on the right-hand sides of the equations in a Taylor series about the steady state solution, and insert an arbitrary parameter μ in front of the nonlinear terms

$$\frac{dy_1}{d\tau} = -a_{11}y_1 - b_{11}v + \mu(c_{11}y_1^2 + c_{12}y_1 v + c_{13}v^2) \qquad 10.1\text{–}28$$

$$\frac{dy_2}{d\tau} = a_{21}y_1 - y_2 + b_{22}v + \mu(c_{21}y_1^2 + c_{22}y_1 v) \qquad 10.1\text{–}29$$

Now we assume that we can find a solution having the form

$$y_1 = z_{10} + \mu z_{11} + \mu^2 z_{12} + \cdots \qquad 10.1\text{–}30$$

$$y_2 = z_{20} + \mu z_{21} + \mu^2 z_{22} + \cdots \qquad 10.1\text{–}31$$

substitute these assumed solutions into Eqs. 10.1–28 and 10.1–29, and equate terms having like coefficients of μ, to obtain the set of equations

$$\frac{dz_{10}}{d\tau} = -a_{11}z_{10} - b_{11}v \qquad 10.1\text{–}32$$

$$\frac{dz_{20}}{d\tau} = a_{21}z_{10} - z_{20} + b_{22}v \qquad 10.1\text{–}33$$

$$\frac{dz_{11}}{d\tau} = -a_{11}z_{11} + c_{11}z_{10}^2 + c_{12}z_{10}v + c_{13}v^2 \qquad 10.1\text{–}34$$

$$\frac{dz_{21}}{d\tau} = a_{21}z_{11} - z_{21} + c_{21}z_{10}^2 + c_{22}z_{10}v \qquad 10.1\text{–}35$$

$$\vdots$$

For the case of a sinusoidal input

$$v = A_0 \sin \omega_0 \tau \qquad 10.1\text{–}36$$

the pseudo-steady state solution of Eq. 10.1–32 is

$$z_{10} = \left(\frac{-b_{11}A_0}{a_{11}^2 + \omega_0^2}\right)(a_{11} \sin \omega_0 \tau - \omega_0 \cos \omega_0 \tau) \qquad 10.1\text{–}37$$

and the corresponding result for Eq. 10.1–33 is

$$z_{20} = A_0 \left[\frac{b_{22}(a_{11}^2 + \omega_0^2) + (\omega_0 - 1)b_{11}}{(a_{11}^2 + \omega_0^2)(1 + \omega_0^2)}\right] \sin \omega_0 \tau$$

$$+ A_0 \left[\frac{2b_{11}\omega_0 - b_{22}(a_{11}^2 + \omega_0^2)}{(a_{11}^2 + \omega_0^2)(1 + \omega_0^2)}\right] \cos \omega_0 \tau \qquad 10.1\text{–}38$$

Now we can eliminate z_{10} from the right-hand side of Eq. 10.1–34, use trigonometric identities to simplify the expression, and solve this equation. Then we can use the result we obtain, along with Eqs. 10.1–36 and 10.1–37, to eliminate z_{11}, v, and z_{10} from the right-hand side of Eq. 10.1–35 and solve for z_{21}. Taking the time average value of this final expression, which, according to Eqs. 10.1–31 and 10.1–38, must be the same as the time average value of y_2, we obtain

$$y_{2\,\text{av}} = x_{2\,\text{av}} - x_{2s} = \frac{1}{2}\left(\frac{A_0^2}{a_{11}^2 + \omega_0^2}\right)[c_{21}b_{11}^2 - c_{22}a_{11}b_{11}$$

$$+ \frac{a_{21}}{a_{11}}(c_{11}b_{11}^2 - c_{12}a_{11}b_{11} + c_{13}a_{11}^2 + c_{13}\omega_0^2)] \qquad 10.1\text{–}39$$

For the particular set of parameters, $\rho = \frac{3}{4}$ and $a = 1$, we find from Eq. 10.1–26 and the steady state form of the system equations, Eqs. 10.1–23 and 10.1–24, that the optimum steady state design conditions are $u_s = 2.52$, $x_{1s} = 0.2718$, $x_{2s} = 0.185$. Also, the time average value of the yield of component B from Eq. 10.1–39 is

$$x_{2\,\text{av}} = 0.185 + A_0^2\left(\frac{-0.189 + 0.00126\,\omega_0^2}{19.1 + \omega_0^2}\right) \qquad 10.1\text{–}40$$

This result shows that for very low frequency disturbances (i, e., as the driving frequency ω_0 approaches zero), the performance of the periodic process is inferior to the optimum steady state design, $x_{2\,\text{av}} < x_{2s}$. However, for high-frequency disturbances (i. e., as the forcing frequency ω_0 approaches infinity), the periodic operation gives a higher time average yield than the optimum steady state value. Since it can be shown that the correction term in Eq. 10.1–40 increases monotonically with frequency, the maximum improvement is obtained by making ω_0 very large. In addition, as a first approximation, we find that the improvement increases as the square of the amplitude of the input signal, so that it would be advantageous to introduce large fluctuations.

The preceding example provides an elementary illustration of a case where periodic operation can be superior to the optimum steady state operation. If we plot the yield versus u, Eq. 10.1–25, we find that the curve is concave downward (see Figure 10.1–3); thus we expect that low-frequency variations in u will lead to a poorer performance. This behavior was confirmed by our dynamic analysis. For this simple problem, it is also possible to provide a qualitative explanation for the improved performance obtained by unsteady state operation. If we plot k_1 versus k_2 or u versus u^ρ, for a case where $\rho < 1$, we get a curve that is concave downward (see the soild line in Figure 10.1–4). Also, if we plot the values of u and u^ρ that give constant values of x_2 in Eq. 10.1–25, we obtain a series of curves that are also concave downward. These are shown as the dashed lines in Figure 10.1–4 and the direction of increasing values of x_2 is indicated by the arrow. The optimum steady state design is obtained when the locus of attainable points (the solid curve) is tangent to the contour line having the largest possible yield. However, if we vary the values

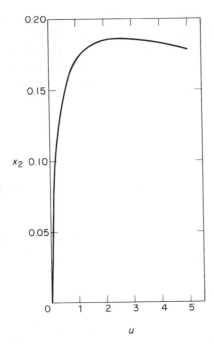

u

Figure 10.1-3. Response surface.

of u and u^ρ as a function of time, the convex hull of the locus of attainable points includes a region of higher yields then the maximum yield on the attainable locus. In other words, we can reach states by dynamic operation that are not possible by steady state operation.

The optimal periodic operation of this process will be discussed in Section 10.4 Perhaps it is worth mentioning at this point that the best operation is obtained when the temperature is switched between its maximum and minimum allowable values as rapidly as possible. It is encouraging to note that our elementary perturbation analysis gives a valid prediction of the direction to

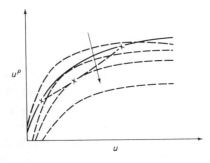

u

Figure 10.1-4. Attainable regions. [Reproduced from F. J. M. Horn and R. C. Lin, *Ind. Eng. Chem. Proc. Design Develop.*, **6**, 21 (1967), by permission of the ACS.]

change the forcing amplitude and frequency in order to obtain an improved performance.

Another study of this type was published by Dorawala ahd Douglas.[4] They considered the effects of sinusoidal oscillations on elementary consecutive and parallel reactions in a stirred-tank reactor. Extensions of the method to polymerization reactors were published by Laurence and Vasudevan,[5] as well as Ray.[6] In the second study, it was observed that the product obtained from the oscillating reactor was more nearly monodispersed than the steady state distribution.

Sinusoidal Inputs to a Nonisothermal Stirred-Tank Reactor

It should be possible to obtain much larger differences between the periodic and optimum steady state operation of a nonisothermal reactor because the equations used to describe the nonisothermal system contain an exponential nonlinearity in the reactor temperature, so that nonlinear behavior often becomes apparent even for small input disturbances. In addition, the system

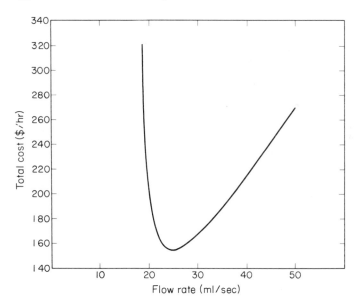

Figure 10.1-5. Total cost versus flow rate. [Reproduced from A. B. Ritter and J. M. Douglas, *Ind. Eng. Chem. Fundamentals*, **9**, 21 (1970), by permission of the ACS.]

[4] T. G. Dorawala and J. M. Douglas, *AIChE Journal*, **17**, 974 (1971).

[5] R. L. Laurence and G. Vasudevan, *Ind. Eng. Chem. Proc. Design Develop.*, **7**, 427 (1968).

[6] W. H., Ray, *Ind. Eng. Chem. Proc. Design Develop.*, **7**, 422 (1968).

equations can have complex conjugate roots; and, as discussed in Section 4.5, if the damping coefficient of the linearized equations is less than 0.707, the system can exhibit resonance. In this case, the reactor tends to amplify the effect of disturbances in the neighborhood of the resonant frequency, and the larger deviations from steady state conditions will cause the nonlinear phenomena to become more pronounced.

A study of the effect of sinusoidal input fluctuations on the operation of a nonisothermal CSTR was undertaken by Ritter and Douglas.[7] The reactor was designed to give the optimum, economic, steady state performance, using the procedure described in Example 2.1–2. A plot of the total cost versus feed rate, for the particular system parameters considered, is given as Figure 10.1–5, This graph provides some indication of the sensitivity of the optimum and makes it apparent that low-frequency variations in flow rate, or other design variables that can be manipulated, will give a poorer time average performance

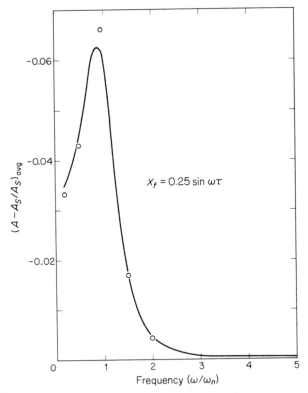

Figure 10.1-6. Time average performance versus forcing frequency. [Reproduced from A. B. Ritter and J. M. Douglas, *Ind. Eng. Chem. Fundamentals*, **9**, 21 (1970), by permission of the ACS.]

[7] A. B. Ritter and J. M. Douglas, *Ind. Eng. Chem. Fundamentals*, **9**, 21 (1970).

of the reactor. The linearized system equations have complex conjugate roots, with a damping coefficient of 0.39.

The perturbation solution described in Example 6.1-3 was used to estimate the response of the reactor to sinusoidal variations in feed composition, a variable assumed to be constant in the design procedure, and flow rate. A plot of the time average effluent composition versus forcing frequency for 25 percent amplitude fluctuations in the feed composition is shown as Figure 10.1-6. The results show that the maximum improvement is obtained at the resonant frequency and that high-frequency fluctuations do not give any improvement. The points on this graph correspond to a numerical solution of the nonlinear equations, whereas the curve was obtained using perturbation theory. According to the analytical solution, the potential improvement should increase with the square of the amplitude of the input fluctuation. However, since only the first-order correction terms were considered in the analysis, we expect the predictions to break down at large amplitudes. This discrepancy is shown in Figure 10.1-7.

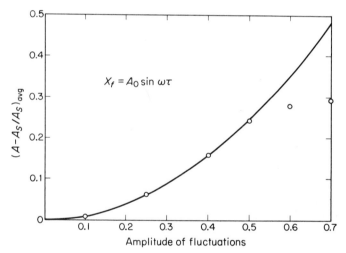

Figure 10.1-7. The effect of forcing amplitude on system performance. [Reproduced from A. B. Ritter and J. M. Douglas, *Ind. Eng. Chem. Fundamentals*, **9**, 21 (1970), by permission of the ACS.]

One interesting feature of the results is that when the forcing frequency is less than the resonant frequency, the output signal exhibits a considerable amount of higher harmonics. This behavior is illustrated in Figure 10.1-8 for a case where the forcing frequency is one-half the natural frequency and the amplitude is 25 percent. This kind of behavior is to be expected, for the system nonlinearities produce higher harmonics that are near the resonant frequency and therefore have high gains. At higher driving frequencies, the effect of the higher harmonics is to cause distortion of the output signal without changing the apparent frequency of the effluent oscillations (see Figure 10.1-9). By in-

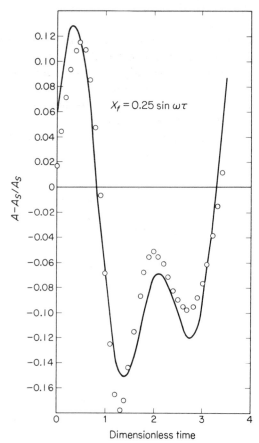

Figure 10.1-8. Effluent composition fluctuations. [Reproduced from A. B. Ritter and J. M. Douglas, *Ind. Eng. Chem. Fundamentals*, **9**, 21 (1970), by permission of the ACS.]

cluding the first-order correction terms in the approximate analytical solutions, it is possible to obtain good estimates of the output signals (see Figures 10.1–8 and 10.1–9). A more comprehensive picture of the effect of the forcing frequency on the distortion is shown in Figure 10.1–10.

It is not surprising that feed composition disturbances can lead to an improved performance, for this variable was not considered in the optimum steady state design procedure. However, the foregoing results indicate that it is advantageous to allow these fluctuations to enter the reactor and that it would be a mistake to install a surge tank, or a control system, to cut down on the effect of these disturbances. In addition, it can be shown that flow rate disturbances sometimes can be used to increase the time average conversion (see Figure 10.1–11). At low, input-frequencies, the periodic performance is inferior to the optimum steady state design, which agrees with the predictions from the steady

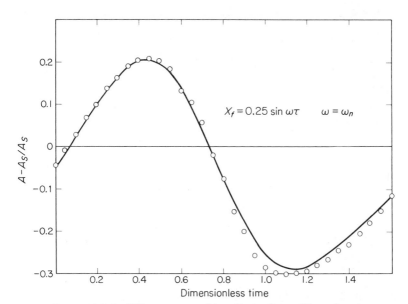

Figure 10.1-9. Effluent composition fluctuations. [Reproduced from A. B. Ritter and J. M Douglas, *Ind. Eng. Chem. Fundamentals,* **9,** 21 (1970), by permission of the ACS.]

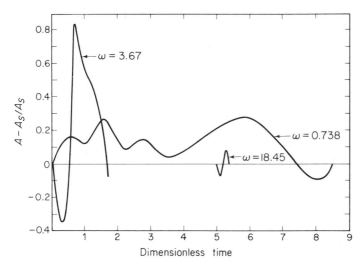

Figure 10.1-10. Distortion of output signal. [Reproduced from A. B. Ritter and J. M. Douglas, *Ind. Eng. Chem. Fundamentals,* **9,** 21 (1970), by permission of the ACS.]

state analysis; but as the frequency is increased, the time average output changes sign and approaches a maximum near the resonant frequency. Further increases in the forcing frequency lead to smaller improvements, until the output approaches the normal steady state value.

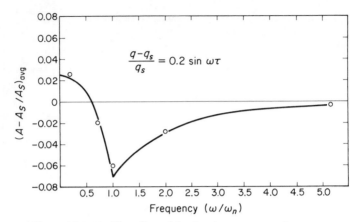

Figure 10.1-11. The effect of forcing frequency on time average composition. [Reproduced from A. B. Ritter and J. M. Douglas, *Ind. Eng. Chem. Fundamentals,* **9,** 21 (1970), by permission of the ACS.]

When feed composition and flow rate fluctuations are introduced into the reactor simultaneously, the shift in the time average value depends on the phase angle between two signals (see Figure 10.1–12). Similar to the case of a single reaction in an isothermal reactor, we find that it is possible to achieve an improvement that is greater than that corresponding to either of the individual forcing functions. In other words, we can use flow rate fluctuations to amplify the effects of feed composition disturbances, instead of using these flow rate fluctuations to damp out the effects of these disturbances.

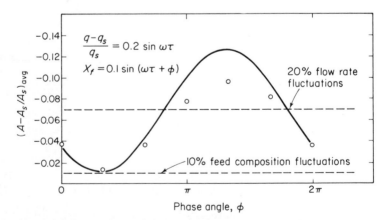

Figure 10.1-12. Variation of average performance with phase angle. [Reproduced from A. B. Ritter and J. M. Douglas, *Ind. Eng. Chem. Fundamentals,* **9,** 21 (1970), by permission of the ACS.]

The same approach can be used to assess the desirability of periodic operation of a nonisothermal CSTR for complex reactions. A few calculations for the parallel reaction system just described were published by Dorawala and Douglas.[4] Sinusoidal variations in flow rate with a 10 percent amplitude gave a 2 percent improvement in the yield of component *B*, but 10 percent changes in feed temperature produced a 15 percent increase in the performance. The corresponding results for a set of consecutive reactions were much lower, for the particular set of parameters considered.

Periodic Operation of a Heat Exchanger

Although the preceding discussion is limited to examples of lumped parameter processes, exactly the same ideas can be applied to distributed parameter plants. As a simple example, we can consider the problem of determining the optimum steady state flow rate of cooling water required to remove a specified amount of heat, Q, from a vapor condensing at a constant temperature T_c. Letting C_A equal the depreciated cost of exchanger area, A_H equal the area

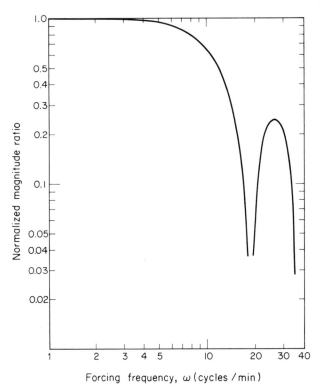

Figure 10.1-13. Bode plot for a steam-heated exchanger.

required, C_W equal the cost per pound of cooling water, and W equal pounds of cooling water required, the total cost is

$$C_T = C_A A_H + C_W W \qquad\qquad 10.1\text{-}41$$

and the system equations are

$$Q = C_p W(T - T_f) = \frac{UA_H(T - T_f)}{\ln\left(\dfrac{T_c - T_f}{T_c - T}\right)} \qquad\qquad 10.1\text{-}42$$

These two expressions can be used to eliminate W and A_H from the cost equation, so that the total cost can be written as an explicit function of the cooling-water outlet temperature T. Then the optimum design can be determined by setting the derivative of this expression with respect to T equal to zero, solving the resulting equation for T, and, finally, using Eqs. 10.1–42 to find the corresponding optimum values of W and A_H.

The equations describing the dynamic behavior of the condenser are identical to those for a steam-heated exchanger. Therefore we can use the perturbation solution developed in Example 6.2–1 to estimate the response to sinusoidal fluctuations in the coolant flow rate W or its equivalent, the velocity. This solution, Eq. 6.2–19, shows that the time average performance of the periodic

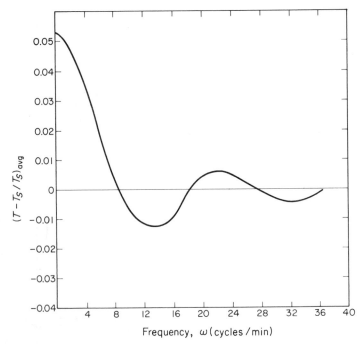

Figure 10.1-14. Effect of forcing frequency on time average output. [Reproduced from A. B. Ritter and J. M. Douglas, *Ind. Eng. Chem. Fundamentals*, **9**, 21 (1970), by permission of the ACS.]

system will not be the same as the steady state case; consequently, it may sometimes be possible to find periodic exchangers that are superior to the optimum
steady state system. Rather than pursue this problem of the design of a condenser, we will review some of the results Ritter and Douglas[7] obtained for a
steam-heated exchanger.

It is well known that resonance peaks are frequently observed when a
distributed parameter process is forced by an input that is distributed in space,
such as velocity or condensate temperature. A Bode plot of the linearized system equations, presented in Figure 10.1-13, illustrates this phenomenon. The
shift in the time average operating level predicted by the analytical solution is
given in Figure 10.1-14, and it is easy to see that improvements are obtained
both at low frequencies and near resonance. Figure 10.1-15 shows a comparison
between the analytical solution, including the first-order correction terms, and
a numerical solution of the partial differential equation describing the system.

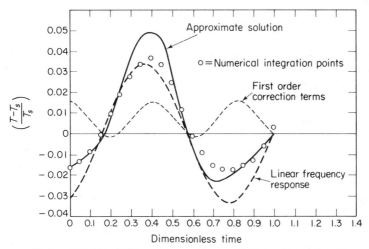

Figure 10.1-15. Effluent temperature oscillations. [Reproduced
from A. B. Ritter and J. M. Douglas, *Ind. Eng. Chem. Fundmentals,*
9, 21 (1970), by permission of the ACS.]

The response of the exchanger to sinusoidal changes in feed temperature
must be symmetrical, because the system equation is linear and has constant
coefficients for this case. In other words, the time average output will be identical to the output at the mean value of the input. However, if both the condensate temperature and the velocity are changed sinusoidally, the shift in the
time average operating level depends on the phase angle between the signals,
and it is possible to obtain improvements that are greater than with flow rate
changes alone (see Figure 10.1-16). This is the same kind of behavior observed
for lumped parameter plants. The improvements for this heat-exchanger
problem are very small because the system equation is linear. More significant

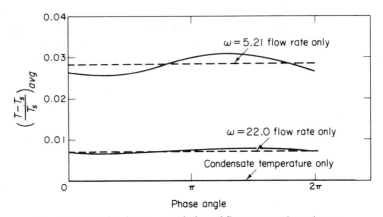

Figure 10.16. Simultaneous variation of flow rate and condensate temperature. [Reproduced from A. B. Ritter and J. M Douglas, *Ind. Eng. Chem. Fundamentals*, **9**, 21 (1970), by permission of the ACS.]

differences between steady state and periodic operation would be expected for a highly nonlinear process, such as a reactor, where the rate constant is an exponential function of temperature.

Parametric Pumping

The term *parametric pumping* was introduced into the chemical engineering literature by Wilhelm.[8] He was able to construct a separating unit that is analogous to some electrical and mechanical systems in that an improved performance is obtained through a coupling of oscillating fields. In his simplest configuration, pistons at each end of a column are used to move a fluid back and forth and heat exchangers near the ends of the column are used to establish a temperature gradient (see Figure 10.1–17). The system equations are written in the form

$$\frac{\partial \theta_f}{\partial t} = -\alpha f(t) \frac{\partial \theta_f}{\partial z} + \beta \gamma (\theta_s - \theta_f) \qquad 10.1\text{--}43$$

$$\frac{\partial \theta_s}{\partial t} = -\gamma (\theta_s - \theta_f) \qquad 10.1\text{--}44$$

$$\frac{\partial \phi_f}{\partial t} = -\alpha f(t) \frac{\partial \phi_f}{\partial z} + \kappa \lambda (\phi^* - \phi_f) \qquad 10.1\text{--}45$$

$$\frac{\partial \phi_s}{\partial t} = -\lambda (\theta^* - \theta_f) \qquad 10.1\text{--}46$$

$$\phi^* = \phi^* (\theta_f, \theta_s, \phi_f, \phi_s) \qquad 10.1\text{--}47$$

[8] R. H. Wilhelm, A. W. Rice, and A. R. Bendelius, *Ind. Eng. Chem. Fundamentals*, **5**, 141 (1966).

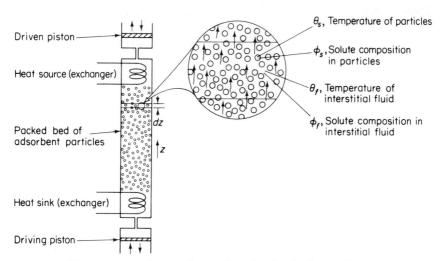

Figure 10.1-17. Parametric pumping of a closed column. [Reproduced from R. H. Wilhelm, A. W Rice, and A. R. Bendelius, *Ind. Eng. Chem. Fundamentals*, **5**, 141 (1966), by permission of the ACS.]

where $\theta_f =$ fluid temperature, $\theta_s =$ solids temperature, $\phi_f =$ solute composition in the fluid phase, $\phi_s =$ solute composition in the solid phase, $\phi^* =$ the equilibrium concentration, $f(t) =$ a periodic velocity, and $\alpha,\ \beta,\ \gamma,\ \kappa,\ \lambda =$ dimensionless constants

In terms of physical considerations, the column operates by an initial downward displacement of hot fluid. This raises the temperature of the adsorbent in the lower region of the column; and as a result of the temperature change, the adsorbent transfers solute to the fluid. Next, the enriched and cooled fluid is displaced upward again to its original position. The fluid cools the adsorbent in this region and transfers solute to it. By repeating this procedure over and over again, it is possible to establish a concentration gradient in the column and to achieve a net separation of binary mixtures. Wilhelm and co-workers carried out a number of experimental and numerical studies demonstrating the validity of his idea, and attempts are being made to find other applications.

One approach that has been used to gain a better understanding of the parametric pumping phenomenon was described by Pigford and co-workers.[9] They considered a greatly simplified version of Wilhelm's model. By assuming that the fluid and solid phases were always in equilibrium, the system could be represented by the equations

$$\epsilon \rho_f \frac{\partial c_f}{\partial t} + \epsilon \rho_f v \frac{\partial c_f}{\partial z} + (1 - \epsilon)\rho_s \frac{\partial c_s}{\partial t} = 0 \qquad 10.1\text{-}48$$

$$c_s = M(t)c_f \qquad 10.1\text{-}49$$

[9] R. L. Pigford, B. Baker, III, and D. E. Blum, *Ind. Eng. Chem. Fundamentals*, **8**, 144 (1969).

These expressions can be combined and put into the form

$$(1 + u)\frac{\partial Y}{\partial t} + v\frac{\partial Y}{\partial z} = - Y\frac{du}{dt}\frac{\partial T}{\partial t} \qquad 10.1-50$$

The authors denote the synchronized square-wave variations in the velocity and the bed temperature by the expressions

$$v = v_0 \text{sq}\,(\omega t) \qquad u = u_0 - a\text{sq}\,(\omega t) \qquad 10.1-51$$

With these input fluctuations, it is possible to use the method of characteristics to develop an analytical solution of Eq. 10.1–50. The results show that an improvement will be obtained if

$$\frac{u_0}{1 - b} < \frac{\omega Z}{\pi} \qquad 10.1-52$$

where

$$b = a/(1 + u_0) \qquad 10.1-53$$

Moreover, the separation factor for n cycles of operation, α_n, is found to be

$$\alpha_n = \left(2 + \frac{2b}{1 + b}\right)\left(\frac{1 + b}{1 - b}\right)^n - \left(\frac{1 + b}{1 - b}\right)^2 \qquad n > 0 \qquad 10.1-54$$

Although this relation indicates that the separation factor approaches infinity as the cycling operation goes on without limit, the actual value will have an upper bound that depends on the mass transfer between phases and the axial diffusion in the bed.

Another interesting extension of Wilhelm's basic work was suggested by Wagle and Denn.[10] They also considered a simplified version of Wilhelm's original system equations but attempted to develop the parametric pumping control policy as the solution of an optimal control problem. In other words, they define a time average separation factor as a performance index and look for the periodic velocity in the column that will give the best performance. Their analysis is described in greater detail in Section 10.4, but it should be noted here that the optimal periodic control resembles Wilhelm's parametric pumping operation.

As Wilhelm points out, most of the electrical and mechanical examples of parametric pumping are second-order differential equations—that is, similar to Mathieu equations. In all these systems, a parameter in the differential equation is forced to become a periodic function of time. In other words, the differential equation describing the system has a periodic coefficient. Of course, our previous studies of the effect on system performance of sinusoidal variations of flow rate, or velocity (which introduces a periodic coefficient into the equations describing most transport problems), also fall into this category. Thus the cases

[10] A. K. Wagle, and M. M. Denn, *Proceedings of the Joint Automatic Control Conference*, p. 286, University of Colorado, Boulder, Colo., August 1969.

where flow rate fluctuations were used to amplify the effects of disturbances can be classified as parametric pumping. Rather than attempt to provide exact definitions of particular phenomena, however, we are more interested in demonstrating that periodic processing can be superior to conventional steady state operation and that more research is needed to gain a better understanding of the dynamic behavior of chemical systems.

Pulsed Operation of Heat Exchangers and Mass Transfer Equipment

A great many experimental studies have been concerned with the pulsed operation of heat and mass transfer equipment, In particular, pulsed extraction[11] has received a considerable amount of attention ever since it was applied to the reprocessing of spent fuel for nuclear reactors. Unfortunately there is a great deal of scatter present in most data, a wide disparity between the results of various investigators, and a wide variety of proposed reasons why an improvement was or was not observed. For example, in heat-transfer studies, it has been claimed that higher heat-transfer rates were due to cavitation, water hammer, secondary flow phenomenon, generation of local turbulence, acoustical streaming, thermoacoustic transduction, boundary layer effects, and so on. Although some or all of these effects may be present in some studies, a unified treatment of the subject is still not available.

One of the earliest and simplest attempts to provide an explanation for the difference between steady state and pulsed operation of heat exchangers, or mass transfer equipment, involved a steady state analysis of the effect of the periodic flow rate on the transport coefficient.[12] It was reasoned that for the case of very low frequency inputs, the system dynamics would not be significant; hence the conventional steady state correlations could be used to describe the system. Thus for a heat transfer process, where it is known that

$$h \simeq c_1 R_e^n = c_2 v^n \qquad \text{10.1-55}$$

the time-average heat-transfer coefficient for a pulsed system,

$$v = v_s + A_0 \sin \omega t \qquad \text{10.1-56}$$

would be

$$h_{\mathrm{av}} = \frac{\omega}{2\pi} \int_0^{2\pi/\omega} c_2 (v_s + A_0 \sin \omega t)^n \, dt \qquad \text{10.1-57}$$

Experimental data indicate that the exponent n is less than unity, however, so that this approach predicts that periodic operation will be inferior to conventional steady state behavior. Similar results are obtained for the mass transfer case.

[11] See Chemical Engineers' Handbook (4th ed.), pp. 21-32, R. H. Perry, C. H. Chilton, and S. D. Kirkpatrick, McGraw-Hill, N, Y., 1963.

[12] A recent treatment is given by M. H. I. Baird, G. J. Duncan, J. I. Smith, and J. Taylor, *Chem. Eng. Sci.*, **21**, 197 (1966).

A very simple model that includes the system dynamics was presented earlier (Eq. 6.2–19), for the case of a steam-heated exchanger. Although this approach reveals the fact that the periodic coefficient has an important effect, it does not consider any effects like cavitation. Therefore, according to the model, it would not make any difference whether the pulsing device was placed upstream or downstream from the exchanger—a factor that Lemlich's[13] experimental study shows can make the difference between obtaining up to an 80 percent increase in the overall heat-transfer coefficient or a decrease in the rate of heat transfer. The model could also be used to predict the rate of mass transfer from a tube wall to a fluid in plug flow, but again it might not provide a complete description of the system behavior.

A more precise technique for determining the effect of a sinusoidally varying pressure gradient on mass transfer from the wall of a circular tube to a fluid with a well-developed laminar velocity profile was proposed by Krasuk and Smith.[14] The momentum equation for this system is

$$\rho \frac{\partial v_z}{\partial t} = -\frac{\partial P}{\partial z} + \eta \left[\frac{1}{r} \frac{\partial}{\partial r} \left(r \frac{\partial v_z}{\partial r} \right) \right] \qquad \text{10.1–58}$$

And in Section 5.2 we showed how to determine the velocity profiles as a function of radius and time. For dilute mixtures, the corresponding mass balance is

$$\frac{\partial c}{\partial t} = -v_z \frac{\partial c}{\partial z} + D \left[\frac{1}{r} \frac{\partial}{\partial r} \left(r \frac{\partial c}{\partial r} \right) \right] \qquad \text{10.1–59}$$

with the boundary conditions (assuming a saturated solution at the wall)

$$\text{at } r = R \qquad c = c_s$$

$$\text{at } r = 0 \qquad \frac{dc}{dr} = 0 \quad \text{for all } z \qquad \text{10.1–60}$$

$$\text{at } z = 0 \qquad c = 0 \quad \text{for all } r$$

In order to determine the composition as a function of radius, distance, and time, it is necessary to substitute the solution we obtained previously for v_z as a function of r and t, Eq. 5.2–23, into Eq. 10.1–59 and then find a solution of the resulting partial differential equation. Although it would be an extremely difficult, if not impossible, task to find an exact analytical solution of this equation, it might be possible to develop an approximate solution using perturbation methods[15] or to solve the equation numerically. Once the concentrations have been determined, the radial gradients at the wall can be calculated, and we can use them to define local mass transfer coefficients. Next, the average mass-transfer coefficient along the tube length can be determined, and, finally,

[13] R. Lemlich, *Chem. Engr.*, **68**, No. 10, 171 (1961).

[14] J. H. Krasuk and J. M. Smith, *Chem. Eng. Sci.*, **18**, 591 (1963).

[15] An analysis of this type was recently published by E. B. Fagela–Alabastro and J. D. Hellums, *AIChE Journal*, **15**, 164 (1969).

we can calculate the time average value of this quantity and compare it with the conventional steady state design condition.

Instead of following this procedure, Krasuk and Smith attempt to develop an analogy between pulsed mass transfer and pulsed momentum transfer. An analysis of the normal state equations is used to relate the steady state mass-transfer coefficient, k_s, to the average velocity and wall shear stress, τ_s, that is, k_s is proportional to $\tau_s^{1/3}$. Then the solution of the momentum equation is used to find the time-average wall shear stress, τ_{av}, for the pulsed case. Finally, it is assumed that the ratios of the pulsed to the steady state transport coefficients must be the same; that is,

$$\frac{k_{av}}{k_s} = \left(\frac{\tau_{av}}{\tau_s}\right)^{1/3} \qquad 10.1\text{-}61$$

From the solution to the unsteady state momentum balance, it can be shown that two dimensionless groups are required to describe the difference between pulsed and steady flow

$$\phi = \frac{A_0 \omega}{u_{av}} \qquad N_p = R\left(\frac{\rho\omega}{\mu}\right)^{1/2} \qquad 10.1\text{-}62$$

where A_0 = the amplitude, ω = the frequency, u_{av} = the time and radial average velocity for pulsed flow, R = the tube radius, ρ = the fluid density, and μ = the fluid viscosity. The improvement in the mass transfer coefficient, predicted by the analogy and the results from the unsteady state momentum balance, is shown in Figure 10.1-18.

Krasuk and Smith used Eq. 10.1-61 to correlate mass transfer data for the solution of β-naphthol into water under steady state and pulsed conditions. Their results, along with some other available data, are shown in Figure

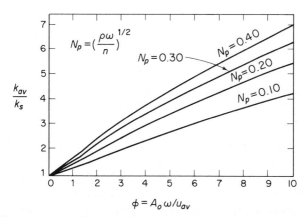

Figure 10.1-18. Mass transfer coefficients for pulsed operation. [Reproduced from J. H. Krasuk and J. M. Smith, *Chem. Eng. Sci.*, **18,** 591 (1963), by permission of Pergamon Press.]

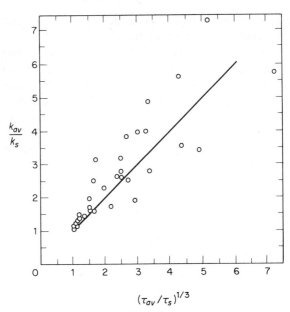

Figure 10.1-19. Correlation of mass transfer coefficients. [Reproduced from J. H. Krasuk and J. M. Smith, *Chem. Eng. Sci.*, **18**, 591 (1963), by permission of Pergamon Press.]

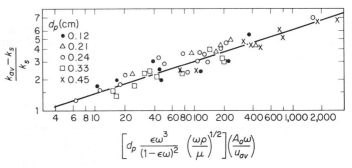

$$\left[d_p \frac{\epsilon \omega^3}{(1-\epsilon \omega)^2} \left(\frac{\omega \rho}{\mu} \right)^{1/2} \right] \left(\frac{A_o \omega}{u_{av}} \right)$$

Figure 10.1-20. Periodic mass transfer coefficient in a packed bed. Reproduced from J. H. Krasuk and J. M. Smith, *AIChE Journal*, **10**, 759 (1964), by permission.]

10.1–19. Although there is about an 11 percent average deviation between the theory and data, this is about the same as the accuracy of the data. Even though this approach neglects any effects caused by the periodic coefficient, v_z, in Eq. 10.1–59, it represents a major extension in our ability to estimate the potential improvement by periodic operation for laminar flow systems. In a later study,[16] the authors extended their results to mass transfer in packed

[16]J. H. Krasuk, and J. M. Smith, *AIChE Journal*, **10**, 759 (1964).

beds by replacing the tube radius in the pulse Reynolds number, given in Eq. 10.1–62, by the appropriate function of particle diameter and porosity. Their results for this case are given in Figure 10.1–20.

SECTION 10.2   CONTROLLED CYCLING

The examples in the previous section show that in some cases periodic operation of a plant can lead to a time average performance that is better than the optimum steady state design. The analytical results often indicate that the improvement increases as the square of the amplitude of the input fluctuations, as a first approximation, which implies that we should make the input oscillations as large as possible. It is obvious that the limiting amplitude will be simply the normal steady state input value; otherwise the input would have to become negative over a certain time interval. However, intuitively we might expect to obtain an additional increase by flattening out the top and bottom of the sine wave signal, so that the input would remain near its maximum and minimum allowable values for a longer period of time. In other words, we expect that a square wave would give us a better average performance than a sine wave.

A similar conclusion can be reached by using the concepts of optimal control theory. For most problems of interest, the state equations and the performance index will be linear in the control variables, such as the flow rate or the condensate temperature in a steam-heated exchanger. Here we are assuming that profit, the production rate, or yield is used as a measure of the plant performance. When the optimal control problem has this structure, we normally find that a "bang-bang" control policy is optimal, providing that there are no singular solutions. Thus we should manipulate the control variable as a square wave between its maximum and minimum allowable values (unless there is a singular solution).

If we apply these concepts to situations where flow rate is used as the control variable, we know that the maximum flow rate is obtained when the valve is wide open and the minimum flow rate is zero. Problems of this type generally cannot be studied by using the perturbation techniques described earlier; for when the flow rate is set equal to zero over some finite time interval, one of the terms often disappears from the steady state equation. The resulting equation is a type different from the former, and it has very different kinds of solutions. For example, if we consider an isothermal, continuous-stirred-tank reactor

$$V\frac{dx}{dt} = q(x_f - x) - kVx \qquad 10.2\text{--}1$$

and set $q = 0$ for a finite time period, we obtain the equations for a batch reactor. The CSTR can be operated at steady state conditions, but this is never possible for a batch system, except for the trivial case where we have already converted all of the reactants—that is, $x = 0$. In order to solve problems of this type, we must use singular perturbation techniques,[1] which are considered beyond the scope of this text.

Nevertheless, we are interested in studying the periodic outputs of plants where the input is a square wave varying between zero and some maximum value. We will refer to this kind of operation as *controlled cycling* of the plant. The term controlled cycling was introduced by Cannon.[2,3] Although the idea is applicable to a wide variety of processes, most of our discussion will be directed toward a plate absorption column. Horn's[4,5] theoretical analysis of this problem will be described in detail, and then some of the experimental results and other applications will be reviewed.

Steady State Behavior of the Column

We consider a plate absorption column, similar to that described in Examples 3.1–2 and 4.6–3, where there is a countercurrent flow of vapor and liquid. A binary mixture is present in both phases, and the compositions of the transferable component of interest are taken as y and x in the vapor and liquid phases, respectively. Initially we assume that the flow rates are constant and that the equilibrium relationship is linear

$$y_n = Kx_n \qquad \qquad 10.2\text{--}2$$

Both assumptions normally are valid in the critical part of the column (where most of the plates are located) for cases of difficult separations. The analysis is also valid for linear equilibrium relationships that have the form

$$Y_n = mX_n + b \qquad \qquad 10.2\text{--}3$$

since the transformation of variables

$$y_n = Y_n - \frac{b}{1 - m} \qquad x_n = X_n - \frac{b}{1 - m} \qquad 10.2\text{--}4$$

reduces Eq. 10.2–3 to Eq. 10.2–2, except for the special case where $m = 1$.

For simplicity, we assume that the liquid on every plate is well mixed and that the Murphree plate efficiency is equal to unity. The method is not limited to this case, however, and Horn's analysis includes the more general case. With this simplification, a material balance for the nth plate in the column becomes

$$L(x_{n+1} - x_n) = V(y_n - y_{n-1}) \qquad \qquad 10.2\text{--}5$$

When the vapor composition is eliminated from this expression, using the equi-

[1] A. Acrivos, *Chem. Eng. Educa.*, **2**, 62 (1968).

[2] M. R. Cannon, *Oil Gas J.*, **54**, 68 (1956).

[3] M. R. Cannon, *Ind. Eng. Chem.*, **53**, 629 (1961).

[4] F. Horn, *Chem. Eng. Tech.*, **36**, 99 (1964).

[5] F. J. M. Horn, *Ind. Eng. Chem. Proc. Design Develop.*, **6**, 30 (1967).

librium relationship, we obtain

$$x_{n+1} - x_n = \frac{KV}{L}(x_n - x_{n-1}) \qquad 10.2\text{-}6$$

or letting

$$\alpha = \frac{KV}{L} \qquad 10.2\text{-}7$$

and rearranging,

$$x_{n+1} - (1 + \alpha)x_n + ax_{n-1} = 0 \qquad 10.2\text{-}8$$

This is a linear, second-order, finite-difference equation with constant coefficients, so we try the solution

$$x_n = c\beta^n \qquad 10.2\text{-}9$$

Substituting this assumed solution and factoring the result gives the expression

$$\beta^{n-1}[(\beta - 1)(\beta - \alpha)] = 0 \qquad 10.2\text{-}10$$

Thus we find that there are two values of interest for β, and the general solution becomes

$$x_n = c_1 + c_2\alpha^n \qquad 10.2\text{-}11$$

where c_1 and c_2 are similar to integration constants. These are chosen so that the solution matches the boundary conditions at the ends of the column

$$\text{at } n = 0, \qquad y = y_0 = Kx_0 \qquad 10.2\text{-}12$$

$$\text{at } n = N + 1, \qquad x_{N+1} = x_e$$

Using these expressions, we find that

$$c_1 = \frac{x_0\alpha^{N+1} - x_e}{\alpha^{N+1} - 1} \qquad c_2 = \frac{x_e - x_0}{\alpha^{N+1} - 1} \qquad 10.2\text{-}13$$

so that, after some manipulation, the solution can be written

$$\frac{x_n - x_0}{x_e - x_0} = \frac{1 - \alpha^n}{1 - \alpha^{N+1}} \qquad 10.2\text{-}14$$

which is known as the Kremser equation.

The outlet liquid composition x_1 can be determined by setting $n = 1$

$$\frac{x_1 - x_0}{x_e - x_0} = \frac{1 - \alpha}{1 - \alpha^{N+1}} \qquad 10.2\text{-}15$$

which can be written as

$$x_e = \frac{\alpha x_0 - x_1}{\alpha - 1} + (x_1 - x_0)\left(\frac{\alpha^{N+1}}{\alpha - 1}\right) \qquad 10.2\text{-}16$$

This result provides a relationship between the liquid leaving the column, x_1, the vapor entering the column, $x_0 = y_0/K$, the liquid entering the column, x_e, the number of plates in the column, N, and a parameter, $\alpha = KV/L$. Hence this expression completely describes the operation of the column as far as the environment is concerned. In other words, any time two columns (steady state or periodic) are described by the same relationship, they will behave in an identical manner, regardless of the way in which the ends of the column are connected to their environment. For example, it can be shown that the steady state behavior of a column for a case where the Murphree efficiency

$$\epsilon = \frac{y_n - y_{n-1}}{y_n^* - y_{n-1}} \qquad y_n^* = Kx_n \qquad\qquad 10.2\text{-}17$$

is not equal to unity, is given by the result

$$x_e = \frac{\alpha x_0 - x_1}{\alpha - 1} + (x_1 - x_0)\left(\frac{\alpha}{\alpha - 1}\right)[1 + \epsilon(\alpha - 1)]^{N_\epsilon} \qquad 10.2\text{-}18$$

Thus two columns, one having N stages with a unit plate efficiency, $\epsilon = 1$, and one having N_ϵ plates with $\epsilon < 1$, will have exactly the same behavior if Eq. 10.2-18 is identical to Eq. 10.2-16. This occurs when

$$[1 + (\alpha - 1)\epsilon]^{N_\epsilon} = \alpha^N \qquad\qquad 10.2\text{-}19$$

or

$$\frac{N}{N_\epsilon} = \frac{\ln[1 + (\alpha - 1)\epsilon]}{\ln \alpha} \qquad\qquad 10.2\text{-}20$$

We call the ratio, N/N_ϵ, the relative stage efficiency, and we are interested in evaluating this quantity for a periodically operated column.

Periodic Operation

Now we consider the behavior of a column where we cycle the vapor flow rate. We assume that liquid enters the column at a constant flow rate, but the vapor rate varies between zero and some maximum value. We make the maximum vapor rate high enough so that during the vapor flow portion of the cycle all the liquid will be held up on each tray. Then when we set the vapor rate equal to zero, the froth will collapse and the liquid will start to flow down the column. As soon as the liquid has dropped one tray, we turn on the vapor again. By repeating this procedure over and over again, eventually we will obtain periodic conditions inside the column.

The analysis of the periodic column operation can be simplified if we neglect the period when the vapor flow is equal to zero, as well as any mass transfer that occurs during this period. With this assumption, we can treat the system as if the vapor flow rate is constant, the liquid flow rate is zero; and at the end of each operating period, we simply move the liquid on any tray to the tray below. Thus a material balance gives the result

$$H\frac{dx_n}{d\theta} = V(y_{n-1} - y_n) \qquad\qquad 10.2\text{-}21$$

where H is the liquid holdup on each tray and θ represents time. Letting τ denote the cycle time, $q = V\tau$ be the total amount of vapor flowing during each cycle, and $t = \theta/\tau$ be the dimensionless operating time during one cycle, the system equation becomes

$$H\frac{dx_n}{dt} = q(y_{n-1} - y_n) \qquad 10.2\text{-}22$$

Also, the periodic boundary condition for the liquid composition is

$$x_n(0) = x_{n+1}(1) \qquad 10.2\text{-}23$$

where the cycle time $t = 1$ corresponds to the conditions just before we move the liquid down one tray. The equation can be simplified further by using the equilibrium relationship to eliminate the vapor composition and letting

$$\alpha = \frac{qK}{H} \qquad 10.2\text{-}24$$

In this way we obtain the state equation

$$\frac{dx_n}{dt} = \alpha(x_{n-1} - x_n) \qquad 10.2\text{-}25$$

This is a linear, difference-differential equation, which we must solve using the periodic boundary condition given by Eq. 10.2–23. We could try to find a solution by taking the Laplace transform of the equation above and the boundary condition, solving the difference equation in terms of the Laplace parameter s, and, finally, looking for the inverse transform of this result. However, we know from Eq. 4.2–16 that the Laplace transforms of periodic functions involve transcendental expressions, and therefore we expect this approach to give complicated results.

A somewhat simpler technique for the problem in question is to recognize that the conditions in a column with N stages must be the same as those in a column with an infinite number of stages, providing that we force the conditions entering the zeroth and Nth trays to be the same. We can represent the state of the infinite column by the sequence of functions

$$x_1(t), x_2(t), \ldots, x_n(t), \ldots$$

Also, we define a characteristic function by the equation

$$u(z,t) = \sum_{n=1}^{\infty} x_n(t)z^n \qquad 10.2\text{-}26$$

This is analogous to a Laplace transform, where z replaces s, $u(z,t)$ replaces $\tilde{f}(s)$, and we sum instead of integrate. Our procedure will be to apply this transformation to the system equation and boundary condition, solve the equation for $u(z,t)$, and then take the inverse transformation.

Multiplying both sides of Eq. 10.2–25 by z^n and summing from $n = 1$ to

∞ gives

$$\sum_{n=1}^{\infty} \frac{dx_n}{dt} z^n = \alpha \sum_{n=1}^{\infty} (x_{n-1} - x_n) z^n = -\alpha \sum_{n=1}^{\infty} x_n z^n + \alpha \sum_{n-1}^{\infty} x_{n-1} z^{n-1} z \qquad 10.2\text{--}27$$

Letting $r = n - 1$ in the last summation and manipulating the results somewhat leads to the expression

$$\frac{d}{dt} \left(\sum_{n=1}^{\infty} x_n z_n \right) = -\alpha \sum_{n=1}^{\infty} x_n z^n + \alpha \sum_{r=1}^{\infty} (x_0 + x_r z^r) z \qquad 10.2\text{--}28$$

Now, letting $r = n$ in the last term and substituting the definition of u, Eq. 10.2–26, we obtain

$$\frac{du}{dt} = \alpha(z - 1)u + \alpha x_0 z \qquad 10.2\text{--}29$$

Thus we have reduced an infinite number of first-order equations to a single equation. We transform the periodic boundary condition, Eq. 10.2–23, in a similar way.

$$\sum_{n=1}^{\infty} x_n(0)z^n = \sum_{n=1}^{\infty} x_{n+1}(1)z^n = \sum_{n=1}^{\infty} x_{n+1}(1)z^{n+1} z^{-1} \qquad 10.2\text{--}30$$

After some manipulation of the indices in the last term, we obtain

$$u(0) = \frac{u(1)}{z} - x_1(1) \qquad 10.2\text{--}31$$

The solution of the differential equation, given by Eq. 10.2–29, is

$$u = c_1 e^{-\alpha(z-1)t} - \frac{x_0 z}{(z - 1)} \qquad 10.2\text{--}32$$

where c_1 is an integration constant. If we evaluate this expression at $t = 0$, we can write

$$c_1 = u(0) + \frac{x_0 z}{z - 1} \qquad 10.2\text{--}33$$

Similarly, at $t = 1$, we have

$$u(1) = c_1 e^{-\alpha(z-1)} + \frac{x_0 z}{z - 1} = \left[u(0) + \frac{x_0 z}{z - 1} \right] e^{-\alpha(z-1)} - \frac{x_0 z}{z - 1} \qquad 10.2\text{--}34$$

By using the transformed periodic boundary condition, Eq. 10.2–31, to eliminate $u(0)$ from this expression, we can show that

$$u(z,1) = \frac{x_0 z}{1 - z} + [x_1(1) - x_0] \left[\frac{z e^{\alpha(z-1)}}{e^{\alpha(z-1)} - z} \right] \qquad 10.2\text{--}35$$

This is the solution of the system equation in terms of $u(z,t)$ that satisfies the boundary condition; consequently, now we want to transform this result back into the composition domain. The first thing we notice is that

$$\frac{z}{1 - z} = z + z^2 + z^3 + \dots = \sum_{n=1}^{\infty} z^n \qquad 10.2\text{--}36$$

Similarly, we find that

$$\frac{ze^{\alpha(z-1)}}{e^{\alpha(z-1)} - z} = z + z^2 e^{-\alpha(z-1)} + z^3 e^{-2\alpha(z-1)} + \cdots$$

$$+ z^{n+1} e^{-n\alpha(z-1)} + \cdots \qquad 10.2\text{-}37$$

We need to write this expression as a power series in z, but this can be accomplished by expanding each of the exponentials in a series

$$e^{-az} = 1 - \frac{ax}{1!} + \frac{a^2 x^2}{2!} - \frac{a^3 x^3}{3!} + \cdots \qquad 10.2\text{-}38$$

and collecting terms. In this way we find that

$$\frac{ze^{\alpha(z-1)}}{e^{\alpha(z-1)} - z} = z + e^\alpha z^2 + e^{2\alpha}(1 - \alpha e^{-\alpha})z^3$$

$$+ e^{3\alpha}\left(1 - 2\alpha e^{-\alpha} + \frac{\alpha^2}{2!} e^{-2\alpha}\right)z^4 + \cdots \qquad 10.2\text{-}39$$

A general expression for these coefficients is

$$A_n = e^{n\alpha} \sum_{j=0}^{n-1} (-1)^j \frac{(n-j)^j}{j!} \alpha^j e^{-j\alpha} \qquad 10.2\text{-}40$$

where $A_0 = 1$, $A_1 = e^\alpha$, $A_2 = e^{2\alpha}(1 - \alpha e^{-\alpha})$, and so forth. Hence our polynomial can be put into the form

$$\frac{ze^{\alpha(z-1)}}{e^{\alpha(z-1)} - z} = \sum_{n=1}^{\infty} A_{n-1} z^n \qquad 10.2\text{-}41$$

Substituting the definition of $u(z,t)$, Eq. 10.2-26, plus Eqs. 10.2-36 and 10.2-41, into the solution of the differential equation, Eq. 10.2-35, and rearranging the results, we find that

$$\sum_{n=1}^{\infty} \{x_n(1) - x_0 - [x_1(1) - x_0]A_{n-1}\}z^n = 0 \qquad 10.2\text{-}42$$

which is always satisfied when

$$x_n(1) = x_0 + [x_1(1) - x_0]A_{n-1} \qquad 10.2\text{-}43$$

This is the inverse transform we were seeking. It gives the solution for a column with an infinite number of plates. In order to make it correspond to our finite column, we require that

$$\text{at } n = N + 1, \qquad x_n = x_{N+1} = x_e \qquad 10.2\text{-}44$$

Hence the finite, periodic column must be described by the equation

$$x_e = x_0 + [x_1(1) - x_0]A_N \qquad 10.2\text{-}45$$

where A_N can be determined from Eq. 10.2-40, x_e is the entering liquid composition, $x_0 = K/y_0$ is proportional to the entering vapor composition, and $x_1(1)$ is the composition of the liquid leaving the bottom of the column.

In order to determine the relative stage efficiency, we need to put the preceding result into the same form as Eq. 10.2-16, which describes the steady state column. Letting N_p refer to the number of plates in a periodic column, and after some manipulation, we find that Eq. 10.2–45 can be written as

$$x_e = \frac{\alpha x_0 - x_1}{\alpha - 1} + (x_1 - x_0)\left(A_{N_p} + \frac{1}{\alpha - 1}\right) \qquad 10.2\text{–}46$$

Hence the steady state and periodic columns will be equivalent when

$$A_{N_p} + \frac{1}{\alpha - 1} = \frac{\alpha^{N+1}}{\alpha - 1} \qquad 10.2\text{–}47$$

or

$$N \ln \alpha = \ln\left[\frac{A_{N_p}(\alpha - 1) + 1}{\alpha}\right] \qquad 10.2\text{–}48$$

Then, dividing by $N_p \ln \alpha$, we find that the relative stage efficiency is

$$\frac{N}{N_p} = \frac{\ln[A_{N_p}(\alpha - 1) + 1]}{N_p \ln \alpha} - \frac{1}{N_p} \qquad 10.2\text{–}49$$

For the case where $\alpha = 1$, which is a case of particular interest because it corresponds to a very difficult separation problem, it is possible to find a simple approximate expression for this equation. By plotting A_n versus n in Eq. 10.2–40, we find what when $n > 2$ we can approximate the coefficients quite closely by the expression

$$A_n = 2n + \tfrac{2}{3} \qquad 10.2\text{–}50$$

Substituting this result into Eq. 10.2–49, we obtain

$$\frac{N}{N_p} = \frac{1}{N_p} \frac{\ln\left[(2N_p + \tfrac{2}{3})(\alpha - 1) + 1\right]}{\ln \alpha} - \frac{1}{N_p} \qquad 10.2\text{–}51$$

Using L'Hôpital's rule to take the limit as α approaches unity, we see that

$$\frac{N}{N_p} = \frac{1}{N_p}\left(2N_p + \frac{2}{3}\right) - \frac{1}{N_p} = 2 - \frac{1}{3N_p} \qquad 10.2\text{–}52$$

Thus if we consider columns having more than three plates, we find that the relative stage efficiency is

$$\frac{N}{N_p} = 2 \qquad 10.2\text{–}53$$

This means that the relative stage efficiency is 200 percent, or that we can accomplish exactly the same job in a periodic column that has one-half the number of plates as a steady state column.

Horn's original paper discusses the effect of Murphree efficiency on this result and considers cases where only a fraction of the liquid on a plate falls down one tray during the interval when the vapor flow is stopped. Both cases lead to a lower performance than that reported here. Calculated values of the relative stage efficiencies for various values of α are also included in his paper.

Similar results were obtained in an analytical and numerical study by Schrodt[6,7,8] and a number of his co-workers.

Experimental Studies of Controlled Cycling

A number of experimental investigations provide confirmation of theoretical analysis given above. In the early work of Gaska and Cannon[9] and McWhirter and Cannon[10] on distillation columns operating at total reflux, efficiencies as high as 160 percent were obtained. Also, it was shown that the column with controlled cycling could be operated so that it had a significantly higher throughput than a conventional steady state column. Later studies by McWhirter and Lloyd,[11] Schrodt,[12] and Schrodt, Sommerfeld, Martin, Parisot, and Chien[8] demonstrated that both separation and capacity improvements were possible. These investigations cover the range from small laboratory columns to semiplant units.

The controlled cycling technique has also been applied to extraction columns by Cannon,[3] Szabo, Lloyd, Cannon, and Speaker,[13] Schrodt,[12] and Belter and Speaker.[14] For this case, the cycle contains periods for dense phase flow down the column, a coalescence stage with no flow, light phase flow up the column, and, again, a coaelescence stage with no flow. Efficiencies greater than 100 percent were observed, and the column capacities were 2 to 10 times greater than the steady state case.

Many investigators have searched for simple qualitative reasons that explain the avantages of controlled cycling. Much of this material has recently been reviewed by Schrodt.[15] One of the main conclusions is that a higher time-average driving force for mass transfer is obtained with the controlled cycling case. It is well known that when two phases containing a transferable component are brought into contact, initially the driving force for mass transfer is very high, but it falls off and approaches zero as the phases approach equilibrium. For controlled cycling conditions, there will be more contacting of the

[6] J. T. Sommerfeld, V. N. Schrodt, P. E. Parisot, and H. H. Chien, *Separation Science*, **1**, Nos. 2 and 3, 245 (1966).

[7] H. H. Chien, J. T. Sommerfeld, V. N. Schrodt, and P. E. Parisot, *Separation Science.*, **1**, Nos. 2 and 3, 281 (1966).

[8] V. N. Schrodt, J. T. Sommerfeld, O. R. Martin, P. E. Parisot, and H. H. Chien, *Chem. Eng. Sci.*, **22**, 759 (1967).

[9] R. A. Gaska, and M. R. Cannon, *Ind. Eng. Chem.*, **53**, 630 (1961).

[10] J. R. McWhirter, and M. R. Cannon, *Ind. Eng. Chem.*, **53**, 632 (1961).

[11] J. R. McWhirter, and W. A. Lloyd, *Chem. Eng. Prog.*, **59**, No. 6, 58 (1963).

[12] V. N. Schrodt, *Ind. Eng. Chem. Fundamentals*, **4**, 108 (1965).

[13] T. T. Szabo, W. A. Lloyd, M. R. Cannon, and S. M. Speaker, *Chem. Eng. Prog.* **60**, No. 1, 66 (1964).

[14] P. A. Belter and S. M. Speaker, *Ind. Eng. Chem. Proc. Develop.*, **6**, 36 (1967).

[15] V. N. Schrodt, *Ind. Eng. Chem.*, **59**, No. 6, 58 (1967).

fresh phases, thus leading to a higher time-average driving force than the normal steady state design, which must be close to the equilibrium conditions in order to achieve the maximum possible transfer of material in the equipment. Although this argument does not seem to agree with the preceding analysis at first, it has been shown that the equations we used are completely analogous to a conventional column with a transverse concentration gradient across the tray, where time in the unsteady state case is replaced by distance in the steady state case. The fact that a transverse concentration gradient leads to significant improvements was discussed by Lewis in 1936.[16]

Another explanation for improved performance by controlled cycling is that almost any system can be overloaded for small time intervals. For example, the flow rates in a steady state distillation column are usually set at some fraction of the flooding velocity. Satisfactory operation can be obtained in this way even if a flow disturbance, which we do not measure, enters the column and persists for some time interval. However, flooding is not an instantaneous phenomenon, and there is no reason why we should not exceed the flooding velocity, just as long as we make certain that we reduce the flow rate before flooding has a chance to occur. Hence, by periodically overloading the system but then relaxing the overload conditions and giving the unit a chance to recover, we might be able to obtain a better time average performance.

Controlled cycling techniques have been applied to many other kinds of systems, including crystallizers, absorption units, ion exchangers, fluidized beds, reactors, screening units, fuel cells, welding, and film boiling. Most of these studies include experimental verification of the advantages of the cycled operation over the conventional steady state design. An interested reader should refer to Schrodt's review papers.

10.3 CHEMICAL OSCILLATORS

Up to now we have considered only those periodic systems where oscillating outputs are produced because the inputs to the plant were manipulated in a periodic manner. However, in some instances it might be possible for a unit to generate periodic outputs even if the inputs are maintained constant. We call these systems chemical oscillators because they are analogous to electrical oscillators and can be described by using the same mathematical tools. It is somewhat against our intuition, and certainly opposed to the training of most chemical engineers, ever to expect a system having constant inputs to generate fluctuating outputs. Therefore it will be necessary first to establish the conditions for the existence of a chemical oscillator. Next, we need to determine whether or not this type of reactor can ever have a time average performance that is superior to the optimum steady state design. After this information has become available, some other problems that seem of interest can be discussed.

[16] W. K. Lewis, *Ind. Eng. Chem.*, **28**, 399 (1936).

An Unstable Design

In Section 7.1 we presented an example where there was only one possible solution of the equations describing the optimum steady state design of a continuous-stirred-tank reactor, but the optimum design turned out to be unstable; that is, the characteristic roots were complex conjugates with positive real parts. This result means that the reactor will always move away from the desired steady state operating point. Thus we designed simple feedback control systems that would make the operating point become stable. Then even if disturbances enter the reactor, it will always return to the steady state design conditions. However, it is certainly reasonable to wonder what would happen to the reactor if we neglected to install the controller. This question will become of paramount importance if the control system fails due to a mechanical or some other breakdown.

Because of the presence of the positive real parts of the complex conjugate roots, the linearized analysis in the neighborhood of the singular point—that is, the steady state design condition—tells us that the system will tend to grow without a bound. Actually, we know that this linearized analysis is based on the assumption that all the quadratic, cubic, and higher-order terms in a Taylor series expansion of the nonlinear state equations are negligible in comparison with the linear terms. Obviously this assumption is no longer valid if the state variables become very large, and, in fact, the linear terms might become negligible in this region. Hence the unbounded nature of the linearized equations is something we must ignore, although the nonlinear system indeed may be unbounded. In other words, the nature of the system bounds is the first thing break we must establish.

If we write the dynamic equations for the stirred-tank reactor as

$$V\frac{dA}{dt} = q(A_f - A) - kVA \qquad 10.3\text{-}1$$

$$VC_p\rho\frac{dT}{dt} = qC_p\rho(T_f - T) + U_a(T_H - T) + (-\Delta HkVA) \qquad 10.3\text{-}2$$

where

$$k = k_0 e^{-E/RT} \qquad U_a = \frac{2C_{PH}\rho_H UA_H q_H}{(UA_H + 2C_{PH}\rho_H q_H)} \qquad 10.3\text{-}3$$

we know from physical considerations that the composition in the reactor can never exceed the feed composition, nor can it ever become less than zero. From the material balance expression, we see that if A ever becomes equal to zero, the composition must increase because $dA/dt > 0$. Similarly, if A ever becomes equal to the feed composition A_f, the composition must decrease as time progresses because $dA/dt < 0$. Thus we have been able to establish an upper and a lower bound on the composition variable and to show that only operating points that lie between these bounds are permissible according to the material balance equation. Applying the same kind of idea to the energy balance, we know that T_f is the lowest possible temperature in the system and we see that $dT/dt > 0$ when $T = T_f$. An upper bound for the temperature is slightly more difficult to obtain. However, we are certain that the temperature can never exceed the value of the heating temperature plus the adiabatic temperature rise. Letting $T = T_H + (-\Delta H)kVA_f$ on the right-hand side of Eq. 10.3-2, we find that $dT/dt < 0$, so that the temperature must always decrease from this value. Hence the temperature variable is also bounded, and the reactor must operate within these bounds.

After all that has been published about reactor "runaways" and reactor explosions, the foregoing results seem somewhat surprising. They indicate that the nature of the system equations requires that the system always remain within some finite bounds. Of course, the contradiction is more apparent than real, because the upper temperature bound might be above the melting point of the reactor material. Similarly, if the rate of increase of the temperature is extremely rapid—a situation not considered in the argument—then a failure in

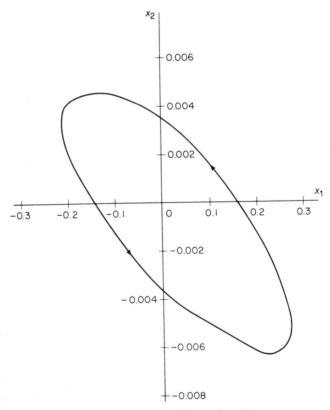

Figure 10.3-1. Phase plane trajectories.

a weld on the reactor might occur. In other words, the bounds for a reactor failure, or explosion, can be within the bounds proposed above, for there are other equations describing the system, such as $T < (T$ melting point), which we have not considered.

Nevertheless, our arguments concerning the system bounds tell us that there might be some systems that will not run away. For these bounded systems, and for cases where there is only one steady state solution that is unstable, we are still uncertain about the nature of the system output. At this point we must refer to a theorem in topology proposed by Bendixson, which tells us that any system trajectory that remains within a finite region and does not approach a singular point must either be a stable limit cycle or else approach a stable limit cycle asymptotically[1,2,3]. This means that if we plot the values of composition versus temperature with time as a parameter, eventually we will obtain a closed curve. Alternately, if we plot composition and temperature versus time, ultimately we will observe a periodic output. In very crude terms, we can say that

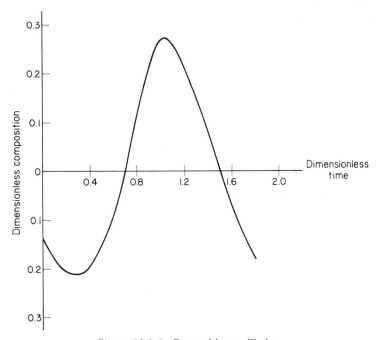

Figure 10.3-2. Composition oscillation.

[1] N. Minorsky, *Nonlinear Oscillations*, Ch. 3, Van Nostrand, Princeton, N. J., 1962.

[2] D. Graham and D. McRuer, *Analysis of Nonlinear Control Systems*, p. 345, Wiley, N. Y., 1961.

[3] J. M. Douglas and D. W. T. Rippin, *Chem. Eng. Sci.*, **21**, 305 (1966).

our system cannot go to a steady state, it cannot run away, and therefore it runs around in circles forever.

In physical terms, the oscillating behavior can be explained as follows: Since the reaction is exothermic, heat is liberated as the reaction proceeds. The heat generation and the heat supplied by the coil cause an increase in the reactor temperature, which increases the rate constant k, and therefore the rate at which reactant is converted. This autocatalytic phenomenon proceeds at an accelerating rate, which becomes much faster than the rate at which reactants are supplied to the system. Hence at some point the effective reactant concentration in the system drops to a very small value, and the reaction rate term, kVA in Eq. 10.3–1, becomes very small. At the same time, the heat generated by the reaction must drop to a low value. Then, for the proper combination of system parameters, the heat loss by convective flow will become greater than the heat supplied through the coil and the heat generated by the reaction, so that the temperature will start to decrease. As the cooling process continues, the effective reactant concentration is building up because the convective flow of reactant into the system is greater than the reaction rate. Although the decreasing

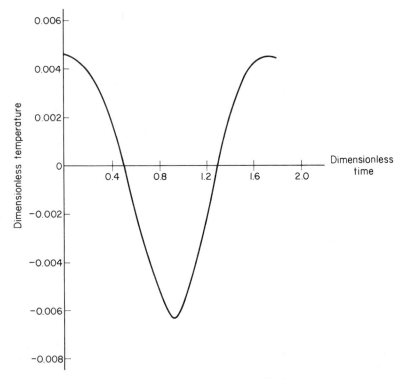

Figure 10.3-3. Temperature oscillation.

temperature tends to reduce the value of the rate, kVA, at some point the increasing value of concentration will become large enough so that the rate term again becomes appreciable. Then the system will follow the autocatalytic path once more, and the reactor outputs will be periodic even though the inputs are maintained constant.

When we numerically integrate Eqs. 10.3–1 through 10.3–3 for the parameters given in Table 7.1–1, we find that after some initial transient period, the system with constant inputs generates periodic composition and temperature outputs (see Figures 10.3–1, 10.3–2, and 10.3–3). Because of the nonlinearities in the system equations, the periodic signals resemble distorted sine waves. It is clear that the fluctuating outputs have no connection with a steady state or a linearized dynamic analysis. Therefore, at present, we have no way of determining the time average behavior of the system, nor how this average performance may compare to the optimum steady state design. For the particular problem under consideration, we can find the time average conversion numerically, and in this way establish that the oscillator performance is slightly better than steady state (see Table 10.3–2). Of course, this means that our whole motivation in Section 7.1 for designing control systems to make unstable plants become stable is not always valid. In fact, as we shall show later, in some cases it seems advantageous to use positive feedback controllers to make stable systems become unstable.

Necessary Conditions for a Chemical Oscillator

The foregoing discussion indicates that whenever we have one unstable singular point lying within a bounded domain, where the trajectories originating on the boundaries are directed inward, we expect to obtain a limit cycle. The existence of a single steady state solution with either a pair of real positive roots or complex conjugate roots with positive real parts is not sufficient to guarantee the existence of a unique periodic output, however, because there might be situations where multiple limit cycles arise. In other words, an unstable singular point might be surrounded by a stable limit cycle, which, in turn, is surrounded by an unstable limit cycle, etc. Thus it is not a simple matter to determine, even on a qualitative basis, what kind of an oscillating output we might expect to observe.

The problem becomes much more complicated if we consider stirred-tank reactors having multiple steady state solutions. For this case, we know that the intermediate equilibrium point is always unstable; that is, in Section 7.1 we proved that it must be a saddlepoint, but it is also possible to find examples where one, or both, of the extreme points is unstable. Then, from topological considerations, we can show that a limit cycle can exist around one of the extreme steady state points, there can be limit cycles around each of these points, or a limit cycle can surround all three. Sketches of the various configurations have been presented by Aris and Amundson[4] and Douglas and Rippin,[3] and an illustration of a limit cycle enclosing three singular points was published by Schmitz and Amundson.[5]

Even though it might be possible to obtain a wide variety of different kinds of limit cycles, experience has shown that they are seldom encountered in practice. Thus, rather than worry about all the difficulties we may encounter, we will assume that the necessary conditions for a chemical oscillator are that we have a single unstable equilibrium point. However, we will carefully check this assumption by numerically integrating the nonlinear equations before we attempt to build a real system. Moreover, we hope that an approximate analytical solution describing the behavior of the oscillator will provide additional information about the existence of limit cycles.

Approximate Analytical Solutions of Oscillator Performance

Throughout this text we have emphasized the fact that it is impossible to find analytical solutions of sets of nonlinear equations. Despite this limitation, we need to establish some technique for predicting the performance of a chemical oscillator, particularly in view of the fact that we gain no useful information from a steady state analysis. Some of the theories of nonlinear oscillations that can be used to develop approximate analytical solutions are described below. Initially the methods are applied to a simple, nonisothermal, stirred-tank reactor problem, but then several extensions of the basic ideas are discussed.

Dimensionless State Equations for a Single Reaction

We consider the dynamic model given by Eq. 10.3–1 through 10.3–3 and include the possibility of having a proportional feedback controller by adding the equation

$$U_a = U_{as} + K(T - T_s) \qquad\qquad 10.3\text{–}4$$

to the set. Here we are assuming that the heating flow rate is manipulated proportional to the difference between the actual reactor temperature and the optimum steady state value, T_s, and that the overall heat transfer group can be approximated by a linear function of the heating flow rate. For negative feedback control systems, we want K, which is proportional to the controller gain, to be negative, so that whenever the reactor temperature exceeds the design value, we decrease the heat transferred to the reactor. Since the optimum steady state design conditions are known, we can introduce the transformation

---

[4] Footnote 2, p. 8.

[5] R. A. Schmitz and N. R. Amundson, *Chem. Eng. Sci.*, **18**, 265, 391, 415, 477 (1965).

of variables

$$x_1 = -\frac{A - A_s}{A_s}, \quad x_2 = -\frac{T - T_s}{T_s}, \quad \tau = \frac{qt}{V}, \quad \beta_1 = \frac{k_s V}{q}, \quad \beta_3 = \frac{E}{RT_s}$$

$$10.3\text{-}5$$

$$\beta_4 = 1 + \frac{U_{as}}{qC_p\rho}, \quad \beta_5 = \frac{(-\Delta H)Vk_sA_s}{qC_p\rho T_s}, \quad \beta_6 = \frac{KT_s}{qC_p\rho}, \quad \beta_7 = 1 - \frac{T_H}{T_s}$$

which translates the singular point to the origin and makes the variables dimensionless. The system equations become

$$\frac{dx_1}{d\tau} = -x_1 - \beta_1\left[1 - (1 - x_1)\exp\left(\frac{\beta_3 x_2}{x_2 - 1}\right)\right] \tag{10.3-6}$$

$$\frac{dx_2}{d\tau} = -\beta_4 x_2 + \beta_5\left[1 - (1 - x_1)\exp\left(\frac{\beta_3 x_2}{x_2 - 1}\right)\right] - \beta_6 x_2(\beta_7 - x_2)$$

$$10.3\text{-}7$$

We want to find a periodic solution of these equations for a case where

$$d_e = 1 + \beta_1 + \beta_4 - \beta_3\beta_5 + \beta_6\beta_7 < 0 \tag{10.3-8}$$

which makes the real part of the roots positive, and

$$d_f = (1 + \beta_1)(\beta_4 + \beta_6\beta_7) - \beta_3\beta_5 > 0 \tag{10.3-9}$$

which ensures that our steady state solution will not be a saddlepoint. Of course, these criteria can be combined to give

$$\beta_1(\beta_4 + \beta_6\beta_7) > \beta_3\beta_5 + \beta_6\beta_7 > 1 + \beta_1 \tag{10.3-10}$$

as a necessary condition to obtain an oscillator.

Although, eventually, we would like to develop a solution for the pair of simultaneous equations given by Eqs. 10.3-6 and 10.3-7 that can be extended to higher-order systems, as a first approach it is somewhat easier if we rewrite the equations as a single, second-order, nonlinear system. Because these state equations are linear in the composition variable x_1, this reduction can be accomplished by differentiating Eq. 10.3-7 with respect to τ, using Eq. 10.3-6 to eliminate $dx_1/d\tau$ from the result, solving Eq. 10.3-7 for x_1 and using this expression to eliminate x_1 from the relation obtained in step 2. In this way we get

$$\ddot{x}_2 = \frac{-\beta_3}{(x_2 - 1)^2} - \dot{x}_2\left\{1 + \beta_4 + \beta_1\exp\left(\frac{\beta_3 x_2}{x_2 - 1}\right) + \beta_6(\beta_7 - x_2)\right.$$

$$-\frac{\beta_3}{(x_2 - 1)^2}[\beta_5 - \beta_4 x_2 - \beta_6 x_2(\beta_7 - x_2)]\right\} - \left\{\beta_5\left[\exp\left(\frac{\beta_3 x_2}{x_2 - 1}\right) - 1\right]\right.$$

$$+ \left[1 + \beta_1\exp\left(\frac{\beta_3 x_2}{x_2 - 1}\right)\right][\beta_4 x_2 + \beta_6 x_2(\beta_7 - x_2)]\right\} \tag{10.3-11}$$

This expression is difficult to deal with; therefore we approximate it by a Taylor series expansion in x_2 and $\dot{x}_2$ about the origin, where we keep all terms up to the third-order

$$\ddot{x}_2 + \omega_0^2 x_2 = K_1 x_2 + K_2 \dot{x}_2 + K_3 x_2^2 + K_4 \dot{x}_2^2 + K_5 x_2 \dot{x}_2 + K_6 x_2^3$$
$$+ K_7 x_2^2 \dot{x}_2 + K_8 x_2 \dot{x}_2^2 + \cdots \qquad 10.3\text{-}12$$

where

$$\omega_0 = \frac{1}{2}(4d_f - d_e^2)^{1/2}, \quad K_1 = -\frac{d_e^2}{4}, \quad K_2 = -d_e$$

$$K_3 = \beta_1 \beta_3 (\beta_4 + \beta_6 \beta_7) + (1 + \beta_1)\beta_6 + \beta_3 \beta_5 - \frac{1}{2}\beta_3^2 \beta_5, \quad K_4 = -\beta_3$$

$$K_5 = \beta_1 \beta_3 + 2\beta_6 + 2\beta_3 \beta_5 - \beta_3(\beta_4 + \beta_6 \beta_7), \quad K_6 = \beta_1 \beta_3 (\beta_4 + \beta_6 \beta_7)$$
$$+ \frac{1}{6}\beta_3^3 \beta_5 - \beta_3^2 \beta_5 + \beta_3 \beta_5 - \beta_1 \beta_3 \beta_6 - \frac{1}{2}\beta_1 \beta_3^2 (\beta_4 + \beta_6 \beta_7)$$

$$K_7 = \beta_1 \beta_3 - \frac{1}{2}\beta_1 \beta_3^2 - 2\beta_3(\beta_4 + \beta_6 \beta_7) + \beta_3 \beta_6 + 3\beta_3 \beta_5$$

$$K_8 = -2\beta_3 \qquad 10.3\text{-}13$$

It is immediately apparent that if all the terms on the right-hand side of Eq. 10.3-12 were equal to zero, the terms on the left will have a periodic solution, for they describe a simple harmonic oscillator. We know that we are looking for a periodic solution of the nonlinear equation, and we might try to obtain it by building up a perturbation solution around the harmonic oscillator result. In other words, we assume that all the terms on the right-hand side are small, which we denote by inserting an artificial, small parameter, μ, into the equation

$$\ddot{x}_2 + \omega_0^2 x_2 = \mu(K_1 x_2 + K_2 \dot{x}_2 + K_3 x_2^2 + K_4 \dot{x}_2^2 + K_5 x_2 \dot{x}_2 + K_6 x_2^3$$
$$+ K_7 x_2^2 \dot{x}_2 + K_8 x_2 \dot{x}_2^2) \qquad 10.3\text{-}14$$

or

$$\ddot{x}_2 + \omega_0^2 x_2 = \mu f(x_2, \dot{x}_2) \qquad 10.3\text{-}15$$

Then we hope that the solution of this nonlinear equation is periodic and not too different from the harmonic oscillator result.

Method of Poincaré and Lindstedt[6]

If our foregoing assumptions are satisfied, we should be able to find an approximate solution having the form

$$x_2 = y_0 + \mu y_1 + \mu^2 y_2 + \cdots \qquad 10.3\text{-}16$$

where $y_0, y_1, y_2, \cdots$ are unknown periodic functions of time that must be determined, y_0 is the generating solution, μy_1 represents a first correction term that should be small in comparison with y_0, $\mu^2 y_2$ represents a second-order

[6] J. M. Douglas and N. Y. Gaitonde, *Ind. Eng. Chem. Fundamentals*, **6**, 265 (1967).

correction term, and so on. In addition, we consider the possibility that the nonlinear terms will produce distortion of the output signal, and therefore we include a frequency correction term

$$\omega^2 = \omega_0^2 + \mu\omega_1^2 + \mu^2\omega_2^2 + \cdots \qquad 10.3\text{-}17$$

Substituting these assumed solutions into Eq. 10.3–15 gives the result

$$\ddot{y}_0 + \mu\ddot{y}_1 + \mu^2\ddot{y}_2 + \cdots + (\omega^2 - \mu\omega_1^2 - \mu\omega_2^2 - \cdots)(y_0 + \mu y_1 + \mu^2 y_2 + \cdots)$$

$$= \mu f(y_0 + \mu y_1 + \mu^2 y_2 + \cdots, \dot{y}_0 + \mu\dot{y}_1 + \mu^2\dot{y}_2 + \cdots) \qquad 10.3\text{-}18$$

where the right-hand side can be written as a power series in μ by substituting Eq. 10.3–16 and its derivative into Eq. 10.3–14. Expanding the various expressions and equating terms having like coefficients of powers of μ leads to the set of equations

$$\ddot{y}_0 + \omega^2 y_0 = 0 \qquad 10.3\text{-}19$$

$$\ddot{y}_1 + \omega^2 y_1 = \omega_1^2 y_0 + K_1 y_0 + K_2 \dot{y}_0 + K_3 y_0^2 + K_4 \dot{y}_0^2 + K_5 y_0 \dot{y}_0 + K_6 y_0^3$$

$$+ K_7 y_0^2 \dot{y}_0 + K_8 y_0 \dot{y}_0^2 \qquad 10.3\text{-}20$$

$$\vdots$$

In a similar way, we can obtain the equations for as many correction functions as we please.

The formal solution of the first equation is

$$y_0 = P_0 \sin \omega\tau + Q_0 \cos \omega\tau \qquad 10.3\text{-}21$$

However, we have no boundary conditions to evaluate the integration constants P_0 and Q_0. We know from the numerical study that with any set of initial conditions we always obtain the same periodic output, but we cannot make an a priori prediction of any point on those curves. Nevertheless, we proceed on the basis of faith and use the boundary conditions

$$\text{at } \tau = 0, \quad y_0 = A_0, \quad \dot{y}_0 = 0 \qquad 10.3\text{-}22$$

which simply say that at some time $\tau = 0$ the output will have its maximum value, the unknown amplitude A_0. With these boundary conditions, the solutions become

$$y_0 = A_0 \cos \omega\tau \qquad \dot{y}_0 = -A_0 \omega \sin \omega\tau \qquad 10.3\text{-}23$$

Now we can substitute these solutions into the right-hand side of Eq. 10.3–20. After using the trigonometric identities

$$\sin^2 \omega\tau = \tfrac{1}{2}(1 - \cos 2\omega\tau) \qquad \cos^2 \omega\tau = \tfrac{1}{2}(1 + \cos 2\omega\tau)$$

$$\sin \omega\tau \cos \omega\tau = \tfrac{1}{2} \sin 2\omega\tau \qquad \cos^3 \omega\tau = \tfrac{1}{4}(\cos 3\omega\tau + 3 \cos \omega\tau)$$

$$\sin^2 \omega\tau \cos \omega\tau = \tfrac{1}{4}(\cos \omega\tau - \cos 3\omega\tau),$$

$$\sin \omega\tau \cos^2 \omega\tau = \tfrac{1}{4}(\sin \omega\tau + \sin 3\omega\tau) \qquad 10.3\text{-}24$$

to simplify the result, we obtain

$$\ddot{y}_1 + \omega^2 y_1 = A_0 \left(\omega_1^2 + K_1 + \frac{3}{4} K_6 A_0^2 + \frac{1}{4} K_8 \omega^2 A_0^2 \right) \cos \omega\tau$$

$$+ \frac{A_0^2}{2} (K_3 + K_4 \omega^2) - \omega A_0 \left(K_2 + \frac{1}{4} K_7 A_0^2 \right) \sin \omega\tau$$

$$+ \frac{1}{2} A_0^2 (K_3 - K_4 \omega^2) \cos 2\omega\tau - \frac{\omega A_0^2}{2} K_5 \sin 2\omega\tau$$

$$+ \frac{A_0^3}{6} (K_6 - K_8 \omega^2) \cos 3\omega\tau - \frac{\omega A_0^3}{4} K_7 \sin 3\omega\tau \qquad 10.3\text{-}25$$

The complementary solution of this equation is

$$y_1 = P_1 \cos \omega\tau + Q_1 \sin \omega\tau \qquad 10.3\text{-}26$$

so that the particular integral will contain terms having the form $\tau \cos \omega\tau$ and $\tau \sin \omega\tau$. It is clear that these so-called secular terms will become unbounded as time increases, and therefore our solution will not be of much value unless they can be eliminated. However, this step can be accomplished by letting

$$A_0^2 = -\frac{4K_2}{K_7} \qquad 10.3\text{-}27$$

and

$$\omega_1^2 = -\tfrac{1}{4} [4K_1 + (3K_6 + K_8 \omega^2) A_0^2] \qquad 10.3\text{-}28$$

Thus the removal of the secular terms, to ensure that our solution will be periodic, gives us an estimate of the unknown amplitude of the oscillation, A_0, and the frequency correction factor.

The complete solution for the first-order correction function can be shown to be

$$y_1 = \frac{A_0^2}{2\omega^2} (K_3 + K_4 \omega^2) + P_1 \cos \omega\tau + Q_1 \sin \omega\tau$$

$$- \frac{A_0^2}{6\omega^2} (K_3 - K_4 \omega^2) \cos \omega\tau + \frac{A_0^2}{6\omega} K_5 \sin 2\omega\tau$$

$$- \frac{A_0^3}{32\omega^2} (K_6 - K_8 \omega^2) \cos 3\omega\tau + \frac{A_0^2}{32\omega} K_7 \sin 3\omega\tau \qquad 10.3\text{-}29$$

The integration constants are obtained by using the boundary conditions

$$\text{at } \tau = 0, \qquad y_1 = 0, \quad \dot{y}_1 = 0 \qquad 10.3\text{-}30$$

These boundary conditions are used for every correction function, for we have used the boundary conditions on x_2 to establish y_0, and from Eq. 10.3–16 we see that all the corrections must be equal to zero at $\tau = 0$. In this way we find that

$$P_1 = \frac{A_0^2}{3\omega^2} (K_3 + 2K_4 \omega^2) + \frac{A_0^3}{32\omega^2} (K_6 - K_8 \omega^2)$$

$$\qquad\qquad\qquad\qquad\qquad 10.3\text{-}31$$

$$Q_1 = -\frac{A_0^2}{3\omega} K_5 - \frac{3A_0^3}{32\omega} K_3$$

Now that we have completely determined the generating solution and the first correction function, y_1, we could proceed to evaluate the second correction function, y_2. However, the algebra involved becomes somewhat tedious. Therefore, as a first approximation, we write

$$x_2 = y_0 + \mu y_1 + \cdots \qquad 10.3\text{-}32$$

or

$$x_2 = \frac{\mu A_0^2}{2\omega^2}(K_3 + K_4\omega^2) + (A_0 + \mu P_1)\cos \omega\tau$$

$$+ \mu Q_1 \sin \omega\tau - \frac{\mu A_0^2}{6\omega^2}(K_3 - K_4\omega^2)\cos 2\omega\tau$$

$$+ \frac{\mu A_0^2}{6\omega}K_5 \sin 2\omega\tau - \frac{\mu A_0^3}{32\omega^2}(K_6 - K_8\omega^2)\cos 3\omega\tau$$

$$+ \frac{\mu A_0^3}{32\omega}K_7 \sin 3\omega\tau \qquad 10.3\text{-}33$$

where

$$A_0^2 = -\frac{4K_2}{K_7} \qquad 10.3\text{-}27$$

$$\omega^2 = \omega_0^2 + \mu\omega_1^2 = \omega_0^2 - \frac{\mu}{4}[4K_1 + (3K_6 + K_8\omega^2)A_0^2] \qquad 10.3\text{-}28$$

and P_1 and Q_1 are given by Eqs. 10.3-31. The parameter μ was introduced into the equation in an arbitrary way; and after comparing Eqs. 10.3-12 and 10.3-14, it becomes apparent that we must set

$$\mu = 1 \qquad 10.3\text{-}34$$

However, the reasoning we used in the analysis will be valid even for this condition if we demand that all the terms in Eq. 10.3-33 that contain μ are small in comparison with the others; for example,

$$P_1 \ll A_0, \quad Q_1 \ll A_0, \quad \tfrac{1}{8}[4K_1 + (3K_6 + K_8\omega_0^2)A_0^2] \ll \omega_0^2 \qquad 10.3\text{-}35$$

If these restrictions are not met, our analytical solution will not be valid, for the series solution given by Eq. 10.3-16 will not be convergent.

It is interesting to note that the approximate solution does indeed supply an additional criterion for the existence of a chemical oscillator. Clearly, the solution only applies to cases where Eq. 10.3-27 gives a real amplitude, or

$$-4\frac{K_2}{K_7} > 0 \qquad 10.3\text{-}36$$

However, one of our previous conditions for an oscillator was that $d_e < 0$, Eq. 10.3-8, and from Eq. 10.3-13 we know that $K_2 = -d_e$. Thus K_2 must be positive, and from the preceding expression we see that we must have

$$K_7 < 0 \qquad 10.3\text{-}37$$

This criterion might be particularly helpful for determining the existence of limit cycles when there are multiple singular points.

Even though we obtain this additional existence criterion as a fringe benefit, the most important feature of the solution for the reactor temperature is that it contains a constant term (d-c component), together with simple harmonic terms and higher-order harmonics. When we take the time average value of the solution, only the constant term remains

$$x_{2av} = \frac{A_0^2}{2\omega^2}(K_3 + K_4\omega^2) \qquad 10.3\text{-}38$$

This quantity will be positive providing that $(K_3 + K_4\omega^2) > 0$. By linearizing Eq. 10.3-7 and solving it for x_1, we obtain

$$x_1 = \frac{1}{\beta_5}[\dot{x}_2 + (\beta_4 - \beta_3\beta_5 + \beta_6\beta_7)x_2] \qquad 10.3\text{-}39$$

Then if we substitute our solution for x_2, Eq. 10.3-33, and calculate the time average value for x_1, we find that

$$x_{1\,av} = \frac{\beta_4 - \beta_3\beta_5 + \beta_6\beta_7}{\beta_5} x_{2av} \qquad 10.3\text{-}40$$

Hence the time-average composition output of the oscillator will be superior to the steady state output $(A_{av} < A_s)$ if $x_{1\,av} > 0$.

Method of Krylov and Bogoliubov

The analytical solution just described only provides information about the ultimate periodic output of the system. However, it is possible to estimate the transient portion of the solution by applying the method of Krylov and Bogoliubov.[6] This approach resembles the technique of variation of parameters. We again start with Eq. 10.3-14 and assume we can find a solution like

$$x_2 = A\cos\psi + \mu y_1(A, \psi) + \mu^2 y_2(A, \psi) + \cdots \qquad 10.3\text{-}41$$

where ψ is the total phase

$$\psi = \omega_0\tau + \theta(\tau) \qquad 10.3\text{-}42$$

Now, we require that the amplitude and phase dependence on time satisfy the relationships

$$\dot{A} = \mu A_1(A) + \mu^2 A_2(A) + \cdots$$
$$\dot{\psi} = \omega_0 + \mu\beta_1(A) + \mu^2\beta_2(A) + \cdots \qquad 10.3\text{-}43$$

After substituting the assumed solution into the system equation and after a considerable amount of manipulation, we find that the solution is almost exactly the same as Eq. 10.3-33 except that the amplitude of the limit cycle is given by the expression

$$A = \frac{(-4K_2/K_7)^{1/2}}{[1 - (1 + 4K_2/K_7 A_i^2)e^{-K_2\tau}]^{1/2}} \qquad 10.3\text{-}44$$

where A_i is any initial guess at the amplitude. It is clear from this result that whenever $K_2 > 0$, which we know to be satisfied, the amplitude will approach the value

$$A = \left(-\frac{4K_2}{K_7}\right)^{1/2} \qquad 15.3\text{-}45$$

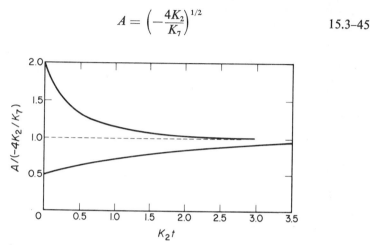

Figure 10.3-4. Approach of system to stable oscillating condition. [Reproduced from J. M. Douglas and N. Y. Gaitonde, *Ind. Eng. Chem. Fundamentals,* **6**, 265 (1967), by permission of the ACS.]

as time approaches infinity. The time dependence of this approach is shown in Figure 10.3–4. Additional details of the method, as well as criteria that can be used to assess the validity of the solution, are available in a paper by Douglas and Gaitonde.[6]

Example 10.3–1   Analytical estimates of the performance of a chemical oscillator

In order to verify the analytical solutions just discussed, Douglas and Gaitonde[6] compared the approximate solutions with a numerical integration of the original nonlinear equations for the case of a single reaction in a CSTR containing a cooling coil. The presence of a cooling coil means that the system can never correspond to an optimum steady state design (see Chapter 2), but this is a unit that has been discussed extensively in the literature. Two sets of parameters were considered. The first set, originally published by Aris and Amundson,[4] corresponds to a case where the oscillator performance was inferior to the arbitrary steady state design, while the second set, taken from a paper by Douglas and Rippin,[3] corresponds to a case where the oscillator performance is superior. Both sets of parameters, plus the results of some of the calculations, are given in Table 10.3–1. The tabulated values indicate that the approximate

TABLE 10.3-1
SYSTEM PARAMETERS*

β_6	d_e	d_f	A_0 Eq. 10.3-27	P_1 Eq. 10.3-31	Q_1 Eq. 10.3-31	ω_0^2 Eq. 10.3-13	ω_1^2 Eq. 10.3-28	Inequality Criteria, Similar to Eq. 10.3-35

Case A Parameters of Aris and Amundson†
$\beta_1 = 1.0$, $\beta_3 = 25.0$, $\beta_4 = 2.0$, $\beta_5 = 0.25$, $\beta_7 = 0.125$

β_6	d_e	d_f	A_0	P_1	Q_1	ω_0^2	ω_1^2	Inequality
17	−0.125	2.0	0.100	0.148	0.110	1.99	8.56	0.157 ≪ 0.480, 34.8 ≪ 16.0
16	−0.25	1.75	0.120	0.224	0.159	1.73	11.78	0.062 ≪ 0.420, 47.1 ≪ 14.0
15	−0.375	1.5	0.127	0.250	0.176	1.46	12.26	0.141 ≪ 0.362, 51.8 ≪ 12.0
14	−0.50	1.25	0.137	0.294	0.204	1.19	13.34	0.250 ≪ 0.301, 52.9 ≪ 10.0
13	−0.625	1.00	0.141	0.322	0.212	0.90	0.95	0.392 ≪ 0.241, 51.8 ≪ 8.0

Case B Parameters Similar to those used by Douglas and Rippin‡
$\beta_1 = 2.0$, $\beta_3 = 33.45$, $\beta_4 = 6.0$, $\beta_5 = 0.228$, $\beta_7 = 0.1$

β_6	d_e	d_f	A_0	P_1	Q_1	ω_0^2	ω_1^2	Inequality
6.0	−0.033	10.17	1.02×10^{-2}	0.03×10^{-2}	0.41×10^{-2}	10.17	0.448	0.001 ≪ 2.4, 1.9 ≪ 81.3
5.5	−0.083	10.02	1.61×10^{-2}	0.37×10^{-2}	1.05×10^{-2}	10.01	1.17	0.007 ≪ 2.4, 4.6 ≪ 80.1
5.0	−0.133	9.87	2.01×10^{-2}	0.58×10^{-2}	1.75×10^{-2}	9.87	1.82	0.018 ≪ 2.38, 7.3 ≪ 78.9
0	−0.633	8.37	4.2×10^{-2}	2.93×10^{-2}	7.23×10^{-2}	8.27	6.40	0.400 ≪ 2.02, 28.0 ≪ 68.9

*Reproduced from J. M. Douglas and N. Y. Gaitonde, *Ind. Eng. Chem. Fundamentals*, 6, 265 (1967), by permission.
†Footnote 2, p. 8.
‡Footnote 3, p. 395.

TABLE 10.3-2
COMPARISON OF ANALYTICAL AND NUMERICAL SOLUTIONS.* SYSTEM PARAMETERS SIMILAR TO THOSE USED BY DOUGLAS AND RIPPIN,† $\beta_1 = 1.0$, $\beta_3 = 33.45$, $\beta_4 = 6.0$, $\beta_5 = 0.228$, $\beta_7 = 0.1$

β_6	Period of Output		Temperature Shift in Mean Value		Composition Shift in Mean Value	
	Analytical Solution	Numerical Solution	Analytical Solution	Numerical Solution	Analytical Solution	Numerical Solution
6.0	1.93	1.95	-0.224×10^{-3}	-0.252×10^{-3}	0.236×10^{-2}	0.201×10^{-2}
5.5	1.89	1.99	-0.875×10^{-3}	-0.909×10^{-3}	0.935×10^{-2}	1.15×10^{-2}
5.0	1.85	2.03	-1.71×10^{-3}	-1.38×10^{-3}	1.86×10^{-2}	2.59×10^{-2}
0	1.65	2.13	-14.6×10^{-3}	-4.47×10^{-3}	18.4×10^{-2}	13.3×10^{-2}

*Reproduced from J. M. Douglas and N. Y. Gaitonde, *Ind. Eng. Chem. Fundamentals*, 6, 265 (1967), by permission.
†Footnote 3, p. 395.

analytical solution does not apply to Aris and Amundson's parameters because the constraints on the solution are not satisfied. The solutions for the other set of parameters are valid at high values of the negative feedback, controller gain, β_6. This corresponds to the smallest limit cycles and systems having the smallest values for the positive real part of the characteristic roots. Comparisons of the amplitudes and periods of the oscillations, as well as the time average outputs, are given in Table 10.3–2.

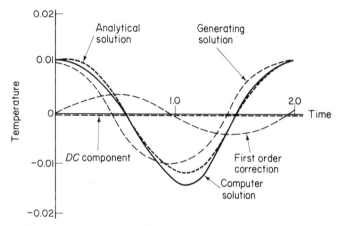

Figure 10.3-5. Periodic effluent temperature, $\beta_6 = 6.0$. [Reproduced from J. M. Douglas and N. Y. Gaitonde, *Ind. Eng. Chem. Fundamentals*, **6**, 265 (1967), by permission of the ACS.]

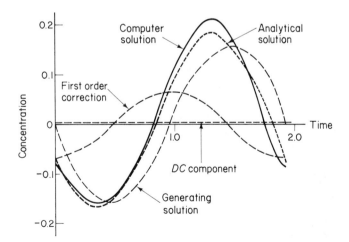

Figure 10.3-6. Periodic effluent composition $\beta_6 = 6.0$. [Reproduced from J. M. Douglas and N. Y. Gaitonde, *Ind. Eng. Chem. Fundamentals*, **6**, 265 (1967), by permission of the ACS.]

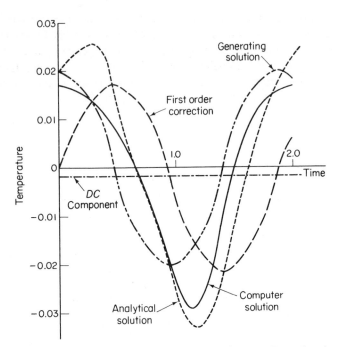

Figure 10.3-7. Periodic effluent temperature, $\beta_6 = 5.0$. [Reproduced from J. M. Douglas and N. Y. Gaitonde, *Ind. Eng. Chem. Fundamentals,* **6,** 265 (1967), by permission of the ACS.]

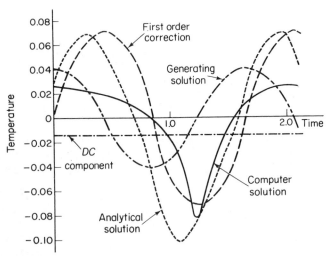

Figure 10.3-8. Periodic effluent temperature, $\beta_6 = 0$. [Reproduced from J. M. Douglas and N. Y. Gaitonde, *Ind. Eng. Chem. Fundamentals,* **6,** 265 (1967), by permission of the ACS.]

The behavior of the approximate solutions becomes more apparent by studying Figures 10.3–5 through 10.3–8. The analytical solutions for these graphs were taken as the sum of the generating solution, the d-c component, and the first harmonic terms of the first correction function. From the graphs we see that as the controller gain β_6 is decreased, so that the system becomes more unstable and the size and distortion of the limit cycle grow, the contribution of the first-order correction term finally becomes larger than the generating solution. This implies that our assumed series solution, Eq. 10.3–16, probably will not be convergent; thus our analytical results break down. This is unfortunate, for the largest deviation between the oscillator and steady state performance is obtained when there is a large distortion (see Figure 10.3–8). Nevertheless, the analytical results are still very useful for design purposes because they give us an initial estimate of the direction of the shift in the average operating level from steady state operation. Using this first estimate, we can then pursue a numerical study if desirable.

Extensions of the Methods

Although the approximate analytical solutions developed above are very helpful, they are limited to the case of a first-order chemical reaction (where the equations are linear in one of the state variables) and to cases where there are only two equations describing the system. Consequently, Gaitonde and Douglas[7] and Dorawala and Douglas[8] used matrix techniques to extend the results to arbitrarily large sets of nonlinear, first-order differential equations having a single unstable singular point; that is, there is one pair of complex conjugate roots having positive real parts, but all the remaining eigenvalues correspond to a stable plant. This solution should be applicable to most lumped parameter, chemical engineering processes, and, in particular, it should describe complex reaction mechanisms in perfectly mixed reactors.

The study by Dorawala and Douglas[8] specifically considered simple, consecutive and parallel, reaction schemes in chemical oscillators. The steady state design of the reactor was chosen to correspond to a maximum amount of the intermediate component, but in one case the performance of the oscillating reactor was 20 percent higher than the steady state value. This appears to be a fruitful area for additional research, for there is a large economic incentive for finding methods that will alter the selectivity and yield of complex reactions. Various control systems were used to change the nature of the self-generated oscillations, and it was noted that the improvement passed through a maximum as the controller gain increased. However, discrepancies were sometimes observed between the approximate analytical solutions and the numerical solutions of the system equations; therefore a better theoretical approach is needed.

[7] N. Y. Gaitonde and J. M. Douglas, *AIChE Journal*, **15**, 902 (1969).
[8] Footnote 4, p. 367.

By modifying the perturbation solution and including up to fifth-order terms in the Taylor series expansion of the process nonlinearities (see Eqs. 10.3–11 and 10.3–12), it is possible to develop a solution for multiple limit cycles.[9] In a similar way to the simple procedure we described earlier, the analysis should also provide criteria for the existence of hard oscillators. Unfortunately, however, some sets of system parameters do not satisfy the constraints on the solution, and thus the results are often qualitative in nature. Nevertheless, this approach was used to demonstrate that multiple limit cycles would be observed for a previously published example of the optimum, economic, steady state design of a CSTR, where there was only one stable singular point. It was interesting to note that the oscillating reactor had about a 5 percent better performance than the steady state value.

Of course, it is always dangerous to rely on theoretical predictions of an unusual kind of behavior without experimental verification of the results. Although there have been a number of studies of the dynamic behavior of CSTRs, they usually considered the step or impulse response of a stable system. Even though a good agreement was obtained between the experimental data and the model predictions,[10,11,12] this kind of a test really is not adequate because an asymptotic approach to a steady state is very different from self-generated oscillations. However, continuous oscillations in laboratory reactors have been produced by Hoffman,[13] who studied the decomposition of hydrogen peroxide in an acetic acid solution with ferric ions as a catalyst, by Bush,[14] who studied the vapor phase chlorination of methyl chloride to carbon tetrachloride, and by Baccaro, Gaitonde, and Douglas,[15] who studied the hydrolysis of acetyl chloride in an acetone solvent. These studies help to confirm the validity of the CSTR model.

Positive Feedback Control Systems

Thus far we have limited our attention to chemical oscillators that arise when the steady state design of a plant turns out to be unstable, and we have shown that in some cases these oscillating reactors can yield more profit than the optimum steady state system. As an extension of these results, it seems reasonable to speculate that it also might be possible to improve on the performance of a stable reactor if we could only make it become unstable. Of course, we spent a considerable amount of effort in Chapter 7 learning how to

[9] P. V. Heberling, N. Y. Gaitonde, and J. M. Douglas, *AIChE Journal*, **17**, 1506 (1971).
[10] Footnote 2, p. 43.
[11] Footnote 5, p. 313.
[12] R. I. Kermode and W. F. Stevens, *Can. J. Chem. Eng.*, **43**, No. 2, 68 (1965).
[13] H. Hoffman, *Proceedings of the Third Symposium on Chemical Reaction Engineering*, p. 283, Pergamon Press, London, 1965.
[14] S. F. Bush, *Proc, Roy. Soc.*, **A309**, 1 (1969).
[15] G. P. Baccaro, N. Y. Gaitonde, and J. M. Douglas, *AIChE Journal*, **16**, 249 (1970).

use control systems to make unstable plants become stable, so that now we might examine what would happen if we reversed all the "rules of thumb" we developed previously. For example, suppose that we installed a positive feedback controller on a CSTR, which would increase the flow rate of heating fluid whenever the reactor temperature exceeded the desired value. In other words, whenever the reactor appears to be too hot, we add additional heat and vice versa. Certainly we will be able to make a stable reactor become unstable with this procedure. Then, providing that the system remains bounded, we expect that the reactor will generate periodic outputs, and the time average behavior of the reactor might be superior to the steady state design conditions.

A study of this type was undertaken by Gaitonde and Douglas.[7] They used the procedure described in Example 2.1–2 to establish the optimum steady state design of a stirred-tank reactor. Next, they installed a positive feedback control system, which had a gain sufficiently high that the originally stable reactor became unstable. Both an approximate analytical solution and a numerical solution of the nonlinear system equations were used to demonstrate that the reactor would generate periodic outputs. Shifts in the time average value of the fraction unconverted of 20 to 40 percent were observed, and the optimum operating cost was lowered by an additional 5 percent. Some of the effluent composition oscillations are shown in Figure 10.3–9.

It should be recognized that it might also be possible to make stable systems act like chemical oscillators by installing negative feedback control systems with

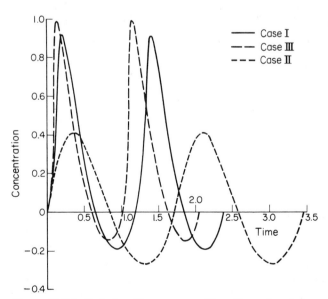

Figure 10.3-9. Periodic effluent composition. [Reproduced from J. M. Douglas and N. Y. Gaitonde, *AIChE Journal*, **15,** 902 (1969), by permission.]

very high gains. In Chapter 7 we noted that a sufficiently high gain on a negative feedback controller would always cause a plant to become unstable, and we made the claim that this would be an undesirable situation. However, now we realize that if the system remains bounded, the unstable condition simply means that the plant will generate periodic outputs. In some cases the time average value of these outputs may be better than the conventional steady state design, although in other situations they may give a worse performance. More research is needed on problems of this type.

Pulsing Chemical Oscillators

In Section 10.1 we found that oscillating the inputs to a plant sometimes would lead to an improved performance (pulsed extraction). The results indicated that we seemed to get the largest effects when the frequency of the forcing signal was close to the resonant frequency of the unit, because in this case the plant tends to amplify the input signal so that the nonlinear phenomena become more pronounced. Of course, we should be able to extend this idea to chemical oscillators. By varying some of the oscillator inputs sinusoidally at frequencies close to the oscillator frequency, it might be possible to obtain an additional resonance effect and in this way achieve an increase in the performance over normal oscillator operation.

Gaitonde[16] developed analytical solutions for this problem. The analysis is quite similar to that discussed previously; therefore the details will be omitted. Sinusoidal variations in feed composition, feed temperature, and feed flow rate were considered, and in almost every case the pulsations caused the size of the limit cycle and the conversion to increase, as compared with the unforced oscillator. The greatest improvement observed was with 5 percent fluctuations in the feed composition, and this gave about a 4 percent change in the effluent composition, as compared to 0.8 percent for the unforced case. As expected, the analytical predictions become very poor as the amplitudes of the input signal and the limit cycle increase.

Self-generated Oscillations in Distributed Parameter Systems

Although the foregoing discussion has been limited to simple lumped parameter processes that act like chemical oscillators, the same kind of behavior can be obtained for distributed parameter plants. For example, Reilly and Schmitz[17] showed that a tubular reactor with a recycle stream could give rise to unique steady state composition and temperature profiles that were unstable (see Section 7.10 and Figure 7.10–8). From physical considerations, we again know that the reactant composition and temperature must be bounded; that is,

[16] N. Y. Gaitonde, Ph.D. Thesis, University of Rochester, Rochester, N.Y., June 1968.
[17] See footnote 5, p. 156.

their values must always lie within the ranges $0 < A < A_f$ and $T_f < T < T_{ad}$ where T_{ad} is the adiabatic temperature rise. Hence it seems reasonable to expect to obtain periodic outputs for this case. This phenomenon was demonstrated[18] when they numerically solved the partial differential equations describing the system for a case where $U_r = 0.79$ and $\theta_f = 2.9$. Their results are shown in Figure 10.3–10. It is important to note that any set of initial conditions will lead to a discontinuous but closed curve of this type. However, the actual curve obtained does depend on the initial conditions, because of the discontinuities present in the system equations and the recycle boundary conditions.

Another example of self-generated oscillations in a distributed parameter process occurs for exothermic reactions on the surface of a spherical particle. Winegardner and Schmitz[19] showed that it was possible to find unique solutions of the steady state equations describing a nonadiabatic system that were unstable. Hence a numerical solution of the dynamic system equations should give a periodic output. In addition, the results can be extended to porous particles,[20] so that we could obtain periodic "effectiveness factors" in certain situations.

Similarly, Sherwin, Shinnar, and Katz[21] used a linear stability analysis to show that a mixed-suspension, mixed-product-removal crystallizer could also generate limit cycles. Then they numerically solved the set of nonlinear

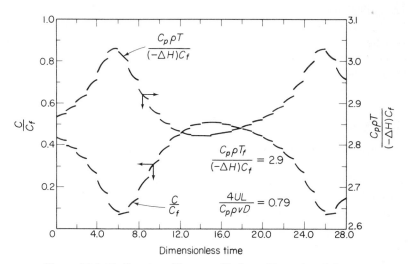

Figure 10.3-10. Reactor effluent oscillations. [Reproduced from M. J. Reilly and R. A. Schmitz, *AIChE Journal*, **13**, 519 (1967), by permission.]

[18] M. J. Reilly and R. A. Schmitz, *AIChE Journal*, **13**, 519 (1967).

[19] D. K. Winegardner and R. A. Schmitz, *AIChE Journal*, **14**, 301 (1968).

[20] J. C. M. Lee and D. Luss, *AIChE Journal*, **16**, 620 (1970).

[21] M. B. Sherwin, R. Shinnar, and S. Katz, *AIChE Journal*, **13**, 1141 (1967) and *Chem. Eng. Prog. Symposium Ser.*, **65**, No. 95, 75 (1969).

dlfferential equations describing the moments of the crystal-size distribution function and the amount of solid in solution. Their results indicate that the period of the oscillation is quite long, of the order of six to eight times greater that the solid drawdown time; thus, in practice, the oscillations might be obscured by other operating changes. Nevertheless, continuous oscillations have been observed in industrial units. Normally the oscillations are considered detrimental, for the product quality specifications often include the uniformity of the crystal product. However, in a case where the total amount of solid produced was of paramount importance, there might be some advantage to periodic operation.

The preceding studies have considered only the stability analysis of distributed parameter plants, and no attempt has been made as yet to compare the performance of the oscillating system with the conventional steady state operation. This should be an active area of research for the next several years.

10.4 OPTIMAL PERIODIC PROCESSES

Now that we have been able to show that periodic processes are sometimes superior to an optimum steady state design, the natural thing to do is look for the optimal periodic process. Of course, it would be even better if we could develop a criterion that would tell us whether or not we could expect an optimal periodic system to be superior to the optimum steady state design, because if an improvement is not possible, we do not want to waste our time studying periodic phenomenon. Fortunately Horn[1] was able to establish a criterion of this type, and therefore his analysis is detailed in this section.

System Equations

Horn considered the problem of finding the optimum steady state temperature and the optimal periodic temperature signal that maximized the yield of component B, produced by the parallel reaction scheme $\alpha A \rightarrow B$ and $A \rightarrow C$ in a stirred-tank reactor. His problem was previously discussed in Section 10.1, where the state equations were written as

$$\frac{dc_A}{dt} = q(c_{Af} - c_A) - Vk_{10}e^{-E_1/RT}c_A^\alpha - Vk_{20}e^{-E_2/RT}c_A \qquad 10.4\text{--}1$$

$$\frac{dc_B}{dt} = -qc_B + Vk_{10}e^{-E_1/RT}c_A^\alpha \qquad 10.4\text{--}2$$

The equations were put into dimensionless form by letting

$$x_1 = \frac{c_A}{c_{Af}}, \quad x_2 = \frac{c_B}{c_{Af}}, \quad u = \frac{(Vk_{10}e^{-E_1/RT}c_{Af}^{\alpha-1})}{q}$$

$$\rho = \frac{E_2}{E_1}, \quad a = \frac{Vk_{20}}{q}\left[\frac{(Vk_{10}c_{Af}^{\alpha-1})}{q}\right]^{-\rho}, \quad \tau = \frac{qt}{V} \qquad 10.4\text{--}3$$

[1] See footnote 3, p. 363.

so that we obtained

$$\frac{dx_1}{d\tau} = 1 - x_1 - ux_1^\alpha - au^p x_1 \qquad 10.4\text{-}4$$

$$\frac{dx_2}{d\tau} = -x_2 + ux_1^\alpha \qquad 10.4\text{-}5$$

The control variable, u, is taken as the rate constant, but it is a monotonic function of temperature.

At steady state conditions, both time derivatives are equal to zero, and we want to find the value of the reactor temperature, or u, that maximizes the yield of component B, or x_2. For the dynamic problem, we look for $u(t)$ as a periodic function of time that maximizes the time average value of $x_2(t)$.

Optimum Steady State Design

In order to generalize the problem somewhat, Horn considered the steady state performance index

$$P = F(x_1, x_2) \qquad 10.4\text{-}6$$

(where for his problem $F = x_2$) and the two state equations

$$f_1(x_1, x_2, u) = 0 \qquad f_2(x_1, x_2, u) = 0 \qquad 10.4\text{-}7$$

which correspond to Eqs. 10.4–4 and 10.4–5 with the time derivatives set equal to zero. He used an implicit differentiation technique to determine the optimum steady state design conditions.

$$\frac{\partial F}{\partial u} = 0 = \frac{\partial F}{\partial x_1}\frac{\partial x_1}{\partial u} + \frac{\partial F}{\partial x_2}\frac{\partial x_2}{\partial u} \qquad 10.4\text{-}8$$

Again, we are going to have to use summation notation in this problem to keep the algebra from becoming unmanageable, and with this approach we write Eq. 10.4–8 as

$$\frac{\partial F}{\partial x_i}\frac{\partial x_i}{\partial u} = 0 \qquad 10.4\text{-}9$$

The terms $\partial x_1/\partial u$ and $\partial x_2/\partial u$ in this expression are unknown; thus we need to develop equations relating these quantities to other known values. This step is accomplished by differentiating the state equations with respect to u. Using the chain rule, we obtain

$$\frac{\partial f_1}{\partial x_1}\frac{\partial x_1}{\partial u} + \frac{\partial f_1}{\partial x_2}\frac{\partial x_2}{\partial u} + \frac{\partial f_1}{\partial u} = 0 \qquad 10.4\text{-}10$$

$$\frac{\partial f_2}{\partial x_1}\frac{\partial x_1}{\partial u} + \frac{\partial f_2}{\partial x_2}\frac{\partial x_2}{\partial u} + \frac{\partial f_2}{\partial u} = 0 \qquad 10.4\text{-}11$$

or, in summation notation, we can write both equations as

$$\frac{\partial f_i}{\partial x_j}\frac{\partial x_j}{\partial u} + \frac{\partial f_i}{\partial u} = 0 \qquad 10.4\text{-}12$$

From Eqs. 10.4–10 and 10.4–11 it is clear that we have two equations for the two unknowns $\partial x_1/\partial u$ and $\partial x_2/\partial u$. In matrix form, these become

$$
\begin{pmatrix}
\dfrac{\partial f_1}{\partial x_1} & \dfrac{\partial f_1}{\partial x_2} \\[2mm]
\dfrac{\partial f_2}{\partial x_1} & \dfrac{\partial f_2}{\partial x_2}
\end{pmatrix}
\begin{pmatrix}
\dfrac{\partial x_1}{\partial u} \\[2mm]
\dfrac{\partial x_2}{\partial u}
\end{pmatrix}
= -
\begin{pmatrix}
\dfrac{\partial f_1}{\partial u} \\[2mm]
\dfrac{\partial f_2}{\partial u}
\end{pmatrix}
\qquad 10.4\text{–}13
$$

Providing that the coefficient matrix is not singular, we can find its inverse

$$
\mathbf{A} =
\begin{pmatrix}
\dfrac{\partial f_1}{\partial x_1} & \dfrac{\partial f_1}{\partial x_2} \\[2mm]
\dfrac{\partial f_2}{\partial x_1} & \dfrac{\partial f_2}{\partial x_2}
\end{pmatrix}^{-1}
= \frac{1}{(\partial f_1/\partial x_1)(\partial f_2/\partial x_2) - (\partial f_2/\partial x_1)(\partial f_1/\partial x_2)}
\begin{pmatrix}
\dfrac{\partial f_2}{\partial x_2} & -\dfrac{\partial f_1}{\partial x_2} \\[2mm]
-\dfrac{\partial f_2}{\partial x_1} & \dfrac{\partial f_1}{\partial x_1}
\end{pmatrix}
$$

$$
=
\begin{pmatrix}
A_{11} & A_{12} \\
A_{21} & A_{22}
\end{pmatrix}
\qquad 10.4\text{–}14
$$

so that the solution of the simultaneous equations in matrix form is

$$
\begin{pmatrix}
\dfrac{\partial x_1}{\partial u} \\[2mm]
\dfrac{\partial x_2}{\partial u}
\end{pmatrix}
= -
\begin{pmatrix}
A_{11} & A_{12} \\
A_{21} & A_{22}
\end{pmatrix}
\begin{pmatrix}
\dfrac{\partial f_1}{\partial u} \\[2mm]
\dfrac{\partial f_2}{\partial u}
\end{pmatrix}
\qquad 10.4\text{–}15
$$

Thus we see that the term $\partial f_i/\partial x_j$ in Eq. 10.4–12 represents a matrix and that when we solve this equation for $\partial x_j/\partial u$ we obtain

$$
\frac{\partial x_j}{\partial u} = -A_{ji}\,\frac{\partial f_i}{\partial u}
$$

or since i and j are interchangeable

$$
\frac{\partial x_i}{\partial u} = -A_{ij}\,\frac{\partial f_j}{\partial u}
\qquad 10.4\text{–}16
$$

where A_{ij} are the elements of the matrix $\mathbf{A}$ given by Eq. 10.4–14. Substituting this result into Eq. 10.4–9, we obtain the necessary conditions for a maximum —that is, an equation we can solve for the optimum temperature.

$$
\frac{\partial F}{\partial x_i}\frac{\partial x_i}{\partial u} = -\frac{\partial F}{\partial x_i}A_{ij}\frac{\partial f_j}{\partial u} = 0
\qquad 10.4\text{–}17
$$

In order to ensure that this solution actually corresponds to a maximum, we require that

$$
\frac{\partial^2 F}{\partial u^2} \le 0
$$

or

$$
\frac{\partial}{\partial u}\!\left(\frac{\partial F}{\partial u}\right) = \frac{\partial^2 F}{\partial x_i \partial x_j}\frac{\partial x_i}{\partial u}\frac{\partial x_j}{\partial u} + \frac{\partial F}{\partial x_i}\frac{\partial^2 x_i}{\partial u^2} \le 0
\qquad 10.4\text{–}18
$$

where we substituted Eq. 10.4–8 for $\partial F/\partial u$ in the equation above. All the terms in this expression are available except $\partial^2 x_i/\partial u^2$, and in order to develop relationships for these terms, we must take the second derivatives of the state equations with respect to u.

$$\frac{\partial^2 f_i}{\partial u^2} = \frac{\partial}{\partial u}\left(\frac{\partial f_i}{\partial u}\right) = \frac{\partial}{\partial u}\left(\frac{\partial f_i}{\partial x_j}\frac{\partial x_j}{\partial u} + \frac{\partial f_i}{\partial u}\right)$$

$$= \frac{\partial^2 f_i}{\partial x_j \partial x_k}\frac{\partial x_j}{\partial u}\frac{\partial x_k}{\partial u} + 2\frac{\partial^2 f_i}{\partial x_j \partial u}\frac{\partial x_j}{\partial u} + \frac{\partial f_i}{\partial x_j}\frac{\partial^2 x_j}{\partial u^2} + \frac{\partial^2 f_i}{\partial u^2} = 0 \qquad 10.4\text{-}19$$

As we let $i = 1$ and $i = 2$, this expression will generate two equations in the two unknowns $\partial^2 x_1/\partial u^2$ and $\partial^2 x_2/\partial u^2$. However, if we arrange the equation in matrix form, the coefficient matrix will have elements $\partial f_i/\partial x_j$, which is similar to the result we obtained for Eq. 10.4-12. Therefore we can immediately write the solution of this equation as

$$\frac{\partial^2 x_j}{\partial u^2} = -A_{ji}\left(\frac{\partial^2 f_i}{\partial u^2} + 2\frac{\partial^2 f_i}{\partial x_k \partial u}\frac{\partial x_k}{\partial u} + \frac{\partial^2 f_i}{\partial x_k \partial x_l}\frac{\partial x_k}{\partial u}\frac{\partial x_l}{\partial u}\right) \qquad 10.4\text{-}20$$

Now we can substitute this expression into Eq. 10.4-18 and use Eq. 10.4-16 to eliminate the first derivative terms. Thus the condition for a maximum becomes

$$\frac{\partial F}{\partial x_k}A_{ki}\frac{\partial^2 f_i}{\partial u^2} - 2\frac{\partial F}{\partial x_i}\frac{\partial^2 f_j}{\partial x_k \partial u}\frac{\partial f_l}{\partial u}A_{ij}A_{kl} + \frac{\partial F}{\partial x_i}\frac{\partial^2 f_j}{\partial x_k \partial x_l}\frac{\partial f_m}{\partial u}$$

$$\times \frac{\partial f_n}{\partial u}A_{ij}A_{km}A_{ln} - \frac{\partial^2 F}{\partial x_i \partial x_j}\frac{\partial f_k}{\partial u}\frac{\partial f_l}{\partial u}A_{ik}A_{jl} \geq 0 \qquad 10.4\text{-}21$$

where we must sum from one to two over all repeated subscripts. This result makes it painfully obvious why we needed summation notation.

In certain cases the foregoing expression can be greatly simplified. For example, whenever

1.

$$f_i(x_1, x_2, u) = g_{i1}(x_1, x_2) + g_{i2}(u)$$

so that the state and control variables are separable in the state equation, then

$$\frac{\partial F}{\partial x_i}\frac{\partial^2 f_j}{\partial x_k \partial u}\frac{\partial f_l}{\partial u}A_{ij}A_{kl} = 0$$

2.

$$f_i(x_1, x_2, u) = g_{i1}(u)x_1 + g_{i2}(u)x_2$$

so that the state equations are linear in the state variables, then

$$\frac{\partial F}{\partial x_i}\frac{\partial^2 f_j}{\partial x_k \partial x_l}\frac{\partial f_m}{\partial u}\frac{\partial f_n}{\partial u}A_{ij}A_{km}A_{ln} = 0$$

3.

$$F(x_1, x_2) = c_1 x_1 + c_2 x_2$$

so that the performance index is a linear function of the state variables, then

$$\frac{\partial^2 F}{\partial x_i \partial x_j}\frac{\partial f_k}{\partial u}\frac{\partial f_l}{\partial u}A_{ik}A_{jl} = 0$$

Example 10.4–1   Optimum operating temperature for parallel reactions

If we apply the preceding method to the problem of parallel reactions in a CSTR, we have

$$F = x_2, \quad f_1 = 1 - x_1 - ux_1^\alpha - au^\rho x_1 = 0, \quad f_2 = -x_2 + ux_1^\alpha = 0$$

Thus

$$\frac{\partial F}{\partial x_1} = 0, \quad \frac{\partial F}{\partial x_2} = 0, \quad \frac{\partial f_1}{\partial x_1} = -(1 + \alpha ux_1^{\alpha-1} + au^\rho)$$

$$\frac{\partial f_1}{\partial x_2} = 0, \quad \frac{\partial f_1}{\partial u} = -(x_1^\alpha + a\rho u^{\rho-1}x_1)$$

$$\frac{\partial f_2}{\partial x_1} = \alpha ux_1^{\alpha-1}, \quad \frac{\partial f_2}{\partial x_2} = -1, \quad \frac{\partial f_2}{\partial u} = x_1^\alpha$$

and the inverse matrix becomes

$$\mathbf{A} = \frac{1}{1 + \alpha ux_1^{\alpha-1} + au^\rho}\begin{pmatrix} -1 & 0 \\ -\alpha ux_1^{\alpha-1} & -(1 + \alpha ux_1^{\alpha-1} + au^\rho) \end{pmatrix}$$

Expanding Eq. 10.4–17 gives the expression

$$0 = \frac{\partial F}{\partial x_1} A_{11} \frac{\partial f_1}{\partial u} + \frac{\partial F}{\partial x_2} A_{21} \frac{\partial f_1}{\partial u} + \frac{\partial F}{\partial x_1} A_{12} \frac{\partial f_2}{\partial u} + \frac{\partial F}{\partial x_2} A_{22} \frac{\partial f_2}{\partial u}$$

or after substituting the relationships above, we obtain

$$\frac{\alpha ux_1^{\alpha-1}(x_1^\alpha + a\rho u^{\rho-1}x_1) - (1 + \alpha ux_1^{\alpha-1} + au^\rho)x_1^\alpha}{1 + \alpha ux_1^{\alpha-1} + au^\rho} = 0$$

or

$$x_1^\alpha[1 + (1 - \alpha\rho)au^\rho] = 0$$

Since, in general, $x_1 \neq 0$ (otherwise we would have complete conversion of the reactant), the optimum value of u, which is monotonically related to the reactor temperature, is given by the solution of the equation

$$u^\rho = \frac{1}{(\alpha\rho - 1)a}$$

It is clear from this expression that an optimum temperature can never exist unless $\alpha\rho > 1$.

In order to show that this solution corresponds to a maximum, we must evaluate the terms in Eq. 10.4–21.

$$\frac{\partial F}{\partial x_k} A_{ki} \frac{\partial^2 f_i}{\partial u^2} = \frac{(\alpha ux_1^{\alpha-1})[a\rho(\rho-1)u^{\rho-2}x_1]}{1 + \alpha ux_1^{\alpha-1} + au^\rho}$$

$$-2\frac{\partial F}{\partial x_i} \frac{\partial^2 f_j}{\partial x_k \partial u} \frac{\partial f_l}{\partial u} A_{ij} A_{kl} = \frac{2(\alpha x_1^{\alpha-1})(x_1^\alpha + a\rho u^{\rho-1}x_1)[1 - (\rho-1)au^\rho]}{(1 + \alpha ux_1^{\alpha-1} + au^\rho)^2}$$

$$\frac{\partial F}{\partial x_i} \frac{\partial^2 f_j}{\partial x_k \partial x_i} \frac{\partial f_m}{\partial u} \frac{\partial f_n}{\partial u} A_{ij} A_{km} A_{ln} = \frac{-[\alpha(\alpha-1)ux_1^{\alpha-1}](x_1^\alpha + a\rho u^{\rho-1}x_1)^2(1 + au^\rho)}{(1 + \alpha ux_1^{\alpha-1} + au^\rho)^3}$$

$$\frac{\partial^2 F}{\partial x_i \partial x_j} \frac{\partial f_k}{\partial u} \frac{\partial f_l}{\partial u} A_{ik} A_{jl} = 0$$

From the solution of the optimization equation, we know that

$$1 + au^p = \alpha a p u^p$$

so that we can write

$$1 + \alpha u x_1^{\alpha-1} + au^p = \alpha u(x_1^{\alpha-1} + apu^{p-1})$$

Summing the three terms, using the expressions above to simplify the results, and dividing through by common factors, we obtain

$$\alpha p + \alpha - 2 - (1 - \alpha)x_1 \geq 0$$

as the criterion for the optimum valve of u to correspond to a maximum. For a case where $\alpha = 2.0$, $p = \frac{3}{4}$, and $0 < x_1 < 1$, we see that the inequality is satisfied.

Optimal Periodic Process

From our previous experience, we expect that if we vary the control variables periodically, the state variables might go through an initial transient, but eventually they will become periodic and have the same period as the driving signal. Thus we define a periodic process as one where we find that

$$x(0) = x(\tau) \qquad u(0) = u(\tau) \qquad\qquad 10.4\text{–}22$$

where τ is the period of the oscillation, or some integral multiple of the period. The system equations are taken as

$$\frac{dx_1}{dt} = f_1(x_1, x_2, u) \qquad \frac{dx_2}{dt} = f_2(x_1, x_2, u) \qquad 10.4\text{–}23$$

and we consider a time-average performance index

$$P = \frac{1}{\tau} \int_0^\tau F(x_1, x_2)\, dt \qquad\qquad 10.4\text{–}24$$

For a case where we have steady state operation, $F(x_1, x_2)$ will be independent of time, and Eq. 10.4–24 reduces to the form $P = F(x_1, x_2)$, which is simply the result we used in the steady analysis (see Eq. 10.4–6).

Now, what we are looking for is the periodic function of time, $u(t)$, which maximizes the performance index given by Eq. 10.4–24. The procedure we use is exactly the same as before: selecting some periodic control function $\bar{u}(t)$, solving the state equations for $\bar{x}_1$ and $\bar{x}_2$, and using the results to evaluate the performance index. Then we make small changes in $u(t)$ to $u = \bar{u} + \delta u$, repeat the preceding steps, and, finally, compare the two cases. Since it turns out that we will need a result for the second derivative, we will give some of the details of the derivation.

When we compare the state equations for the two control functions and expand the difference in a Taylor series, keeping up to second-order terms, we obtain

$$\frac{d\delta x_i}{dt} = \frac{\partial f_i}{\partial x_j}\,\delta x_j + \frac{\partial f_i}{\partial u}\,\delta u + \frac{1}{2}\frac{\partial^2 f_i}{\partial x_j \partial x_k}\,\delta x_j \delta x_k$$

$$+ \frac{\partial^2 f_i}{\partial u\, \partial x_k}\,\delta u\,\delta x_k + \frac{1}{2}\frac{\partial^2 f_i}{\partial u^2}\,\delta u^2 + \cdots \qquad 10.4\text{--}25$$

The difference between the performance indices for the two cases, where we consider that the period of $\bar{u} + \delta u$ is $\tau + \delta\tau$, becomes

$$\delta P = \frac{1}{\tau}\int_0^\tau \left(\frac{\partial F}{\partial x_j}\,\partial x_j\right)dt + \frac{1}{\tau}\left[F(\tau) - \frac{1}{\tau}\int_0^\tau F\,dt\right]\delta\tau$$

$$+ \frac{1}{2}\frac{1}{\tau}\int_0^\tau \left(\frac{\partial^2 F}{\partial x_i \partial x_j}\,\delta x_i\,\delta x_j\right)dt + \frac{1}{\tau^2}\left[\frac{1}{\tau}\int_0^\tau F\,dt - F(\tau)\right]\delta\tau^2$$

$$+ \frac{1}{\tau}\left[\left(\frac{\partial F}{\partial x_i}\,\delta x_i\right)\Big|_\tau - \frac{1}{\tau}\int_0^\tau \frac{\partial F}{\partial x_j}\,\delta x_j\,dt\right]\delta\tau \qquad 10.4\text{--}26$$

Multiplying the state equations by $\lambda_i(t)$, integrating the results from zero to τ, and adding them to Eq. 10.4–26 gives an expression that is similar to those we obtained before. Integration by parts can be used to show that

$$\frac{1}{\tau}\int_0^\tau \lambda_i \frac{d\delta x_i}{dt}\,dt = \frac{1}{\tau}(\lambda_i\,\delta x_i)\Big|_0^\tau - \frac{1}{\tau}\int_0^\tau \delta x_i \frac{d\lambda_i}{dt}\,dt \qquad 10.4\text{--}27$$

For convenience, we let

$$H = F + \lambda_i f_i \qquad 10.4\text{--}28$$

so that the equation for δP becomes

$$\delta P = \frac{1}{\tau}[\lambda_i(0)\,\delta x_i(0) - \lambda_i(\tau)\,\delta x_i(\tau)] + \frac{1}{\tau}\int_0^\tau \left(\frac{d\lambda_i}{dt} + \frac{\partial H}{\partial x_i}\right)\delta x_i\,dt$$

$$+ \frac{1}{\tau}\int_0^\tau \frac{\partial H}{\partial u_i}\,\delta u_i\,dt + \frac{1}{2}\frac{1}{\tau}\int_0^\tau \left(\frac{\partial^2 H}{\partial x_i \partial x_j}\,\delta x_i\,\delta x_j + \frac{\partial^2 H}{\partial u^2}\,\delta u^2\right.$$

$$\left.+ \frac{\partial^2 H}{\partial x_i \partial u}\,\delta x_i\,\delta u\right)dt + \frac{1}{\tau}[F(\tau) - P] + \frac{1}{\tau^2}[P - F(\tau)]\,\delta\tau^2$$

$$+ \frac{1}{\tau}\left[\frac{\partial F(\tau)}{\partial x_i}\,\delta x_i\right]\delta\tau + \frac{1}{\tau^2}[\lambda_i(\tau)\,\delta x_i(\tau) - \lambda_i(0)\,\delta x_i(0)]\,\delta\tau$$

$$- \frac{1}{\tau^2}\left[\int_0^\tau \left(\frac{d\lambda_i}{dt} + \frac{\partial H}{\partial x_i}\right)\delta x_i\,dt + \int_0^\tau \frac{\partial H}{\partial u}\,\delta u\,dt\right]\delta\tau \qquad 10.4\text{--}29$$

If we define the adjoint variables, except for their boundary conditions, so that

$$\frac{d\lambda_i}{dt} = -\frac{\partial H}{\partial x_i} \qquad 10.4\text{--}30$$

we can simplify Eq. 10.4–29.

We want the process to be periodic after we make the change from $\bar{u}(t)$, where $\bar{u}(0) = \bar{u}(\tau)$, to the new control $\bar{u} + \delta u$, where $(\bar{u} + \delta u)|_0 = (\bar{u} + \delta u)|_{\tau + \delta\tau}$. This means that we require that the change from $\bar{x}_i(t)$, where $\bar{x}_i(0) = \bar{x}_i(\tau)$, to

$\bar{x}_i + \delta x_i$ is also periodic—that is, that

$$(\bar{x}_i + \delta x_i)|_0 = (\bar{x}_i + \delta x_i)|_{\tau + \delta \tau} \qquad 10.4\text{-}31$$

Using a Taylor series expansion, we see that

$$x_i(t + \delta t) \simeq x_i(t) + \left(\frac{dx_i}{dt}\Big|_t\right) \delta t + \frac{1}{2}\left(\frac{d^2 x_i}{dt^2}\Big|_t\right) \delta \tau^2 \qquad 10.4\text{-}32$$

or for the case of interest where $t = \tau$ and $x_i = \bar{x}_i + \delta x_i$,

$$x_i(\tau + \delta \tau) = \bar{x}_i(\tau) + \delta x_i(\tau) + \left(f_i + \frac{d\delta x_i}{dt}\right)\Big|_\tau \delta t + \frac{1}{2}\frac{d^2(\bar{x}_i + \delta x_i)}{dt^2}\Big|_\tau \delta \tau^2$$

$$10.4\text{-}33$$

We can use Eq. 10.4–25 to approximate the time derivatives in this expression and then substitute the result into the periodic boundary condition, Eq. 10.4–31. This leads to a relationship between $\delta x_i(0)$ and $\delta x_i(\tau)$, since $\bar{x}_i(0) = \bar{x}_i(\tau)$; thus Eq. 10.4–29 can be written

$$\delta P = \frac{1}{\tau} [\lambda_i(0) - \lambda_i(\tau)] \, \delta x_i(0) + \frac{1}{\tau} \int_0^\tau \frac{\partial H}{\partial u_j} \delta u_j \, dt + \frac{1}{\tau} [H(\tau) - P] \, \delta \tau$$

$$+ \frac{1}{2}\frac{1}{\tau}\int_0^\tau \left(\frac{\partial^2 H}{\partial x_i \partial x_j} \delta x_i \, \delta x_j + \frac{\partial^2 H}{\partial u^2} \delta u^2 + 2\frac{\partial^2 H}{\partial x_i \partial u} \delta x_i \delta u\right) dt$$

$$+ \frac{1}{\tau}\left[\frac{\partial H(\tau)}{\partial x_i} \delta x_i \, \delta \tau + \frac{\partial H(\tau)}{\partial u} \delta u \, \delta \tau\right]$$

$$- \frac{1}{2}\frac{1}{\tau}\lambda_i(\tau)\left[\left(\frac{\partial f_i}{\partial x_j} f_j + \frac{\partial f_i}{\partial u}\right)\Big|_\tau\right] \delta \tau^2$$

$$- \frac{1}{\tau^2}[H(\tau) - P(\tau)] \, \delta \tau^2 + \frac{1}{\tau^2}\left(\int_0^\tau \frac{\partial H}{\partial u} \delta u \, dt\right) \delta \tau \qquad 10.4\text{-}34$$

By picking the boundary conditions on the adjoint variables as

$$\lambda_i(0) = \lambda_i(\tau) \qquad 10.4\text{-}35$$

so that the adjoint variables become periodic functions of time, we can remove the first term in the equation for δP. Also, by considering only the first variations and looking for the necessary conditions for a stationary value of P, we find that

$$\frac{\partial H}{\partial u} = 0 \qquad H(\tau) = P \qquad 10.4\text{-}36$$

Hence the expression for δP in the neighborhood of a stationary point becomes

$$\delta P = \frac{1}{2}\frac{1}{\tau}\int_0^\tau \left(\frac{\partial^2 H}{\partial x_i \partial x_j} \delta x_i \, \delta x_j + \frac{\partial^2 H}{\partial u^2} \delta u^2 + 2\frac{\partial^2 H}{\partial x_i \partial u} \delta x_i \, \delta u\right) dt$$

$$+ \frac{1}{\tau}\left[\frac{\partial H(\tau)}{\partial x_i} \delta x_i \, \delta \tau\right] - \frac{1}{2}\frac{1}{\tau}\lambda_i(\tau)\left[\left(\frac{\partial f_i}{\partial x_j} f_j + \frac{\partial f_i}{\partial u} \dot{u}\right)\Big|_\tau\right] \delta \tau^2$$

$$10.4\text{-}37$$

For the case where δP is a maximum, we require that $\delta P \leq 0$. However, we are only free to vary $u(t)$ and the period. Unfortunately the preceding equation contains the variations in x_i as well. Therefore it is necessary to develop an expression for the order of magnitude of δx_i. This can be accomplished by using a Lipschitz condition (see Rozenoer[2]) to show that

$$|\delta x_i(t)| \leq Kne^{Knt} \sum_1^r |\delta u^*| \, \Delta t \qquad\qquad 10.4\text{--}38$$

where K is the maximum of the Lipschitz constants for the functions, $n = 2$ for our problem, δu^* refers to a special constant variation in the control variable, and Δt is an arbitrary time interval over which we impose the control δu^*. In other words, we find that δx_i is of the order Δt, so that $\delta x_i \, \delta x_j$ will be of the order Δt^2 and $\delta x_i \, \delta u_j$ will be of the order Δt. Hence if we make Δt small enough, these terms will be negligible in comparison to δu^2 and they can be neglected. Similarly, $\delta x \, \delta \tau$ will be negligible in comparison to $\delta u \, \delta \tau$. With these simplifications, our expression for δP becomes

$$\delta P = \frac{1}{\tau} \int_0^\tau \frac{1}{2} \frac{\partial^2 H}{\partial u^2} \delta u^2 dt - \frac{1}{2} \frac{1}{\tau} \lambda_i(\tau) \left[\frac{\partial f_i}{\partial x_j} f_j + \frac{\partial f_i}{\partial u} \dot{u} \right)\Big|_\tau \right] \delta \tau^2 \qquad 10.4\text{--}39$$

Now, clearly, in order for P to be a maximum, $\delta P \leq 0$, we must have

$$\frac{\partial^2 H}{\partial u^2} \leq 0 \qquad\qquad 10.4\text{--}40$$

and

$$\lambda_i \left(\frac{\partial f_i}{\partial x_j} f_j + \frac{\partial f_i}{\partial u} \dot{u} \right) \geq 0, \qquad \text{at } t = \tau \qquad 10.4\text{--}41$$

The analogous result for a multivariable control system is that the Hessian of the Hamiltonian is negative definite, which, again, is similar to the result obtained in classical calculus. The second expression can be simplified somewhat, because we know that

$$\frac{\partial H}{\partial u} = 0$$

and

$$\frac{d}{dt}\left(\frac{\partial H}{\partial u} \right) = 0 = \frac{\partial}{\partial x_i}\left(\frac{\partial H}{\partial u} \right)\dot{x}_i + \frac{\partial}{\partial \lambda_i}\left(\frac{\partial H}{\partial u} \right)\dot{\lambda}_i + \frac{\partial}{\partial u}\left(\frac{\partial H}{\partial u} \right)\dot{u}$$

so that

$$\dot{u} = \frac{-[\partial(\partial H/\partial u)/\partial x_i] - [\partial(\partial H/\partial u)/\partial \lambda_i]\partial H/\partial x_i}{\partial^2 H/\partial u^2} \qquad 10.4\text{--}42$$

Also, since the starting time is arbitrary, Eq. 10.4–41 must be valid for every value of t and not just $t = \tau$.

[2] L. Rozenoer, *Autom. Remote Control*, **20**, 1288, 1405, 1517 (1959).

Application to Steady State Processes

In the foregoing development we have established the conditions for an optimal periodic process. However, it is apparent that we can consider any steady state process to be a periodic process simply by taking an arbitrary time interval of the constant input and output signals as the period. Therefore the preceding equations must also apply to steady state systems, if they are truly optimal from both a steady state and a periodic point of view. For this case, the optimization equations are

$$\frac{\partial H}{\partial u} = 0 = \frac{\partial H}{\partial u} + \lambda_1 \frac{\partial f_1}{\partial u} + \lambda_2 \frac{\partial f_2}{\partial u} \qquad 10.4\text{--}43$$

$$\frac{d\lambda_j}{dt} = -\frac{\partial H}{\partial x_j} = -\left(\frac{\partial F}{\partial x_j} + \lambda_i \frac{\partial f_i}{\partial x_j}\right) = 0 \qquad 10.4\text{--}44$$

where we have recognized that $\lambda_i(t)$ must be independent of time for a steady state system, a condition that automatically satisfies the boundary conditions on the adjoint variables, Eq. 10.4–35. With our formulation of the problem, F is independent of u so that $\partial F/\partial u = 0$.

Equation 10.4–44 provides two equations for the two unknown values of λ_1 and λ_2. If we write the equation in matrix form, then the elements of the coefficients matrix for the unknown λ vector will be $\partial f_i/\partial x_j$, so that the inverse of this matrix is just the matrix A given by Eq. 10.4–14. Thus, in summation notation, the solution of Eq. 10.4–44 is

$$\lambda_i = -\frac{\partial F}{\partial x_j} A_{ji} \qquad 10.4\text{--}45$$

Now if we write Eq. 10.4–43 in summation notation

$$\lambda_i \frac{\partial f_i}{\partial u} = 0 \qquad 10.4\text{--}46$$

and substitute Eq. 10.4–45, we obtain

$$-\frac{\partial F}{\partial x_i} A_{ij} \frac{\partial f_j}{\partial u} = 0 \qquad 10.4\text{--}47$$

where we have interchanged the subscripts. This result is identical to Eq. 10.4–17; therefore the necessary conditions obtained by using an optimum steady state analysis and those obtained by considering the steady state case of an optimal periodic analysis are the same.

Fortunately the results based on the second derivative condition are different, so that we can learn something about steady state versus periodic systems. According to Eq. 10.4–40, the condition for the performance index to be a maximum for a periodic process is

$$\frac{\partial^2 H}{\partial u^2} = \frac{\partial^2 F}{\partial u^2} + \lambda_i \frac{\partial^2 f_i}{\partial u^2} = \lambda_i \frac{\partial^2 f_i}{\partial u^2} \leq 0 \qquad 10.4\text{--}48$$

For the steady state case of a periodic process, we can substitute Eq. 10.4-45 to eliminate λ_i and obtain

$$- \frac{\partial F}{\partial x_i} A_{ij} \frac{\partial^2 f_j}{\partial u^2} \leq 0 \qquad\qquad 10.4\text{-}49$$

where, again, we have interchanged the subscripts i and j. This result must be compared with Eq. 10.4-21, and we find that it is merely the first term in the more complicated expression.

Suppose that we now consider a problem where Eq. 10.4-21 is satisfied but Eq. 10.4-49 is not. This means that the optimum steady state design gives its maximum performance, but this steady state system is not the one that gives the maximum periodic performance. Thus it might be possible to find a periodic system, the one that satisfies Eq. 10.4-49 as well as Eq. 10.4-47, which is superior to the optimum steady state design. Of course, it is possible to extend the results to cases where the instantaneous performance index depends on the control variable, as well as the state variables, and to problems where more than one control variable present.

Example 10.4-2   Optimal periodic temperature variations for parallel reactions

If we return to the problem discussed in Example 10.4-1, we find that the condition for a steady state process to be an optimum periodic process

$$+ \frac{\partial F}{\partial x_i} A_{ij} \frac{\partial^2 f_j}{\partial u^2} \geq 0 \qquad\qquad 10.4\text{-}49$$

becomes

$$\frac{(\alpha u x_1^{\alpha-1})[a\rho(\rho-1)u^{\rho-2}x_1]}{1 + \alpha u x_1^{\alpha-1} + a u^\rho} \geq 0$$

This condition will be violated if $\rho < 1$. Hence if we consider a problem where $\alpha\rho > 1$ and $\rho < 1$, we should be able to find a periodic system that is superior to the optimum steady state design.

Horn and Lin consider a case where $\alpha = 2.0$, $\rho = 0.75$, and $a = 1.0$, and used a general computing algorithm to solve the optimization equations describing the dynamic[1] periodic system, Eqs. 10.4-22, 10.4-23, 10.4-24, 10.4-30, 10.4-35, and 10.4-36. Some of their results are shown in Figure 10.4-1. It is clear from this graph that the time average performance of the periodic system is superior to the optimum steady state conditions. However, it should be noted that the optimal periodic control is a "bang-bang" policy with infinitely fast switching between the maximum and minimum values of the control variables; that is, the optimal period is zero. Although this result is somewhat unrealistic,

[3] J. E. Bailey, F. J. M. Horn, and R. C. Lin. Paper presented at the 63rd National AIChE Meeting, St. Louis, Mo., February 1968.

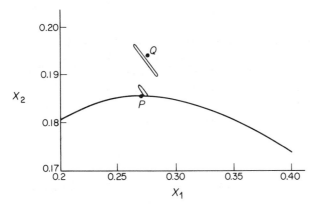

Figure 10.4-1. Process yield. [Reproduced from F. J. M. Horn and R. C. Lin, *Ind. Eng. Chem. Proc. Design Develop.*, **6**, 21 (1967), by permission of the ACS.]

a later study by Bailey, Horn, and Lin[3] showed that if the heat flux to the reactor is used as the control variable, rather than the reactor temperature, the periodic process is still superior to the optimum steady state case, but the addition of the energy balance leads to a solution with a finite period.

A later investigation by Fjeld[4] demonstrated that phase plane techniques are very useful for determining the optimal periodic control of second-order systems. Using Horn's parameters and restricting the control variable to the range $1 \leq u \leq 5$, he found that the optimal periodic yield was 0.195, compared to the maximum steady state yield of 0.185, an improvement of about 5 percent. However, for a case where $0.2 < u < 5$, a 12 percent improvement is possible, and for the same bounds on the control but a case where $a = 2.0$, the improvement is 25 percent. Fjeld also considered a pair of first-order consecutive reactions with $p = 0.75$, $a = 1$, $0.2 < u < 10$, and obtained a 17 percent improvement over the optimum steady state case. He studied a few other cases for different reaction orders and other parameters and observed an improvement in every case, although they were smaller than the 17 percent figure.

It should be noted that Horn's criterion repesents a kind of sufficient condition for obtaining an improvement by periodic operation. This is due to the fact that Eq. 10.4-49 is based on Eq. 10.4-44, and this last expression does not apply to the actual dynamic system. Thus there might be cases where Eq. 10.4-49 is satisfied, but some periodic system still is superior to the optimum steady state design. In fact, a few examples of this type have been described by Fjeld. Nevertheless, the criterion provides a useful starting point for gaining a better understanding of the differences between steady state and periodic operation of processes.

[4] M. Fjeld, "On Periodic Processes," *Report 67-54-D*, Division of Automatic Control, The Technical University of Norway, 1967.

Distributed Parameter Systems

Unfortunately, a general criterion that can be used to predict whether or not the optimal periodic performance of a distributed parameter plant will exceed the optimum steady state design is not available at the present time. However, specific problems have been studied which indicate that periodic operation is sometimes advantageous. Two examples of this type are described below. The necessary conditions for the optimal periodic control are obtained by combining the analysis for periodic systems described earlier with the analysis for distributed parameter systems presented in Section 9.8.

Example 10.4–3   Periodic operation of a jacketed tubular reactor

Chang and Bankoff[5] considered the periodic operation of a pair of first-order consecutive reactions in a jacketed tubular reactor. The system equations were taken as

$$\frac{\partial c_A}{\partial t} = -u\frac{\partial c_A}{\partial z} - k_1 c_A \qquad\qquad 10.4\text{-}50$$

$$\frac{\partial c_B}{\partial t} = -u\frac{\partial c_B}{\partial t} + k_1 c_A - k_2 c_B \qquad\qquad 10.4\text{-}51$$

$$\frac{\partial T}{\partial t} = -u\frac{\partial T}{\partial z} + \frac{(-\Delta H_1)}{C_p \rho} k_1 c_A + \frac{(-\Delta H_2)}{C_p \rho} k_2 c_A + \frac{2U}{C_p \rho r}(T_c - T) \qquad 10.4\text{-}52$$

where r is the tube radius, and the condensate temperature, which only depends upon time, is considered to be the control variable—that is, $u = $ constant and $T_c = v(t)$. For the optimum steady state design problem, they considered that the feed compositions were constant, and they determined the value of the condensate temperature that maximized the yield of component B. Then for the periodic process, they assumed that the feed compositions were fluctuating sinusoidally

$$c_{Af} = c_{Afs} + A\sin \omega t, \quad c_{Rf} = c_{Bfs} - (c_{Afs} + A\sin \omega t), \quad T_f = T_{fs}$$

$$10.4\text{-}53$$

and looked for the jacket temperature as a function of time which would maximize the amount of component B produced at the reactor outlet during one period. Thus

$$P = \int_0^\tau - c_B(t, L)dt = \int_0^\tau \int_0^L -\frac{\partial c_B}{\partial z} dz \, dt \qquad\qquad 10.4\text{-}54$$

where L is the total reactor length and the negative sign is included in the performance index because they were looking for the control that minimized P.

[5] K. S. Chang and S. G. Bankoff, *Ind. Eng. Chem. Fundamentals*, **7**, 633 (1968).

With their formulation of the optimization equations, the Hamiltonian became

$$H = \frac{\partial c_B}{\partial z} + \lambda_1\left(-u\frac{\partial c_A}{\partial z} - k_1 c_A\right) + \lambda_2\left(-u\frac{\partial c_B}{\partial z} + k_1 c_A - k_2 c_B\right)$$

$$+ \lambda_3\left\{-u\frac{\partial T}{\partial z} + \frac{(-\Delta H_1)}{C_p \rho}k_1 c_A + \frac{(-\Delta H_2)}{C_p \rho}k_2 c_B + \frac{2U}{C_p \rho r}[v(t) - T]\right\}$$

$$\text{10.4--55}$$

so that the adjoint equations were

$$\frac{\partial \lambda_i}{\partial t} = -u\frac{\partial \lambda_1}{\partial z} + k_1 \lambda_1 + k_1 \lambda_2 - \frac{(-\Delta H_1)}{C_p \rho}k_1 \lambda_3 \qquad \text{10.4--56}$$

$$\frac{\partial \lambda_2}{\partial t} = -u\frac{\partial \lambda_2}{\partial z} + k_2 \lambda_2 - \frac{(-\Delta H_2)}{C_p \rho}k_2 \lambda_3 \qquad \text{10.4--57}$$

$$\frac{\partial \lambda_3}{\partial t} = -u\frac{\partial \lambda_3}{\partial z} + \frac{E_1}{RT^2}k_1 c_A \lambda_1 - \frac{1}{RT^2}(E_1 k_1 c_A - E_2 k_2 c_B)\lambda_2$$

$$- \frac{1}{RT^2}\left[\frac{(-\Delta H_1)}{C_p \rho}E_1 k_1 c_A + \frac{(-\Delta H_2)}{C_p \rho}E_2 k_2 c_B\right]\lambda_3 + \frac{2U}{C_p \rho r}\lambda_3 \qquad \text{10.4--58}$$

with the boundary conditions

$$\lambda_i(\tau, z) = 0, \quad \lambda_i(t, L) = \frac{\delta_{i2}}{u} \qquad \text{for } i = 1, 2, 3 \qquad \text{10.4--59}$$

where δ_{i2} is the Kronecker delta.

Chang and Bankoff solved these equations numerically for a particular set of system parameters. Their results show that if the condensate temperature is maintained constant at the optimum steady state value, the feed composition fluctuations lead to a lower time average yield than the optimum steady state case. However, when the optimal periodic control is imposed on the system, in some instances the time average yield is superior to the optimum steady state operation. For the parameters considered in the study, this kind of behavior was observed only for low-frequency fluctuations; that is, when the period of the oscillations was three to four times greater than the residence time of the reactor.

Example 10.4–4   Optimal periodic feed rate for parametric pumping

The phenomenon of parametric pumping was described in Section 10.1. One of the mathematical models studied by Wilhelm and his co-workers is presented in Eqs. 10.1–43 through 10.1–47. These equations are sufficiently complicated that no attempt has been made to solve them analytically, although an analytical solution is available for Pigford's simplified model. Wilhelm's experimental and numerical studies cover a fairly wide range of system parameters, so that it is possible to estimate the optimum operating conditions. However,

Wagle and Denn[6] attempted to show that this same information could be obtained by solving an optimal control problem. They simplified Wilhelm's original set of equations by assuming that there was no axial dispersion of heat or mass and that the fluid and solid temperatures were always equal. Thus the system equations they considered were

$$(1 + \beta K)\frac{\partial T}{\partial t} + u\frac{\partial T}{\partial z} = 0 \qquad\qquad 10.4\text{-}60$$

$$\frac{\partial C_f}{\partial t} + u\frac{\partial C_f}{\partial z} + K\frac{\partial C_s}{\partial t} = 0 \qquad\qquad 10.4\text{-}61$$

$$\frac{\partial C_s}{\partial t} + \sigma(C^* - C_s) = 0 \qquad\qquad 10.4\text{-}62$$

$$C^* = C^*(C_s, C_f, T) \qquad\qquad 10.4\text{-}63$$

and the boundary conditions over the two operating intervals were

$$\begin{aligned}
\text{during } \tau_1(u \le 0): & \quad T = 1, \quad C_f = 1 \text{ at } z = 1 \\
\text{during } \tau_2(u \ge 0): & \quad T = 0, \quad C_f = 1 \text{ at } z = 0
\end{aligned} \qquad 10.4\text{-}64$$

All the variables are considered periodic over 2π. With this simplified model, it is possible to develop a formal solution to Eq. 10.4–60

$$T\left(z - \frac{1}{1 + \beta K}\int_0^t u(\tau)\, d\tau\right) = \text{constant} \qquad 10.4\text{-}65$$

which implies that periodic operation is limited to velocities satisfying the condition

$$\int_0^{2\pi} u(t)\,dt = 0 \qquad\qquad 10.4\text{-}66$$

The material balance relationships for the amount of dissolved solute in each stream can be written as

$$\text{Rich product stream} \qquad \int_{\tau_2} u(t)C_f(t, 1)\,dt \qquad\qquad 10.4\text{-}67$$

$$\text{Lean product stream} \qquad -\int_{\tau_1} u(t)C_f(t, 0)\,dt \qquad\qquad 10.4\text{-}68$$

Now it is possible to define the net separation per cycle as

$$S = \int_{\tau_2} u(t)C_f(t, 1)\,dt + \int_{\tau_1} u(t)C_f(t, 0)\,dt \qquad 10.4\text{-}69$$

which is the performance index considered by Wagle and Denn. An alternate approach would be to use Wilhelm's definition of the separation factor

$$SF = -\int_{\tau_2} u(t)C_f(t, 1)\,dt \bigg/ \int_{\tau_1} u(t)C_f(t, 0)\,dt \qquad 10.4\text{-}70$$

[6] A. K. Wagle and M. M. Denn, *Proceedings of the Joint Automatic Control Conference*, p. 286. University of Colorado, Boulder, Colo., August 1969.

as a measure of performance. Somewhat different answers might be obtained
for these two problems, just as a change in the performance criterion led to
different optimal control policies in Section 9.7. However, the purpose of the
study was to gain some experience with problems of this type, rather than developing a complete understanding of optimal parametric pumping, so that
only the simplest case was considered.

The total feed rate during one cycle is

$$F = \int_0^{2\pi} |u(t)| dt = \int_{\tau_2} u(t) dt - \int_{\tau_1} u(t) dt \qquad 10.4\text{-}71$$

which can be combined with Eq. 10.4–66 to give

$$\int_{\tau_2} u(t) dt = -\int_{\tau_1} u(t) dt = \frac{1}{2} F \qquad 10.4\text{-}72$$

Then the optimal periodic control problem is to choose $u(t)$ to maximize the
separation S, Eq. 10.4–69, over one period for some specified amount of feed F.

They derived the necessary conditions for the optimal control action and
showed that the appropriate adjoint equations were

$$(1 + \beta K)\frac{\partial \lambda_1}{\partial t} + u\frac{\partial \lambda_1}{\partial z} - \sigma \lambda_3 \frac{\partial C^*}{\partial T} = 0 \qquad 10.4\text{-}73$$

$$\frac{\partial \lambda_2}{\partial t} + u\frac{\partial \lambda_2}{\partial z} + \sigma \lambda_3 = 0 \qquad 10.4\text{-}74$$

$$\frac{\partial \lambda_3}{\partial t} - \sigma \lambda_3 \frac{\partial C^*}{\partial C_s} + K\frac{\partial \lambda_2}{\partial t} = 0 \qquad 19.4\text{-}75$$

With the boundary conditions

$$\begin{array}{ll} \text{during } \tau_1(u \leq 0): & \lambda_1 = 0, \quad \lambda_2 = -1 \text{ at } z = 0 \\ \text{during } \tau_2(u \geq 0): & \lambda_1 = 0, \quad \lambda_2 = +1 \text{ at } z = 1 \end{array} \qquad 10.4\text{-}76$$

Also, they found that all three adjoint variables are periodic over an interval
2π and that very small changes in the control action, δu, produced a change in
the performance index

$$\delta S = \int_{\tau_1} \left[\Lambda_1 + C_f(t, 0) - \int_0^1 \left(\lambda_1 \frac{\partial T}{\partial z} + \lambda_2 \frac{\partial C_f}{\partial z} \right) dz \right] \delta u(t) dt$$

$$+ \int_{\tau_2} \left[\Lambda_2 + C_f(t, 1) - \int_0^1 \left(\lambda_1 \frac{\partial T}{\partial z} + \lambda_2 \frac{\partial C_f}{\partial z} \right) dz \right] \delta u(t) dt \qquad 10.4\text{-}77$$

The quantities Λ_1 and Λ_2 in this expression are constant Lagrange multipliers,
which are needed to account for the constraint on the total feed rate, Eq.
10.4–72. Using the same kind of arguments described in Chapter 9, we know
that the necessary conditions for the optimal control policy are that the
quantities in square brackets in Eq. 10.4–77 must be equal to zero. Unfortunately this information is not very helpful, for the complete set of equations describing the optimal control policy is too complicated to solve at present.

Wagle and Denn state that an extremely small grid size would be necessary to obtain numerical solutions of the adjoint equations without encountering serious difficulties with stability problems. However, they used an approximate technique to develop a partial solution. Their results indicate that the optimal control policy appears to approach a square wave variation in the feed rate, which is similar to the parametric pumping policy studied by Wilhelm.

10.5   SUMMARY

In this chapter we have attempted to present a number of examples illustrating that the performance of periodic processes is sometimes superior to the optimum steady state design. In the initial studies, we allowed sinusoidal disturbances to enter a system designed to operate at steady state conditions; and we showed that instead of making the performance deteriorate, they actually improved the profitability of the plant. Moreover, we found that it was sometimes advantageous to manipulate one of the inputs periodically in order to amplify the effects of other disturbances and that in certain frequency ranges periodic variations of design variables around their optimum steady state values would also correspond to higher profits. All these results contradict the normal assumptions used in steady state design and the conventional design of control systems.

Since the results for sinusoidal inputs indicate that the improvement often increases with the square of the amplitude of the fluctuation, we might expect to obtain the greatest increase by using square wave inputs. Situations where the flow rate is varied as a square wave, particularly if the flow rate is equal to zero over a portion of the operating cycle, are often referred to as controlled cycling. It was shown that up to 200 percent efficiencies could be obtained by applying this technique to a plate separating unit.

Although it is well known that the outputs of a chemical plant can be forced to become periodic functions of time by manipulating the inputs periodically, it is also possible to design systems that generate periodic outputs even when the inputs are maintained constant. This type of chemical oscillator can also have a time average output that exceeds the optimum steady state value. Both lumped and distributed parameter systems can behave in this manner. Another interesting result we found in our study is that it is sometimes a good idea to use positive feedback control systems to make stable plants become unstable, so that they become oscillators.

With this evidence of the potential advantages of periodic processing, we then described a procedure for determining the optimal periodic process. In addition, we found that it was possible to find a criterion that can often be used to assess whether or not a periodic process might be superior to the best steady state plant.

Of course, our analysis is by no means complete, for we have neglected to evaluate the effects on the units downstream from the system under consideration. However, in cases where these oscillations are detrimental, it should be possible to use surge tanks or control systems to provide damping on the downstream, rather than the upstream, side. Also, there should be situations where the fluctuations are advantageous. For example, if we use the fluctuating temperature obtained as the output from a chemical oscillator to preheat the feed stream to the reactor, we will be forcing the oscillator with feed temperature oscillations; and, in some cases, this might provide an additional, beneficial, resonance effect. Similarly, by the proper coupling of periodic inputs to other units, it might be possible to obtain improvements. In other words, there might be cases where complete plant operation in the unsteady state is better than its steady state counterpart.

The main purpose of this chapter is to show that our understanding of process dynamics and control theory is still in its infancy. Thus even though past practice and "rules of thumb" are extremely important when solving engineering problems, in some cases a bold and imaginative approach might lead to a significantly better system.

Appendix

Contents

for Volume 1

*The asterisk indicates sections or chapters considered to provide a basis for an undergraduate course.

Author Index

Subject Index